BIOPHYSICS OF THE PANCREATIC β-CELL

ADVANCES IN EXPERIMENTAL MEDICINE AND BIOLOGY

Recent Volumes in this Series

BIOPHYSICS OF THE PANCREATIC β-CELL

Edited by

Illani Atwater
and Eduardo Rojas

National Institutes of Health
Bethesda, Maryland

and

Bernat Soria

University of Alicante
Alicante, Spain

PLENUM PRESS • NEW YORK AND LONDON

Library of Congress Cataloging in Publication Data

Biophysics of the pancreatic [beta]-cell.

(Advances in experimental medicine and biology; v. 211)
Proceedings of a workshop held Sept. 30–Oct. 1, 1985, in Alicante, Spain.
Includes bibliographies and index.
1. Pancreas—Congresses. 2. Islands of Langerhans—Congresses. 3. Biophysics—
Congresses. I. Atwater, Illani. II. Rojas, Eduardo, 1936-   III. Soria, Bernat. IV.
Series. [DNLM: 1. Biophysics—congresses. 2. Islands of Langerhans—congresses. W1
AD559 v.211 / WK 800 B6158 1985]
QP188.P26B56   1987   59940132   87-7713

Proceedings of a workshop on the Biophysics of the Pancreatic β-Cell,
held September 30–October 1, 1985, in Alicante, Spain

DOI 10.1007/978-1-4684-5314-0

© 1986 Plenum Press, New York
MyCopy version of the original edition 1986

A Division of Plenum Publishing Corporation
233 Spring Street, New York, N.Y. 10013

Pancreatic β-cell biophysics has undergone a veritable information explosion in the past two years. Single channel and macroscopic
currents have become easily accessible following the introduction of
the patch clamp technique. In addition to this new approach, further
development of techniques for optical measurements, ion-sensitive
microelectrodes, permeabilized cells and mathematical modelling have
recently added to the now classical techniques of membrane potential
recording and tracer flux measurement. The International Workshop on
Biophysics of the Pancreatic β-Cell held in Alicante (Spain) on September 30 - October 1, 1985, has now given us the opportunity to share
experiences with these new techniques applied to the β-cell. Furthermore this was the first occasion for most of the groups doing patch-
clamp studies of the β-cell to decide on appropriate nomenclature and
to debate the different characteristics of the β-cell ionic channels.

To make this information available to the larger scientific
community a record of the meeting has been assembled in this book. It
is a collection of research papers by leading scientists at the meeting working on biophysical, biochemical and physiological aspects of
secretion. We grouped their contributions in seven sections, including new experimental approaches, K-channels, Ca-channels, role of
ionic channels, intracellular ionized calcium, neural regulation and
mechanisms of insulin release. Each section gives an account of the
state of the problem at the time of the meeting, and the subjects are
analyzed from the different perspectives of the various contributors.
For example, the general problem of modulation of membrane ionic
channels by glucose metabolism is approached by using not only the
patch clamp technique, but also by radioactive tracer methods, intracellular recordings using microelectrode techniques and cation-sensitive electrode methods. For the sake of generality various sources of
pancreatic β-cells were considered, ranging from small rodents, to

fish as well as insulin secreting cell lines. Finally, information
from other cell types having relevance to the problem of β-cell function were considered. These systems included mast cells (intracellular dialysis using patch pipettes), adrenocortical and
parathyroid cells (patch clamp electrophysiology with innovative
experimental design) and chromaffin cells from the adrenal medulla
(on-line measurement of hormone release with a time resolution in the
millisecond range).

This book therefore provides an overview of the present status of
the research on the mechanisms underlying the glucose recognition by
the β-cell and presents the latest results on β-cell membrane channels, their regulation and the specific role they might play in stimulus-secretion coupling. The control mechanisms of exocytosis are
explored by experts in various fields, including molecular biochemistry and protein chemistry.

We wish to thank the participants for their lively and sustained
discussion throughout the Workshop and hope that this display of open
enthusiasm will set the stage and tone for future research efforts in
β-cell biophysics.

The Organizing Committee included an enthusiastic group of Ph.D.
students from the Department of Physiology of the University of
Alicante. Cristina Ripoll was the Workshop Secretary, while Salvador
Sala and Juan Vicente Sanchez Andres provided invaluable administra-
tive support in all phases of the conference. Finally, the contributions of Dr. Rosa Ferrer to organization of the banquet and social
events during the meeting as well as to the actual success of the
conference as a whole is gratefully acknowledged.

We are in deep gratitude to our generous sponsors: Universidad
de Alicante, Conselleria de Cultura, Educacio i Ciencia de la
Generalitat Valenciana, Diputacion Provincial de Alicante and Comision
Asesora de Investigacion Cientifica y Tecnica of Spain (CAICYT).

Illani Atwater

Eduardo Rojas

Bernat Soria

CONTENTS

NEW EXPERIMENTAL APPROACHES

Patch Pipettes Used for Loading Small Cells
        with Flourescent Indicator Dyes............................   1
    E. Neher and W. Almers

Optical Detection of Calcium-Dependent ATP Release
        From Stimulated Medullary Chromaffin Cells..................   7
    E. Rojas, E. Forsberg and H.B. Pollard

Glucose-evoked Changes in $[K^+]$ and $[Ca^{2+}]$ in the
        Intercellular Spaces of the Mouse
        Islet of Langerhans...   31
    E. Perez-Armendariz and I. Atwater

K-CHANNELS

A Potassium Channel Modulated by Glucose Metabolism
        in Rat Pancreatic β-cells...................................   53
    F. M. Ashcroft, D.E. Harrison, and S.J.H. Ashcroft

Glucose Suppresses ATP-Inhibited K-channels
        in Pancreatic β-cells.......................................   63
    D.L. Cook, C.N. Hales, and L.S. Satin

K-Channels in an Insulin-secreting Cell Line:
        Effects of ATP and Sulphonylureas...........................   69
    M.L.J. Ashford, N.C. Sturgess, D.L. Cook, and C.N. Hales

Inhibition of K-channels in Insulin Secreting Cells................ 77
    O.H. Petersen, I. Findlay, and M.J. Dunne

Pharmacological Control of ^{86}Rb Efflux from
    Mouse Pancreatic Islets.................................... 83
  J.C. Henquin, M.G. Garrino, M. Nenquin,
  G. Paolisso and M. Hermans

Electrophysiological Measurements Show Marked Differences
    in the Properties of the Pancreatic β-cell K-channels
    from Albino Mice and a Strain of ob/ob (Obese) Mice.......... 95
  L.M. Rosario

Single K-channel Activity in Fish Islet Cells.................... 109
  R.M. Santos, H. Finol, and E. Rojas

Potassium Channels in Adrenocortical and Parathyroid Cells........ 125
  J. Lopez-Barneo, L. Tabares, and A. Castellano

Ca-CHANNELS

Pharmacological Properties of the Chromaffin Cell
    Calcium Channel.. 139
  A.G. Garcia, C.R. Artalejo, R. Borges,
  J.A. Reig, and F. Sala

Insulin Release and K$^+$-induced Depolarization in Mouse
    Pancreatic β-cells... 159
  S. Joost and I. Atwater

Calcium and Potassium Currents Recorded from Pancreatic
    β-cells Under Voltage Clamp Control 167
  G. Trube and P. Rorsman

Voltage-activated Ca^{2+} and K$^+$ Currents in an Insulin-
    secreting Cell Line (RINm 5F).............................. 177
  I. Findlay and M.J. Dunne

Voltage-gated Ca Current in Pancreatic Islet β-cells.............. 189
  L.S. Satin and D.L. Cook

Effects of Varapamil and Nifedipine on Glucose-Induced
    Electrical Activity in Pancreatic β-cells...................... 195
  M. Vasseur, A. Debuyser, and M. Joffre

Stimulation of Insulin Release by Organic Calcium-agonists......... 201
  F. Malaisse-Lagae, A. Sener, and W.J. Malaisse

ROLE OF IONIC CHANNELS

Contribution of Isotope Flux Studies to Understanding
    the Mechanism of the β-cell Membrane......................... 207
  P.C. Croghan, C.M. Dawson, A.M. Scott, and J.A. Bangham

$^{22}Na^+$ Efflux from Normal and ob/ob Mouse Islets
    of Langerhans.. 225
  C.M. Dawson and P.C. Croghan

The Role of Anions in the Regulation of Insulin Secretion.......... 227
  J. Sehlin

Graded Spike Electrogenesis in Mouse Pancreatic β-cell............. 235
  B. Soria and R. Ferrer

Prediction of the Glucose-induced Changes in Membrane Ionic
    Permeability and Cytosolic Ca^{2+} by Mathematical Modeling..... 247
  J. Rinzel, T.R. Chay, D. Himmel, and I. Atwater

Modelling the β-cell Electrical Activity........................... 265
  J.A. Bangham, P.A. Smith, and P.C. Croghan

INTRACELLULAR IONIZED CALCIUM

Insulin Secretion Studied in Islets Permeabilised
    by High Voltage Discharge.................................... 279
  P.M. Jones and S.L. Howell

Regulation of Insulin Release Independent of Changes of
    Cytosolic Ca^{2+} Concentration............................... 293
  T. Tamagawa, H. Niki, A. Niki, and I. Niki

The Role of Cytosolic Calcium in Insulin Secretion
    from a Hamster Beta Cell Line.................................   305
  A.E. Boyd III., R.S. Hill, T.Y. Nelson,
    J.M. Oberwetter, and M. Berg

Differential Effect of Nutrient and Non-nutrient
    Secretagogues on Cytosolic Free Ca^{2+}
    in Pancreatic Islet Cells...................................   317
  A. Herchuelz, M. Juvent, E. Van Ganse, and P. Gobbe

Relationship Between Extracellular Na^+ and the Total
    Ionized Ca^{2+} Content of Rat Pancreatic Islets................   319
  G.H.J. Wolters, M. Vonk, and A. Pasma

NEURAL REGULATION

Mobilization of Different Pools of Glucose-incorporated
    Calcium in Pancreatic β-cells After Muscarinic
    Receptor Activation...   325
  B. Hellman, E. Gylfe, and P. Bergsten

Effect of the Order of Application of Neural Inputs
    on Insulin Secretion..   343
  L.A. Campfield, F.J. Smith, J.E. Settle,
    and R. Sohaey

Muscarinic Receptors and the Control of Glucose Induced
    Electrical Activity in the Pancreatic β-cell................   351
  I. Palafox, J.V. Sanchez-Andres, S. Sala,
    R. Ferrer, and B. Soria

Electrophysiological Evidence for Histaminergic Modulation
    of Pancreatic β-cell Function...............................   359
  A. Marques, R. Ferrer, C. Ripoll, and B. Soria

Effect of Melatonin on Insulin Secretion from Isolated Rat
    Islets of Langerhans..   367
  J.M. Pou, T. Cervera, M. Codina, and A. de Leiva

Calcium Regulation of Membrane Fusion During
   Hormone Secretion.. 369
   H.B. Pollard, K.W. Brocklehurst, E.J. Forsberg,
      A. Stutzin, G. Lee, and A.L. Burns

The Insulin Secretory Granule: Featurees and Functions
   in Common with Other Endocrine Granules...................... 385
   J.C. Hutton, M. Peshavaria, H.W. Davidson,
      K. Grimaldi, and R. Pogge Von Strandmann, and K. Siddle

Effects of Monensin on Glucose-induced Insulin Release
   and $^{45}Ca^{2+}$ Outflow... 397
   A. Kanatusuka, H. Makino, N. Hashimoto,
      M. Sakurada, and S. Yoshida

Interdependency of Ca^{2+} Availability and Cyclic AMP
   Generation in the Pancreatic β-cell........................... 403
   I. Valverde, P. Garcia-Morales, M. Saceda,
      and W.J. Malaisse

Cyclic AMP Antagonist, A Second Messenger for
   Insulin Action, Inhibits Glucose-stimulated
   Insulin Secretion in Isolated Islets
   of Chinese Hamsters... 409
   H. J. Partke and H.K. Wasner

Pulsatile Insulin Release and Electrical Activity
   from Single ob/ob Mouse Islets of Langerhans................ 413
   L.M. Rosario, I. Atwater, and A.M. Scott

Comparison of Stimulus-secretion Coupling in Normal
   and ob/ob (Norwich Colony) Mouse Islets
   of Langerhans... 427
   A.M. Scott and C.M. Dawson

Insulin Release, Ca^{2+} Fluxes and Calmodulin Content
   of Pancreatic Islet in Aging Rats............................ 429
   J.J. Osuna, R. Rubio, E. Rodriguez,
      and C. Osorio

Protein Carboxyl Methylation in Rat Pancreatic Islets.
Possible Role in β-cell Function....................................... 431
J.E. Campillo, P. Mena, S. Alejo, and C. Barriga

Role of Transglutaminase in Proinsulin Conversion
and Insulin Release... 443
R. Gomis, C. Alarcon, I. Valverde,
and W.J. Malaisse

Induction of the Glucokinase-glucose Sensor
in Pancreatic Islets of Insulinoma-bearing
Rats Following Tumor Removal................................ 447
F.J. Bedoya, M.C. Appel, R. Goberna,
and F.M. Matschinsky

Biochemical Design Features of the Pancreatic Islet
Cell Glucose-sensory System................................ 459
F.M. Matschinsky, M. Meglasson, A. Gosh,
M.C. Appel, F.S. Bedoya, M. Prentki,
B. Corkey, T. Shimizu, D. Bernar,
H. Najafi, and C. Manning

Contributors....................................... 471

Index....................................... 485

PATCH PIPETTES USED FOR LOADING SMALL CELLS WITH FLUORESCENT INDICATOR
DYES

Erwin Neher and Wolf Almers[+]

Max-Planck-Institut fur biophysikalische Chemie
D-3400 Gottingen, F.R.G.
[+]University of Washington
Dept. of Physiology and Biophysics, SJ-40
Seattle, Washington 98195, USA

In modern cell physiology, much important information has come from
the intracellular application of membrane-impermeant substances, such as
ion indicator dyes, second messengers, regulatory proteins, antibodies and
messenger RNA. Such substances have been either injected into cells
through microelectrodes or loaded into lipid vesicles and erythrocyte
ghosts that were later allowed to fuse with the cell membrane. We have
monitored cytoplasmic Ca^{2+} in mast cells with intracellularly applied
fura-2, a new Ca-indicator dye.

Fura-2 belongs to a new family of indicator dyes developed by Tsien
and collaborators[2]. These dyes are extremely useful for studies of small
cells, because the cells may be loaded by external application of the
membrane-permeant ester of the dye. When intracellular esterases split off
a lipophilic moiety, the dye becomes impermeant and trapped in the cell.
This convenient approach was tried in our initial experiments until it
appeared that membrane-permeant fura-2 becomes trapped not only into the
cytoplasm but also into mast cell granules, and reports not cytoplasmic
Ca^{2+} but instead a mixed signal originating partly from the cytosol and
partly from the interior of secretory vesicles[1]. To obtain a purely
cytoplasmic signal from mast cells, the dye must evidently be applied to
the cytoplasm in its impermeant form, e.g. by microinjection. However,
while microelectrode injection works well with large cells, such as

oocytes, most cells of the mammalian organism are much smaller, and suffer too much damage from impalement with microelectrodes. In establishing contact with the cytoplasm, firepolished "patch pipettes" inflict less damage and are more efficient[3]. Here, we explore the use of such pipettes to load small secretory cells (mast cells) with fura-2.

METHODS AND RESULTS

From measurements of Na-current equilibrium potential it was inferred[5] that Na^+ reaches diffusional equilibrium with the contents of a patch pipette within 5 to 10 seconds after establishment of whole-cell contact. Another way to study diffusion from a patch pipette into a cell is by monitoring the fluorescence of a dye diffusing into the cell. Fig. 1 shows an experiment on a single rat peritoneal mast cell. The fluorescent dye fura-2 was present in the pipette at a concentration of 100 µM. Fluorescence was measured at 500 nm, with excitation at two wavelengths as described by Almers and Neher[1]. Trace a) (Fig. 1) shows the fluorescence excited at a wavelength (350-360 nm) close to the isosbestic point; this signal reports the total dye concentration. Trace b) shows the fluorescence excited at 390 nm. At this wavelength the fluorescence depends strongly on the concentration of free calcium, since the Ca-bound form has no fluorescence at this wavelength[2].

At the beginning of the traces in fig. 1 the pipette has sealed onto the cell, but the patch of membrane beneath the pipette tip still separates the contents of pipette from the cytoplasm. The weak fluorescence recorded originates mainly from the dye in the pipette tip. At the arrow, the membrane patch is ruptured by a pulse of suction, so that pipette interior and cytoplasm become continuous (whole-cell configuration). As the dye diffuses into the cell, fluorescence is seen to increase and reach a steady level within 20 to 30 seconds. From the ratio of fluorescence intensities at the two wavelengths, one can calculate the cytoplasmic concentration of ionized Ca[1,2]. Cytoplasmic $[Ca^{2+}]$ approaches a steady value of approximately 0.25 µM which is imposed by the mixture of 9 mM EGTA and 6 mM $CaCl_2$ in the pipette.

The rate at which the fluorescence dye inside the cell reaches its steady state value is strongly correlated with the electric resistance

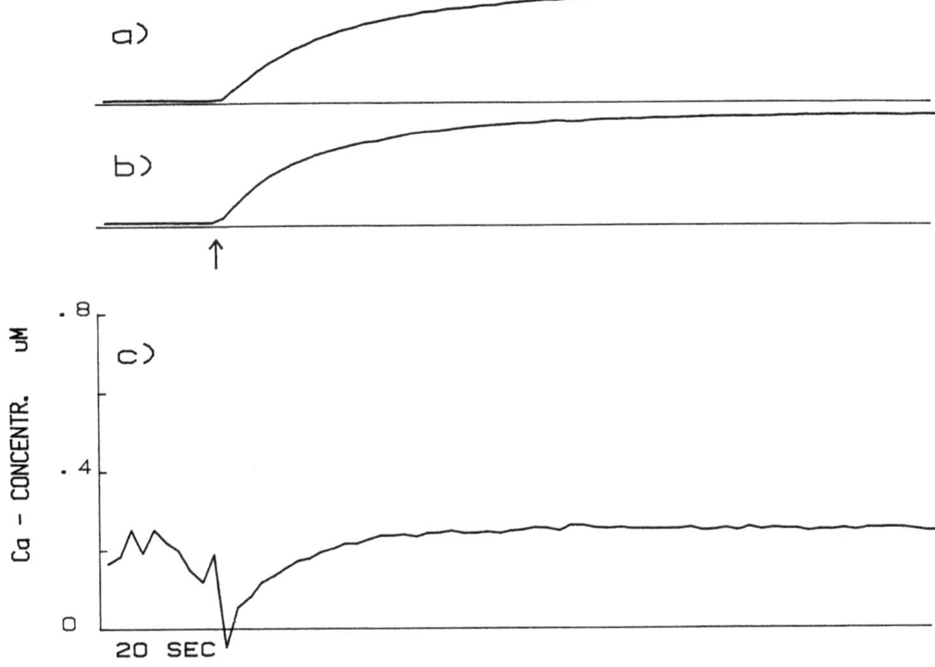

Fig. 1. Changes in fluorescence during the loading process. a) shows the fluorescence at excitation wavelengths 350-360 nm. b) shows fluorescence at 390 nm. c) gives free Ca-concentration calculated from the fluorescence ratio according to Almers and Neher[1]. The erratic portion of the curve before penetration is not significant since it represents the ratio of two very small background fluorescence signals. An exponential fitted to part a) yields a time constant of 22 seconds. Series resistance was 4.4 Mohm; capacitance 7.7 pF. Bath saline (in mM): 140 NaCl, 2.5 KCl, 2 $CaCl_2$, 5 $MgCl_2$, 10 HEPES-NaOH, pH 7.2. Pipette filling solution (in mM): 155 K-glutamate, 4 $MgCl_2$, 10 HEPES-NaOH (pH 7.2), 0.2 ATP, 2.5 ITP, 0.1 fura-2, 9 EGTA, 6 $CaCl_2$. Temperature 25°C.

within the pipette tip. This resistance can be measured with the series resistance cancellation circuit of the patch clamp amplifier (EPC-7, List Elektronik), and is usually two to four times the value of the pipette resistance before a seal is made[5]. In fig. 2 the time constant of the fluorescence increase is plotted versus series resistance. The experimental points are approximated quite well by a straight line which intersects the abscissa close to the origin. Such a relation is expected if the cell is assumed to be a single compartment loaded diffusionally through a rate limiting constriction.

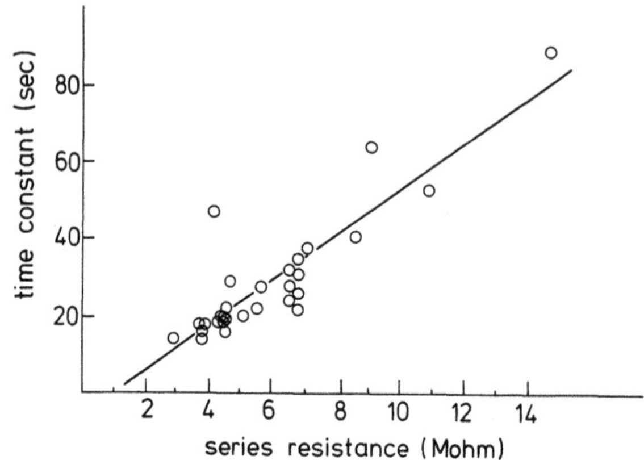

Fig. 2. Time constant of the loading process as a function of series
resistance. Time constant of loading was estimated graphically
from plots like fig. 1a by fitting a tangent to the point of
steepest rise in fluorescence and measuring the time interval
until intersection with a line representing the steady state
level. This method was adopted since in many experiments the
series conductance (and hence the diffusional access) diminished
later in the record so that time courses became nonexponential.
The time constant, τ, was plotted against the series resistance
determined at the beginning of the record. The straight line,
fitted to the data points by unconstrained linear regression is
drawn according to:

$$\tau = -6.44 + 5.965 \, R_s$$

where τ is in seconds and R_s is in Megaohms. The cells had
capacitance values of 7.3 $\pm$ 2.7 pF[mean $\pm$ S.D.] which at 1 $\mu F/cm^2$
would correspond to an average cell diameter of 15 μm.

DISCUSSION

Marty and Neher[5] found an exchange time constant for Na-ions of 5.4
sec at a series resistance of 3.8 Mohm . For this same series resistance
we find an exchange time of 16 seconds for fura-2. The ratio (about 3) of
these two time constants is similar to the ratio of aqueous diffusion
coefficients (about 5) taking for NaCl the value of $1.5 \times 10^{-5} \, cm^2 \, sec^{-1}$,
and for fura-2 (MW 831) a value of $0.3 \times 10^{-5} \, cm^2 \, sec^{-1}$ which is typical
for water soluble compounds of this molecular weight. It is interesting to
extrapolate to molecules of larger size. Based on aqueous diffusion
constants D, a small protein like pituitary growth hormone (MW 49000) would
have an exchange time constant four times longer than fura-2, or about 1
minute (D = $0.07 \times 10^{-5} \, cm^2 \, sec^{-1})$[4].

REFERENCES

1. W. Almer and E. Neher, The Ca-signal from fura-2 loaded mast cells
    depends strongly on the method of dye-loading, _FEBS_ _Lett._, 192:13
    (1985).
2. G. Grynkiewicz, M. Poenie, and Y. Tsien, A new generation
    of Ca^{2+} indicators with grately improved fluorescence properties, _J._
    Biol. _Chem._ 260:3440 (1985).
3. O.P. Hamill, A. Marty, E. Neher, B. Sakmann and F. Sigworth, Improved
    patch-clamp techniques for high-resolution current recording from
    cells and cell-free membrane patches. _Pflügers_ _Arch._ 391:85 (1981).
4. _Handbook_ _of_ _Biochemistry_, 2nd ed. CRC Press (1970).
5. A. Marty and E. Neher, Tight-seal whole-cell recording, in: "Single
    Channel Recording", E. Sakmann and E. Neher, eds., New York: Plenum
    Press, pp 107-122 (1983).
6. R.Y. Tsien, T. Pozzan and T.J. Rink, Measuring and manipulating
    cytosolic Ca^{2+} with trapped indicators, _TIBS_ 9:263 (1984).

OPTICAL DETECTION OF CALCIUM DEPENDENT ATP RELEASE FROM STIMULATED
MEDULLARY CHROMAFFIN CELLS

E. Rojas, E. Forsberg, and H. B. Pollard

Laboratory of Cell Biology and Genetics
NIDDK, National Institutes of Health
Bethesda, Maryland 20892

INTRODUCTION

It is well known that different types of secretory granules
contain adenosine-5'-triphosphate (ATP) in addition to the specific
hormones and neurotransmitters. Examples include secretory granules
from chromaffin cells,[32] pancreatic B-cells,[27] platelets[4,5,30] and
mast cells. We have recently shown[22] that it is possible to obtain
real-time measurements of the kinetics of secretion of vesicle contents by monitoring ATP release from stimulated medullary chromaffin
cells using luciferin-luciferase.

We report here the results of experiments designed to measure
the kinetics and steady-state characteristics of the Ca^{2+}-dependent
acetylcholine-induced ATP release from chromaffin cells. In addition,
the quantal nature of the ATP release together with some of the properties of cholinergic nicotinic and muscarinic receptor stimulation
were determined.

METHODS

The Luminescence Instrument

The instrument used to stimulate the cells and to measure the
light consisted of a reaction chamber mounted vertically resting on
the photocathode of a photomultiplier. This reaction chamber, kept at

constant temperature, consisted of a plastic tube (1000 µl) with
hydrophobic surfaces so as to keep the solution confined to the bottom
of the tube. Solutions, containing secretagogues or drugs, were added
in small volumes (10 µl) to the cell mixture using a microliter
pipette (Gilson) through a light-tight holder. The temperature of the
solution in the tube could be controlled (18-36°C). Emitted light
from the reaction chamber was detected using a photomultiplier with
the anode held at virtual-ground voltage by means of an amplifier with
a negative feedback resistance (either 10^8 or 10^9 Ohm).

The Luciferin-Luciferase Reaction

The efficiency of the luciferin-luciferase reaction, i.e.,
number of photons emitted divided by the number of molecules of substrate oxidized, is close to 1.0.[10]   During the oxidation of
luciferin the energy released (50 to 72 kcal per mole quanta) is
utilized to create an electron excited state either directly or by
energy transfer to fluorescent molecules formed or present during the
reaction. The chemical requirements for light emission can be expressed by the following scheme[6]:

$$\text{Luciferin} + \text{ATP.Mg}^{2+} + O_2 = CO_2 + \text{oxyluciferin} + \text{AMP} + \text{PP}_i.\text{Mg}^{2+} + h\nu.$$

The biochemical mechanism, however, underlying the bioluminescent reaction is

$$\text{Luciferin} + \text{ATP.Mg}^{2+} = \text{luciferyl-adenylate} + \text{PP}_i$$

which is called the activation step; and,

$$\text{Luciferyl-adenylate} + O_2 = \text{AMP} + \text{oxiluciferin} + CO_2 + h\nu$$

which is called the redox step.

The highest energy emission corresponds to a wavelength maximum
of 546 nm. The optimum pH is 7.75 and the optimum temperature is
about $25°C$.[2]

The technique as used here measures the rate at which light is
emitted by the ongoing reaction (not the amount of ATP present).

However, it is easy to infer from the rate of reaction how much ATP is
present. Figure 1 illustrates the calibration of the detector and the
reaction.

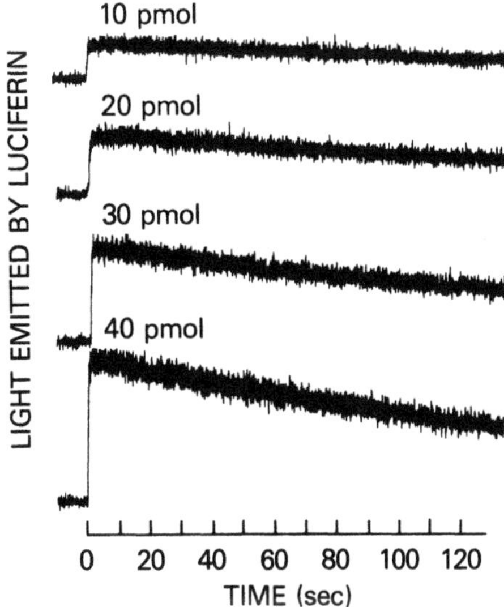

Fig. 1. Calibration of the reaction mixture and luminescent detection.
Four records of the photomultiplier output are shown. At time zero, 10
μl of Krebs solution containing various amounts of ATP (indicated next
to the corresponding record) were added to the reaction mixture.
From[23]

It is clear that the light output follows the amount of ATP
present in the reaction chamber. For this series of calibration
records the concentration of the luciferin-luciferase mixture was kept
constant. Similar levels of emitted light were obtained either by
single additions of ATP or by consecutive additions. Depending on the
concentration of the luciferin-luciferase mixture the order of the
reaction ranges between 1.5 and 2. This implies that the light emitted is proportional to the concentration of ATP added and not to the
amount of substrate present.

Provided the concentration of cells in the reaction mixture was
kept below 100 cells/μl, emitted light levels during the experiments
with chromaffin cells were constant in the initial portion of the
record, and linearly related to ATP levels up to 140 pmol in the
reaction medium. Light levels, although reduced by the dilution of the

ATP during the additions of small volumes, were unaffected by triton
X-100 (up to 1%) and digitonin (up to 200 µM). Furthermore, the
reaction was inhibited by K^+ (25% at 100 mM K^+) and Cd^{2+} (50% at 1
mM). The assays with the cells were performed either at about 18 or
36°C.

Preparation of Chromaffin Cells

Chromaffin cells were prepared from bovine adrenal medulla by
collagenase digestion as described previously in detail.[13] The cells
were placed in Eagles essential medium at a concentration of 50 million cells per flask (30 cm^3 of medium) and allowed to recover from
the isolation procedure for a three day period in a CO_2 incubator at
37°C. After this incubation the cells were collected by centrifugation and resuspended in a modified Krebs solution (135 mM NaCl, 10 mM
Hepes-NaOH at pH 7.2, 5 mM KCl, 2.5 mM $CaCl_2$, 1.1 mM $MgCl_2$, 1.0 mg/ml
bovine serum albumin (crystalline BSA, Calbiochem-Behring) and 10 mM
D-glucose). The cell preparation was kept at room temperature prior
to initiation of experiments at a concentration of 10^7 cells/cm^3 of
Krebs solution.

Luciferin/Luciferase Preparation

The enzmye preparation (from Analytical Luminescence Laboratory,
San Diego, California, USA) consisted of a mixture of highly purified
luciferase, purified bovine serum albumin (Calbiochem-Behring) and
luciferin. The contents of a sealed vial were dissolved in 5 cm^3 of
Krebs solution. The reaction mixture (130 µl Krebs solution, 15 to 25
µl cells and 50 µl of the luciferin-luciferase mixture) was placed in
the plastic tube. Cell density in the reaction mixture was kept below
2.5 x 10^5 cells/cm^3 in order to maintain ecto-ATPases at low activity.
Light output from the reaction mixture containing resting cells was
constant, 0.25 to 2.5% of the maximum signal recorded in each experiment.

RESULTS

Acetylcholine-Stimulated ATP Release

Figure 2 (lower trace) depicts a typical record of the time-
integral of the ATP release process evoked by acetylcholine. It may

be seen that immediately after the application of the acetylcholine (1 µM), the signal increased rapidly and reached a steady-state level after 90 sec. Since the ATP released by the cells is confined to the

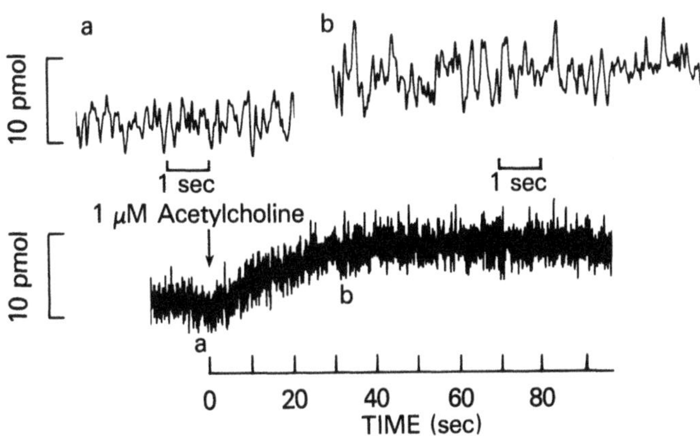

Fig. 2. Acetylcholine-evoked ATP release from freely suspended chromaffin cells. Lower record: photomultiplier output (mV) representing the time-integral of the ATP release in the reaction chamber. About 10,000 cells in the reaction mixture. The arrow indicates the addition of 10 µl of Krebs solution containing 20 µM acetylcholine to 200 µl of reaction medium. Inserts on expanded time-base: a) taken before the application of acetylcholine, b) taken 32 sec after.

reaction chamber, the vertical axis represents the time-integral of the ATP released. Thus, the time-derivative of the record shown in Fig. 2 represents the time course of the rate of ATP release. It is apparent that ATP secretion was almost complete in less than two minutes, consistent with data for the rate of secretion of catecholamines.[21] On the upper part of Fig. 2, the noise in the signal before (a) and during (b) stimulation of the cells with acetylcholine can be compared on an expanded time base. In presence of acetylcholine the amplitude of the fluctuations is larger than in the record taken before stimulation of the cells.

Acetylcholine-Induced ATP Release From Chromaffin Cells Requires External Calcium

In chromaffin cells, the secretion of catecholamines is evoked by stimulation of cholinergic receptors and depends on calcium in the medium.[7,8,16] As shown in Fig. 3, this is also the case for stimulation of ATP release. For the experiment illustrated in Fig. 3A, 30 µM nicotine was applied to cells in the luciferin-luciferase medium in presence of different $[Ca^{2+}]_o$. The concentration of free-calcium, which was measured with a calcium electrode, varied from 0.1 to 1.9 mM. In all instances prompt secretion of ATP was observed. In presence of 1.9 mM Ca^{2+} (part A, top record), nearly 20% of the cellular ATP was released as a result of nicotinic stimulation. In a Ca^{2+}-free

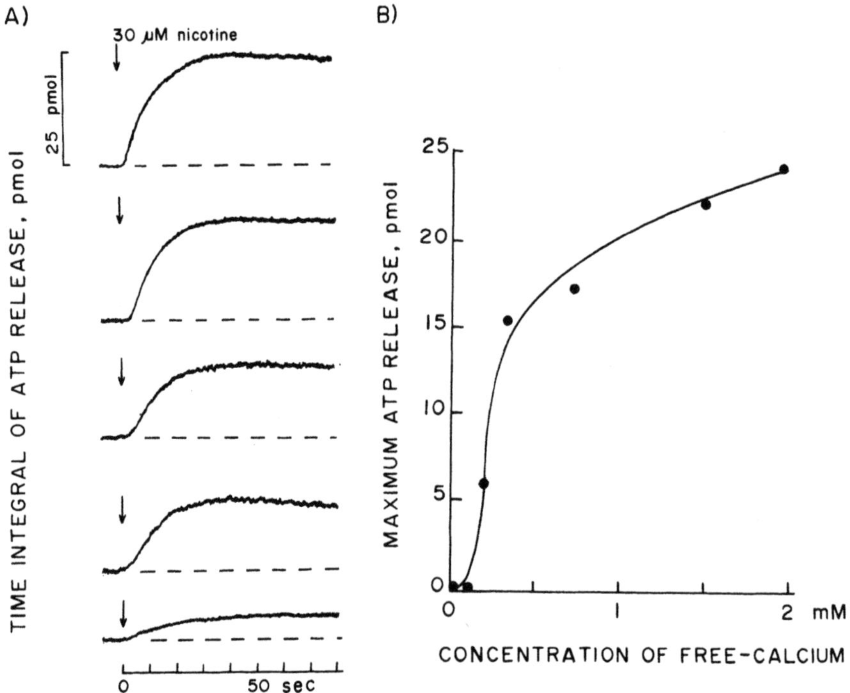

Fig. 3. Calcium dependence of acetylcholine-induced ATP release. (A) Five chart records of the photomultiplier current output representing the time integral of the ATP released by about 10,000 cells in the reaction mixture in response to nicotine. The low pass filter was set at 5 Hz. Arrows indicate the addition of nicotine. Free-calcium concentration (measured with a calcium electrode) was increased 0.1 to 1.9 mM (top record). (B) Maximum ATP released as a function of free-calcium concentration in the reaction medium. From [23]

modified Krebs medium no ATP secretion was observed. The response of the cells to nicotine was gradually recovered by adding increasing amounts of Ca^{2+}, the threshold being near 0.1 mM (not shown). This was also true for higher concentrations of nicotine (up to 240 µM) in a low Ca^{2+} medium (<0.2 mM). Shown on the right side are the maximum ATP levels reached during nicotinic stimulation as a function of free-calcium concentration.

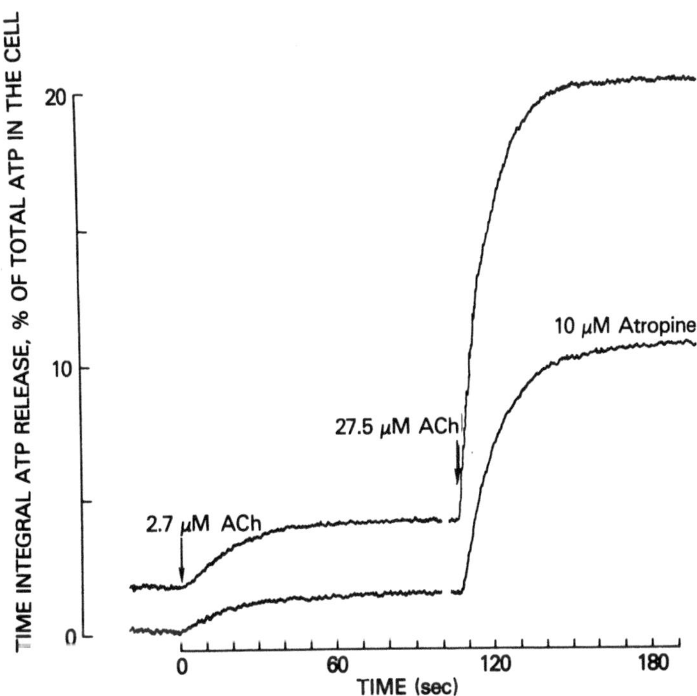

Fig. 4. Effect of atropine on acetylcholine-evoked release. Experiments with about 10,000 cells in the reaction medium. Upper record:
in the absence of atropine two consecutive applications of acetylcholine as indicated by the arrows. Lower record: protocol as in the
upper record but cells were incubated in presence of 10 μM atropine.
Same calibration for both records. Experiment carried out at 18°C.
From [24]

Acetylcholine receptors in chromaffin cells can be desensitized
by prolonged exposure to high concentrations of agonists. This property of the acetylcholine receptors is also expressed in the case of
the ATP release. Submaximal doses of acetylcholine (10 μM) can evoke
substantial ATP release (up to 15% of the cellular ATP). Addition of
a further identical dose five minutes later (20 μM) resulted in no
further ATP release. Control experiments showed that the application
of a single 20 μM dose elicited more ATP release (up to 25%). Thus,
this refractoriness of ATP secretion was not due to the depletion of
ATP.[22] That the nicotinic acetylcholine receptor was primarily involved in the acetylcholine stimulation of ATP release was shown by
pretreatment of the cells with either d-tubocurarine or hexamethonium,

specific blockers of the nicotinic receptor. Both drugs induced a substantial inhibition of the acetylcholine-evoked ATP release. Furthermore, application of muscarine (1-100 µM) alone did not evoke ATP secretion.

On the other hand, the acetylcholine-evoked ATP release in the presence of atropine, a blocker of the muscarinic acetylcholine receptor, was 20 to 45% smaller than in the absence of atropine. These effects of atropine on ATP release are illustrated in Fig. 4. In the absence of atropine (upper record), the ATP release evoked by acetylcholine (2.7 µM) amounted to 2% of the ATP in the cells. Further stimulation with the agonist (27.5 µM) elicited further release of ATP, about 16%. The ATP release in the presence of atropine (10 µM) amounted to 1% for the first dose of acetylcholine (2.7 µM) and, 9% for the second dose (27.5 µM), 56% of the release measured in the absence of the muscarinic receptor blocker. Muscarine alone (up to 50 µM) did not elicit ATP secretion.

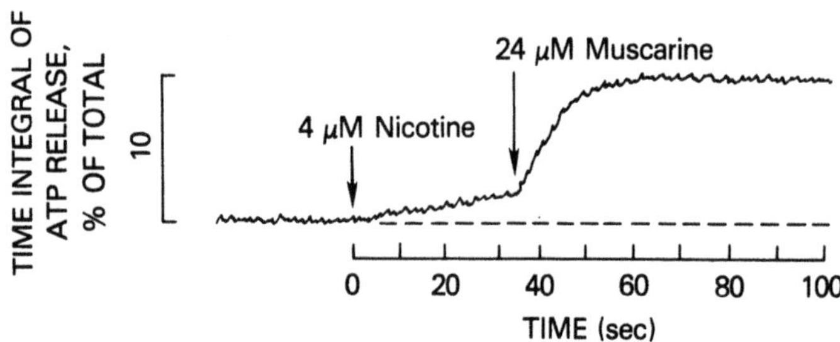

Fig. 5. ATP release induced by sequential stimulation of the nicotinic and muscarinic receptors.

Muscarinic Receptor Activation Potentiates Nicotine-Evoked ATP Release

It has been proposed that muscarinic receptor activation may directly activate a membrane bound phospholipase-C stimulating the hydrolysis of inositol-containing phospholipids and generation of diglyceride and inositol phosphates.[3,14,18] One of the inositol phosphates, inositol 1,4,5 trisphosphate, may then mobilize Ca^{2+} from internal sources and this Ca^{2+} may activate a cytosolic phospholipase-C, which enhances phosphatidylinositol hydrolysis in the cell interior. The relevance of this hypothetical scheme to the chromaffin cell systems has only recently been demonstrated.[11] Inositol phosphates may induce Ca^{2+} release from the endoplasmic reticulum in chromaffin cells,

14

and this additional source of Ca^{2+} may play a fundamental role in stimulus-secretion coupling. Since this sequence of events is extremely rapid, it has been difficult to detect.

However, our technique can be used to establish the temporal relationship between muscarinic receptor activation and ATP release evoked by nicotinic receptor stimulation. Provided the time interval between the two stimuli was less than 120 sec, sequential nicotinic-muscarinic receptor stimulation always gave an enhanced ATP release. The results of a typical experiment are shown in Fig. 5. Initial stimulation with nicotine (4 µM) elicited a moderate release of ATP (about 2%). Stimulation with muscarine induced an additional ATP release (8%). Reversing the sequence of the stimuli to muscarinic-nicotinic, the cells responded only to nicotine (9%), albeit with enhanced release.

Granular Nature of the Secreted ATP

ATP is distributed both in the secretory granules and in the cytosol of the adrenal medullary cells.[28,32] To establish the origin of the secreted ATP, we permeabilized the cells either by application of digitonin (10-40 µM) or by brief exposures (2 µs) to an electrical field (2000 V/cm) in a medium with isethionate in place of chloride.[17]

Chromaffin cells suspended in a medium with low Ca^{2+} and Cl^- plus luciferin-luciferase, were subjected to repetitive high voltage discharges. Fig. 6 shows that 7% of the ATP in the cells was rapidly released to the medium (arrow on the left). Under these conditions, application of acetylcholine (100 µM), failed to evoke further release. Addition of digitonin (20 µM, third arrow from the left), gave rise to a further ATP liberation, 16% of the total ATP in the cells. This ATP may represent a small fraction of the ATP from the granular pool, releasable by Ca^{2+} at 1 µM.[9,31] Finally, treatment of the reaction mixture with 1% triton released the remainder of the ATP in the cells (arrow on the right), probably that within intact chromaffin granules. The ATP level reached with the triton treatment gave the total ATP as 2,000 pmol (Fig. 6). Since the reaction mixture contained about 250,000 cells, the measured ATP content per chromaffin cell was 8 fmol. Under these experimental conditions, namely isethionate in place of chloride and low $[Ca^{2+}]_o$ (1 µM) in the external medium, while 10 to 16% of the ATP in the cells is liberated by

digitonin (10 or 20 μM), the average ATP released after high voltage
permeabilization of the cells amounted to 7% of the total ATP (n=4).
Figure 6 shows that digitonin treatment after high voltage permeabilization induced further ATP release. A careful comparison of ATP
release evoked by different numbers of discharges revealed that, with
our apparatus, no further release could be elicited by more than 5-10
discharges. It seems reasonable to propose that the observed differences are due to differences in the efficiency with which either
treatment removes the permeability barrier.

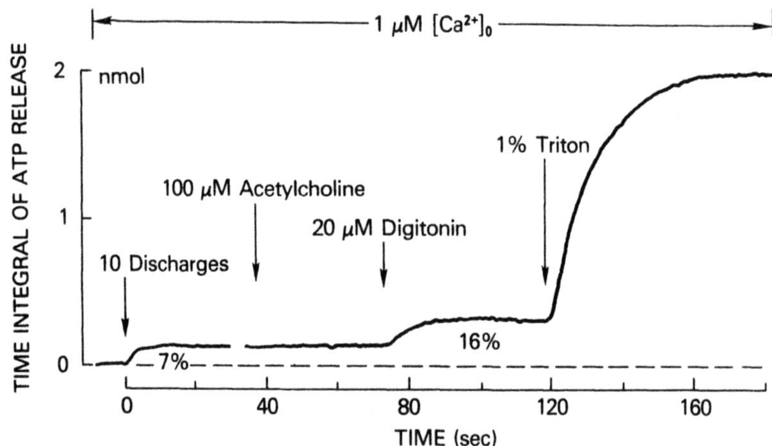

Fig. 6. Identification of the origin of the secreted ATP. Record
represents the time-integral of the ATP released from about 25,000
chromaffin cells in response to: application of 10 high voltage discharges at a frequency of 0.5 sec⁻¹ (arrow on the left); application
of 100 μM acetylcholine (second arrow from left to right); 20 μM
digitonin and, 1% triton X-100. Low [Ca²⁺]₀-reaction medium made with
isethionate in place of chloride ions.

ATP Release Evoked by Membrane Depolarization is Mediated by Activation of Voltage-Gated Calcium Channels

Nifedipine blocks voltage-gated Ca-channels in many excitable
cells. In the case of the chromaffin cells, nifedipine blocks stimulated catecholamine secretion[12] and ATP release elicited by a sudden
depolarization of the membrane (Fig. 7). For example, in the presence
of 1 μM nifedipine (B), an elevation of $[K^+]_0$ from 0 to 100 mM, keeping $[Na^+]_0$ plus $[K^+]_0$ constant in the modified Krebs solution, failed

to induce the ATP release observed in control experiments in the
absence of Ca-channel antagonist. The incomplete inhibition of the
ATP release observed in the presence of 0.1 μM nifedipine (75%, Fig.
7A) indicates that the affinity constant lies in the concentration
range from 0.1 to 1 μM nifedipine as in other preparations.[20]

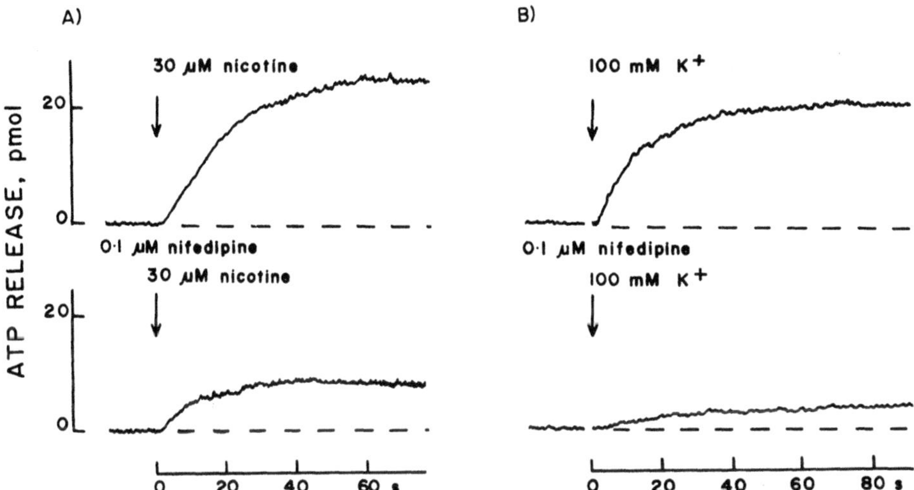

Fig. 7. Blockade of the ATP release induced by membrane depolarization
by nifedipine. Records represent the time-integral of the ATP released from 10,000 chromaffin cells. Upper records in (A) and (B)
represent the responses to 30 μM nicotine and a sudden elevation of
$[K^+]_o$ from 0 to 100 mM, respectively. For the records shown below,
identical medium plus nifedipine (0.1 μM) and about the same number of
cells. From[23]

Consistent with the results obtained with nifedipine were our
studies with Cd^{2+}, a potent blocker of the voltage-gated calcium
channel.[15] In the millimolar concentration range, Cd^{2+} strongly
inhibited ATP release evoked by K^+-induced membrane depolarization.

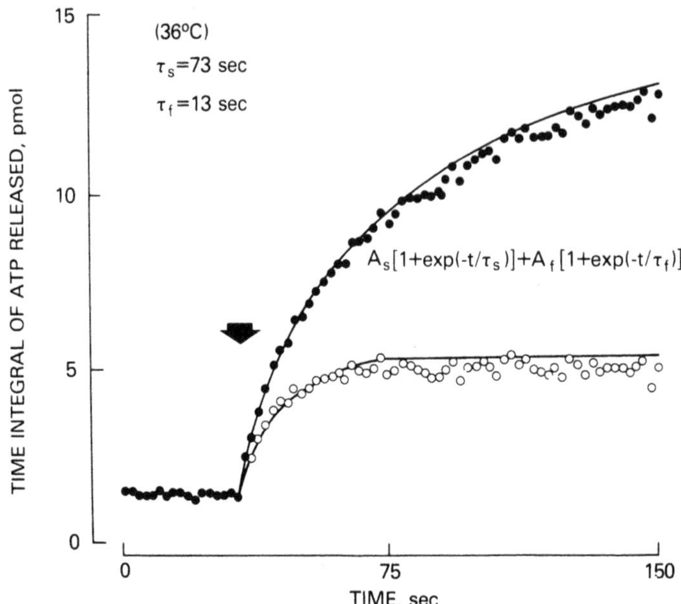

Fig. 8. Kinetic analysis of the acetylcholine-induced ATP release at
36°C. Time integral of the ATP released as a function of time. The
arrow indicates the addition of 10 μM acetylcholine. Filled circles:
mean ATP values. The curve representsthe least squares fit of the
function,

$$A_s[1+\exp(-t/\tau_s)] + A_f[1+\exp(-t/\tau_f)]$$

to the experimental points. Open circles: derived mean values
(measured values minus the corresponding level of slow exponential
component). Curve represents the least squares fit of the exponential
$A_f[1+\exp(-t/\tau_f)]$ to the points. Temperature of the reaction medium was
36°C. Values of the time constant as indicated. From[25]

Analysis of the Time Course of the Stimulated ATP Release

One important feature of our method is the temporal resolution
with which one can follow the release process. Shown in Fig. 8 is the
analysis of the early time course of the ATP release in response to
acetylcholine. Filled circles represent the time integral of the ATP
release. Each filled circle is the average of 1024 samples taken at a

sampling rate of 500 Hz. The curve was fitted to the experimental
points and was calculated as the sum of two exponentials. Open circles were calculated by subtracting from the measured values (filled
circles) the slow exponential.

From the fitted curves one can calculate the rates of ATP release at 36°C as time derivatives. Figure 9 shows the time course of
the rate of ATP release (double exponential curve starting from 0.8
pmol/sec). The temporal course of the slow rate of ATP release
(single exponential starting at 0.23 pmol/sec) is also shown for
comparison.

The simplest interpretation of this result is to consider that

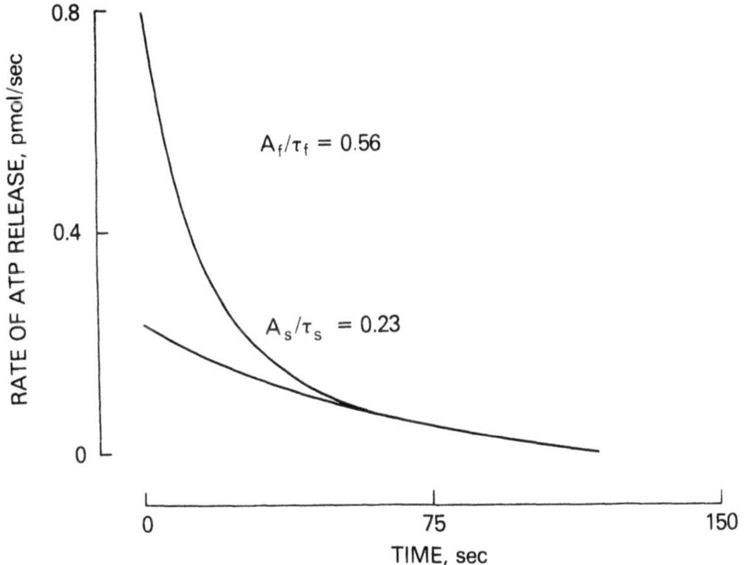

Fig. 9. Calculated rates of ATP release. Data from the fitted functions in Fig. 8. From[25]

the ATP released originates in two different pools. The fastest of the
two (τ_f=13 sec), may correspond to ATP released from a granular pool
close to the membrane. The other (τ_s=37 sec) may correspond to ATP
released by compound exocytosis from another pool, further away from
the membrane.

<u>Effect of Temperature on the Kinetic of the ATP Release</u>

Figure 10 shows the analysis of the time course of the acetylcholine-evoked ATP release at 18°C. The time course could be fitted using two exponentials as before, but the time constants increased

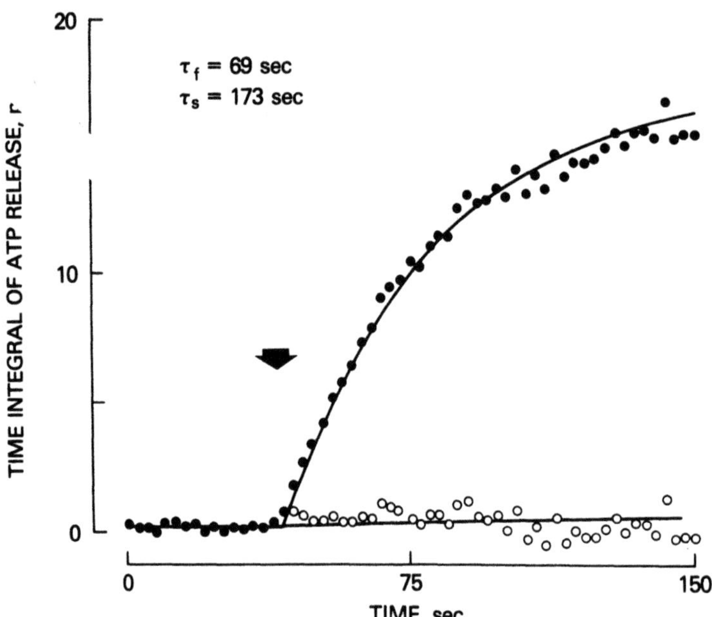

Fig. 10. Kinetic analysis of the acetylcholine-induced ATP release at 18°C. Time integral of the ATP released as a function of time. The arrow indicates the addition of 10 µM acetylcholine. Filled circles: mean ATP values. The curve represents the least squares fit of a double exponential function to the experimental points. Open circles: derived mean values (measured values minus the corresponding level of slow exponential component). Curve represents the least squares fit of a single exponential function to the points. Temperature of the reaction medium was 18°C. Values of the time constant as indicated. From[25]

from 13 to 69 sec for the fast release and, from 73 to 173 sec for the slow release. The activation energy, calculated from the effect of temperature on ATP release, was similar for both processes, about 18 kcal mole^{-1}degree^{-1}.

<u>Microscopic Characteristics of the ATP Release Process</u>

We have seen that the noise content of the optical signal increases during ATP secretion (Fig. 2, compare records a and b). This

excess noise is made up of three components: photomultiplier noise
(mainly shot noise), intrinsic noise of the luciferin-luciferase
reaction and noise due to the discrete nature of the ATP release.

To determine whether or not the excess noise is caused by the
discrete nature of the ATP release, the difference in variance,

$$\sigma^2 \text{ during} - \sigma^2 \text{ before}$$

acetylcholine stimulation, was calculated and plotted as a function of
the rate of ATP release in Fig. 11. The arrows indicate the maximum

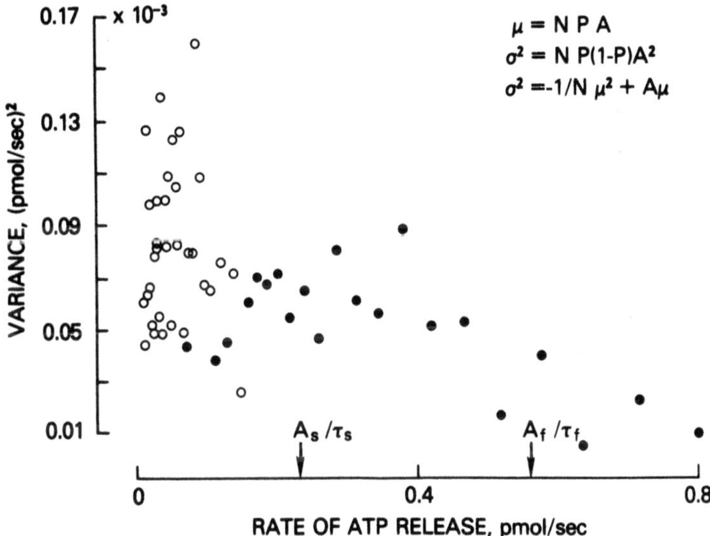

Fig. 11. Variance of the optical signal as a function of the rate of
ATP release. Each symbol represents the variance of the signal calculated from 1024 samples (sampling rate 500 Hz).

rates (A_s/τ_s and A_f/τ_f) and the open symbols were chosen to distinguish two possible parallel processes contributing to the measured
excess noise. After this arbitrary grouping of the data, two parabolic
functions were fitted to the experimental points and the following
parameters were obtained: ($N_s \times A_s$)=0.66 and ($N_f \times A_f$)=0.12 pmol/sec;
N_s=650 and N_f=1,600 events; A_s=20 and A_f=41 fmol/sec.

The frequency spectrum of the excess noise due to acetylcholine-
evoked ATP secretion is shown in Fig. 12. The experimental points

represent the average normalized power as a function of frequency.
Curves were calculated as

$$S(f)/S(f{=}0) = 1/(1 + (f/f_c)^2)$$

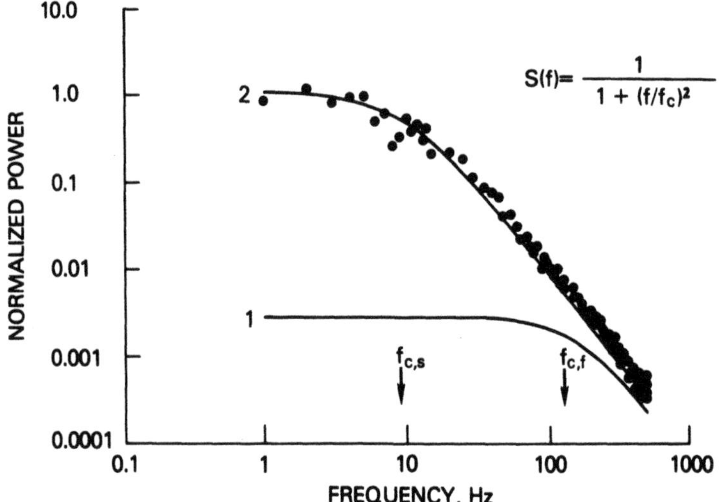

Fig. 12. Single Lorentzian power spectrum of the optical signal.
Filled circles represent the average of 32 points at each frequency up
to 45 Hz. For higher frequencies, four points were condensed. Curves
1 and 2 represent the least squares fit of the experimental averaged
values: 5 min after stimulation of the cells (1) and, 1 min after
stimulation of the cells (2).

$$f_{c,s} = 9.5 \text{ Hz}; \quad f_{c,f} = 150 \text{ Hz}.$$

where

$$S(f{=}0) = 2\, \sigma^2/\Pi\, f_c$$

and σ^2 represents the variance of the noise and f_c represents the
cutoff frequency. This function was used to fit the data obtained 5
min after the application of the agonist, when the rate of ATP release
is unmeasurable (1) and, 1 min after the stimulation of the cells (2).
The arrows indicate the corner frequencies, 9.5 Hz for the record 1
min after stimulation and, 109 Hz for the record 5 min after stimulation. The relaxation time constants for the unitary events $(1/\{2\Pi f_c\})$
were 16.7 and 1.5 msec, respectively.

Origin of the ATP Released by Permeabilization

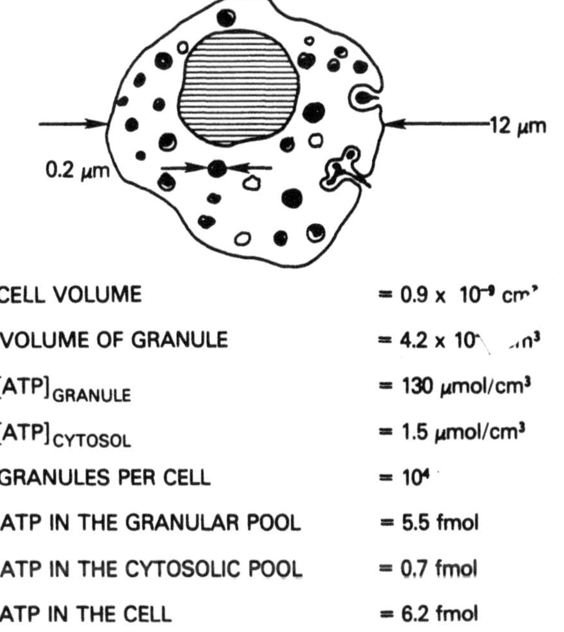

Fig. 13.   ATP distribution in chromaffin cells.

ATP is distributed both in secretory granules and the cytosol. The information provided by the cell permeabilization techniques used here showed that about 11% of the total ATP in the cells is found in the cytosol and, about 88.7% in the granular pool. Taking the volume of the granule as 4.2×10^{-15} cm^3, the ATP concentration in the granule as 130 mmol/cm^3, and 10^4 granules per cell,[28] the granular ATP content per cell is estimated as 6.2 fmol. Taking the cytosolic cell volume as 50% of the cell volume, i.e., 2.5×10^{-10} cm^3 and the cytosolic concentration of ATP as 1.5 μmol/cm^3, the ATP in the cytosol is estimated as 0.7 fmol per chromaffin cell, or less than 10% of the total ATP that can be extracted from the cell.

We have seen that in low Ca^{2+} (≤ 1 μM), close to 11% of the cellular ATP is liberated after treatment of the cells with digitonin. Digitonin solubilizes the plasma membrane of the chromaffin cell, leaving the organelles such as chromaffin granules, mitochondria and endoplasmic reticulum intact.[9,31] The permeabilized cell loses large molecules such as lactate dehydrogenase (LDH), but not DBH or catecholamines which are found in the granules. The high electric

field permeabilization step, on the other hand,[17,33] under the same
experimental conditions induced the release of about 7% of the cellular ATP. Application of high electric field pulses, a few microsec-
onds in duration, makes holes in the plasma membrane estimated to be 4
to 5 A^O in diameter. These holes do not allow LDH to escape.[17] Thus,
it can be assumed that luciferase cannot enter the permeabilized
cells, therefore, the detection of ATP takes place outside the cell
boundary, unlike the situation with digitonin permeabilized cells.
Furthermore, it is likely that in our reaction chamber not all the
cells are located at the critical angle required to break the membrane
dielectric,[17] thus lowering the efficiency of the technique as used
here. It seems logical to assume that the greater percentages obtained
with digitonin, namely 10 to 16% compared with 5 to 7% obtained with
the high voltage pulses, may be attributable to the more efficient
membrane removal by digitonin.

Comparison Between Acetylcholine-Stimulated ATP Release and Catechol-amine Secretion

We have shown that acetylcholine-evoked, Ca^{2+}-dependent ATP
secretion from chromaffin cells can be quantitatively measured using a
highly purified luciferin-luciferase preparation to assay released
ATP. With this new technique it is possible to monitor secretion with
a time resolution in the millisecond range.

Acetylcholine-stimulated ATP release shares many important
properties with the process of catecholamine secretion in chromaffin
cells. For example, i) both secretory processes are evoked by activation of acetylcholine receptors; ii) both processes depend on extracellular Ca^{2+} and are inhibited by Ca-channel blockers such as
nifedipine and, iii) both secretory processes are blocked by trifluoroperazine. The conclusion to be drawn from this comparison is
that ATP is secreted by exocytosis together with the catecholamines.
Thus, the technique can be used to monitor, on real-time, the process
of secretion of granule content.

Discrete Nature of the ATP Release

The simplest way to study the fluctuations in the optical signal
would be to stimulate the cells with a low dose of acetylcholine and
to monitor the fluctuations in the signal around its steady mean value
(Fig. 2). To extract the size of the unit event from a noise record

(inserts in Fig. 2), one can analyse the variance of the fluctuating
light signal. However, it should not be forgotten that the variance of
records, such as those depicted in Fig. 2, includes contributions from
three main sources. Two of them, instrumentation and intrinsic
luciferin-luciferase noise, constitute the background component. The
background noise is irrelevant and can be assumed to be uncorrelated
with the noise resulting from exocytotic activity. If so, its variance can be estimated in the absence of exocytosis (5 min after the
application of the acetylcholine in Fig. 2) and can be substracted
from the total. The time-dependent variance $\sigma^2(t)$ from N identical
exocytotic events is given by the expression,

$$\sigma^2(t) = N \{P(t) (1-P(t))\} A^2$$

where A is the size of the unitary event and, where σ^2 and the probability of the exocytotic event P are written as functions of time to
emphasize that the equation holds even when these quantities are
time-varying. Since the mean rate of release is

$$\mu(t) = N P(t) A,$$

the variance can be rewritten as

$$\sigma^2(t) = -1/N \mu^2 + A \mu,$$

which relates the two experimentally-derived functions $\sigma^2(t)$ and
$\mu(t)$. By curve fitting, therefore, the size of the unitary event A
and the number of events can be estimated.[26]

The analysis of the excess noise from stimulated chromaffin
cells revealed the discrete nature of the ATP release. The simple
spectrum obtained (see Fig. 12), which could be fitted by

$$S(f)/S(f=0) = 1/(1 + (f/f_c)^2,$$

is indicative of the discrete nature of the release process.[19] However, in order to gain useful information from this analysis, it is
necessary to develop a model.

Assuming that ATP release is the result of Poisson-distributed,
independent exocytotic events, the model used before to analyze the
quantal nature of the acetylcholine release at the end-plate[1] can be

applied here with minor modifications. Assuming that the shape of the
unitary event is also similar to that of the miniature end-plate
current, i.e., a fast rise to a peak followed by an exponential decay,
the time constant of the exocytotic event would be of the order of 10
msec. Since the size of the unit event A at 18°C was estimated as 20
pmol/sec and the relaxation time constant as 10 msec, the amount of
ATP released per event is estimated as 0.2 fmol. However, the ATP
content per granule (Fig. 13) calculated from morphological data is
only 0.55×10^{-3} fmol, three orders of magnitude smaller. The explanation for this discrepancy may be that, at the site of release the concentration of ATP is close to the concentration estimated for the core
of the granules. Therefore, the luciferin-luciferase reaction pro-
ceeds at a saturating ATP concentration, leading to local depletion of
luciferin. As ATP diffuses away from the exocytotic site and
luciferin is replenished, a new level of light emission is reached.
This new level, although a step higher than before the unitary addition of ATP, cannot be resolved with the sensitivity available (see
Fig. 1). Alternatively, each release event may mark the simultaneous
release of a substantial portion of the secretory granules in one
cell. Ultrastructural studies show that when a population of
chromaffin cells is stimulated, a given cell is observed to either
have secreted substantially, or not to have secreted at all. If each
"unitary event" actually represented 1000 granules,this would be about
10% of all the granules in a given cell, a reasonable value on both
biochemical and morphological grounds. The length release time (10
msec) is also consistent with this concept.

ACKNOWLEDGMENTS

The authors are pleased to thank Drs. I. Atwater, A. Stutzin,
R. M. Santos and R. Ornberg for comments and discussion. Thanks are
also given to Diane Seaton for preparation of the chromaffin cells,
B. Chidakel and A. Markowitz for electronic design and construction.

REFERENCES

1.   C. L. Anderson and C. F. Stevens, Voltage Clamp analysis of
          acetylcholine produced end-plate current fluctuations of
          frog neuromuscular junction, J. Physiol 235:655 (1973).
2.   W. H. Biggley, J. E. Lloyd, and H. H. Seliger, The spectral
          distribution of firefly light II, J. Gen. Physiol.
          50:1681 (1967).

3.   J. H. Brown and S. Brown-Masters, Does phosphoinositide hydroly-
     sis mediate inhibitory as well as excitatory muscarinic
     receptors? Trends Pharmacol. Sci. 5:417 (1984).

4.   G. V. R. Born, Changes in the distribution of phosphorous in
     platelet-rich plasma during clotting, Biochem. J. 68:695
     (1958).

5.   C. F. Code, Histamine in blood, Physiol. Rev. 32:47 (1952).

6.   M. J. Cormier, J. E. Wampler, and K. Hori, Bioluminescence:
     chemical aspects, Fortschr. Cem. Org. Naturst. 30:1 (1973).

7.   W. W. Douglas, Stimulus-secretion coupling: the concept and
     clues from chromaffin and other cells, Br. J. Pharmacol.
     34:451 (1968).

8.   W. W. Douglas and R. P. Rubin, The role of calcium in the
     secretory response of the adrenal medulla to acetylcholine.
     J. Physiol. 159:40 (1961).

9.   L. A. Dunn and R. W. Holz, Catecholamine secretion from
     digitonin-treated adrenal medullary chromaffin cells, J.
     Biol. Chem. 258:4989 (1983).

10. M. Eley, J. Lee, J. M. Lhoste, C. Y. Lee, M. J. Cormier, and P.
     Hemmerich., Bacterial bioluminescence, comparisons of bio-
     luminescence emission spectra, the fluorescence of
     luciferase reaction mixture and the fluorescence of flavin
     cations, (1970).

11. E. J. Forsberg, E. Rojas, and H. B. Pollard, Receptor
     enhancement of nicotine-induced catecholamine secretion may
     be mediated by phosphoinositide metabolism in bovine adrenal
     chromaffin cells, J. Biol. Chem. 261:4915 (1986).

12. A. G. Carcia, F. Sala, J. A. Reig, S. Viniegra, J. Frias, R.
     Fonteriz, and L. Gandia, Dihydropyridine BAY-K-8644
     activates chromaffin cell calcium channels, Nature 309:69
     (1984).

13. A. Greenberg and O. Zinder, Control of catecholamine secretion
     from isolated adrenal medullat cells, Cell Tissue Res.
     226:655 (1982).

14. H. D. Griffin, J. N. Hawthorne, and M. Sykes, A calcium require-
     ment for the phosphatidylinositol response following activa-
     tion of presynaptic muscarinic receptors, Biochemical
     Pharmacology 28:1143 (1979).

15. S. Hagiwara and L. Byerly, Calcium channel, Ann. Rev. Neurosci.
     4:69 (1981).

16. Y. Kidokoro and A. K. Ritchie, Chromaffin cell action potentials and thier possible role in adrenaline secretion from rat adrenal medulla, J. Physiol. 307:199 (1980).

17. D. E. Knight and P. F. Baker, Calcium-dependence of catecholamine release from bovine adrenal medullary cells after exposure to intense electric fields, J. Membrane Biol. 68:107 (1982).

18. S. B. Masters, T. K. Harden, and J. H. Brown, Relationships between phosphoinositide and calcium responses to muscarinic agonists in Astrocytoma cells, Molecular Pharmacology 26:149 (1984).

19. E. Neher and C. Stevens, Conductance fluctuations and ionic pores in membranes, Ann. Rev. Biophys. Bioeng. 6:345 (1977).

20. W. G. Nayler and J. D. Horowitz, Calcium antagonists: a new class of drugs, Pharmac. Ther. 20:203 (1983).

21. M. Oka, M. Isosaki, and J. Watanabe, Calcium flux and catecholamine release in isolated bovine adrenal medullary cells: effects of nicotinic and muscarinic stimulation, in: "Synthesis, storage and secretion of adrenal catecholamines," Advances in the Biosciences 36:29 (1980).

22. E. Rojas, H. B. Pollard, and Heldman, Real-time measurements of acetylcholine-induced release of ATP from bovine medullary chromaffin cells, FEBS Lett. 185:323 (1985).

23. E. Rojas, H. B. Pollard, and E. Forsberg, Optical detection of acetylcholine-evoked, calcium dependent ATP release from medullary chromaffin cells, Biophys. J. (submitted, 1986A).

24. E. Rojas, E. Forsberg, and H. B. Pollard, Muscarinic receptor activation potentiates nicotine-evoked ATP release from medullary chromaffin cells, FEBS Lett. (submitted, 1986B).

25. E. Rojas, A. Stutzin, H. B. Pollard, Microscopic characteristics of the acetylcholine-evoked ATP release from medullary chromaffin cells, Biophys. J. (submitted, 1986C).

26. F. J. Sigworth, Nonstationary noise analysis of membrane current, in: "Membrane, channels and noise," R. S. Eisenberg, M. Grank, and C. F. Stevens, ed., Plenum Press, N.Y. and London, p. 21 (1984).

27. K. E. Sussman and J. W. Leitner, Conversion of ATP into other adenine nucleotides within isolated islet secretory vesicles. Effect of cyclic AMP on phosphorous translocation, Endocrinology 101:694 (1977).

28. A. Ungar and J. H. Phillips, Regulation of the adrenal medulla,
     Physiol. Revs. 63:787 (1983).

29. B. Uvnas, The molecular basis for the storage and release of
     histamine in rat mast cell granules, Life Sci. 14:2355
     (1974).

30. H. Weil-Malherbe and A. D. Bone, The association of adrenaline
     and noradrenaline with blood platelets, Biochem. J. 70:14
     (1958).

31. S. P. Wilson and N. Kirshner, Calcium-evoked secretion from
     digitonin-permeabilized adrenal medullary chromaffin cells,
     J. Biol. Chem. 258:5994 (1983).

32. H. Winkler and E. Westhead, The molecular organization of adre-
     nal chromaffin granules, Neuroscience 5:1803 (1980).

33. U. Zimmermann, G. Pilwat, and F. Riemann, Dielectric breakdown
     of cell membranes, Biophys. J. 14:881 (1974).

GLUCOSE-EVOKED CHANGES IN $[K^+]$ AND $[Ca^{2+}]$ IN THE INTERCELLULAR

SPACES OF THE MOUSE ISLET OF LANGERHANS

E. Perez-Armendariz and I. Atwater

Laboratory of Cell Biology and Genetics
NIDDK, National Institutes of Health
Bldg 8, Rm. 403
Bethesda, MD 20892

INTRODUCTION

Microelectrode studies of the glucose-evoked burst pattern of
electrical activity in pancreatic β-cells have indicated that the
underlying mechanism involves the cyclic activation of
K-channels.[4,5,9,11,32] Furthermore, it is well documented that the
action potentials during the bursts result from the activation of two
voltage-gated membrane channels, Ca-channels[7,8,25,31,34] and K-channels.

Since β-cells comprise more than 80% of the endocrine part of a
single islet of Langerhans from mouse[13] and electrophysiological
experiments show that β-cells in the islet are electrically coupled,[15]
the extracellular concentrations of the two main ions involved in the
electrical activity, namely, Ca^{2+} and K^+, undergo small changes concomitant with the oscillations in electrical activity.[30]

The gap junctions between β-cells[27,28,29] presumably mediate
cell-to-cell coupling in the islet[22,24] whereas the tight junctions
may restrict diffusion through the tortuous spaces between the cells
in the islet. In this context, an important role has been attributed
to the gap junctions in the homeostasis of the tissue as a whole.[24]
Analogously, tight junctions or very close cell-to-cell contact, by

restricting intercellular diffusion, could play an important part in
the regulation of the composition of the extracellular medium within
the tissue.[30]

We report here the results of experiments designed to measure
the concentration of ionized potassium and calcium in the intercellular spaces of the islet of Langerhans from mouse ($[K^+]_I$ and
$[Ca^{2+}]_I$). Previous results indicate that the diffusion coefficients
for K^+ in the islet space is about one-half of that of K^+ in solution.[30] We have confirmed our earlier observations that in the presence of stimulatory levels of glucose, small oscillations in $[K^+]_I$
occur. These oscillations occurred at the frequency of the bursts of
electrical activity recorded from a nearby cell in the same islet.
Furthermore, maximum $[K^+]_I$ was reached towards the end of the active
phase of the bursts. $[Ca^{2+}]_I$ also changed but in this case the maximum
change was about half of that of $[K^+]_I$ and a minimum level was reached
towards the end of the active phase of each burst. Thus, during the
active phase of the burst there is K^+-accumulation and Ca^{2+}-depletion
in the intercellular spaces of the islets of Langerhans. In this
study, we have also explored the effects of sudden changes in glucose
concentration on $[K^+]_I$ and find that K^+ accumulation is closely correlated with spike frequency. We have also compared $[K^+]_I$ simultaneously in different regions of the same islet and found that while the
amplitude of the oscillations in $[K^+]_I$ varies, the time course is
synchronous throughout a single islet.

METHODS

Membrane Potential Measurements on Single Microdissected Mouse Islets of Langerhans

The electrophysiological methods which were used in this work
have been described before.[7] The volume of the chamber used was 0.04
cm^3 and the rate of perfusion was about 2.5 cm^3/min. When changing
solution at the stopcock,[21] the new solution reached the chamber 2 s
after the switch at the stopcock. The solution in the chamber was
electrically connected to earth potential with Ag/AgCl electrode via a
1000 Ohm resistor.

The modified Krebs solution used during the experiments had the
following composition (mM): 120 NaCl, 25 $NaHCO_3$, 5 KCl, 2.5 $CaCl_2$ and
1.1 $MgCl_2$ and was equilibrated with 95% O_2/ 5% CO_2, pH 0f 7.4 at 37°C.

<u>Cation-Selective Microelectrodes</u>

The method used to prepare these electrodes has been described in
a recent publication.[30] Briefly, microelectrodes were first silanized
with N,N-dimethyl-trimethylsilylamin (Fluka Chemical Corp.) vapor at
200°C. Next, the [K^+]-selective liquid membrane electrode was prepared by back-filling the microelectrode with the reference 100 mM KCl
solution first and, then sucking into the tip of the microelectrode a
column of 100 to 250 μm length of the liquid resin. The resistance of
the final electrode ranged between 0.5 and 1.5 Gohms. The Ca^{2+}-sensi-
tive electrodes were prepared in a similar fashion substituting the
K^+-resin for Ca^{2+}-resin (Fluka Chemical Corp.). The reference solution was 10 mM $CaCl_2$.

To measure the time constant of the ion-sensitive electrode
response to a voltage step, voltage pulses were applied to the solution in the chamber through a Ag/AgCl electrode. For an ion-sensitive
electrode in solution with a tip resistance of a 1.5 Gohms, 63% of the
applied voltage pulse was seen to take place in less than 0.2 s.

<u>Calibration of the Microelectrodes and Controls</u>

The cation-sensitive microelectrodes used here were calibrated
before and after the experiments in which penetration of the islet
tissue with the tip of the microelectrode was involved. To calibrate
the K^+-electrode a modified K-Krebs was prepared (150 mM K^+ in place
of Na^+) and mixed with Na-Krebs to give the desired [K^+]$_o$ keeping the
sum [K^+]$_o$ + [Na^+]$_o$ constant. From the Goldman-Hodgkin-Katz
equation[18,20] it is possible to obtain the following equation,

$$V_K = V_o + RT/zF \ln \{[K]_o + SC\,[Na]_o\} \quad (1)$$

where SC represents the selectivity ratio P_{Na}/P_K for the membrane
electrode; V_o is obtained from (1) when [K^+]$_o$ is 100 mM; R, T, z and F
have their usual meaning. Two parameters are needed to characterize
the ion-sensitive electrode response: V_o and SC. In the case of the
K^+-sensitive microelectrode, these two parameters were slightly affected by the experimental procedures as follows: Increasing the
calcium concentration from 2.5 to 7.5 mM changed V_o of the K^+ electrode from -115.7 to -116.4 mV and SC from 0.012 to 0.014. Another
experimental manipulation used to minimize any possible interferences
due to the process of insulin release, was to perform experiments at

27°C. At this temperature insulin release from mouse islets is
blocked.[6] It was found that lowering the temperature from 37° to 27°C
reversibly changes V_o from -119.9 to -114.6 mV and SC from 0.028 to
0.018. The K^+ selectivity of the electrode was increased in the low
concentration range at 27°C.

For the Ca^{2+}-electrode, $CaCl_2$ solutions in the concentration
range from 1 to 10 mM were used to calibrate the electrode. The slope
in this range was 30 mV for a 10-fold change in Ca^{2+}. The ratio
P_{Mg}/P_{Ca} ranged from 0.001 to 0.002.

Although the cation-sensitive microelectrode potentials are
related to the cation concentration by the Goldman-Hodgkin-Katz equation, for small changes in $[K^+]$ (or $[Ca^{2+}]$) the time course of the
$[K^+]$-sensitive microelectrode response, V_K (or V_{Ca}), is similar to the
time course of $[K^+]$ (or $[Ca^{2+}]$). Whence, the records of the cation-
sensitive electrodes in the figures are good descriptions of the time
course of the changes in concentration of the corresponding cation.

Estimation of the Penetration Depth

A piece of acinar tissue with a semi-dissected islet was pinned
to the silicon rubber base of the chamber. At this time the $[K^+]$-
sensitive electrode was introduced into the islet tissue. Finally, a
β-cell was impaled with a high resistance microelectrode and the
membrane potential (V_m) recorded.

Each electrode was held in a perspex holder by a 3-mm internal
diameter coil spring giving interference fit. The microelectrode
holders were carried by micromanipulators. One manipulator was
mounted vertically (for the $[K^+]$-sensitive electorde), the other (for
the intracellular electrode) at an angle of 7° from vertical to allow
the electrodes to be brought close together without interference
between their carriers. The average penetration depth, estimated using
the calibrated micrometer on the micromanipulator, was 60 μm.

Apparent K⁺-Diffusion Coefficient in the Intercellular Islet Spaces in the Absence of Glucose

If the K^+ exchange between the intercellular space and the external solution is diffusion limited, then an apparent diffusion coefficient can be estimated using a diffusion model to generate the time course of $[K^+]_I$ during a change of $[K^+]$ in the bath, $[K^+]_o$. Although the shape of the islets examined was not spherical (largest islet diameter ranged from 150 to 250 μm; smallest diameter ranged from 100

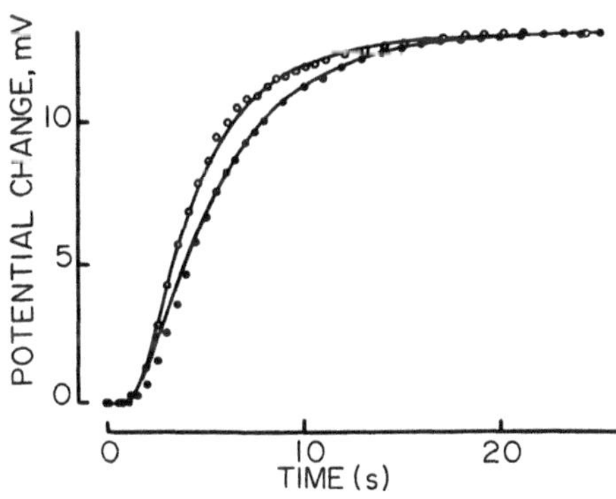

Fig. 1. Comparison between the experimental and theoretical $[K^+]$-sensitive microelectrode responses. Continuous curves represent calculated responses. Points represent experimental data obtained with the electrode on the surface of the islet (open circles) and with electrode at a depth of 28 μm into the islet tissue (filled circles). Experimental points obtained in the islet were normalized to the voltage response of the electrode outside the islet tissue shown in the upper part. Change in $[K^+]_o$ from 5 to 10 mM. Islet in the absence of glucose at 27°C. Parameters for calculated curves: thicknesses of the unstirred layer equal to 100 μm; $D_{K,o} = 1.83 \times 10^{-5}$ cm²/s; $D_{K,I} = 0.9 \times 10^{-5}$ cm²/s.

to 150 μm), the model proposed by Perez-Armendariz et al.[30] considers radial diffusion into a spherical islet (180 μm radius) through a 100 μm thick unstirred layer. Fig. 1 illustrates the analysis of an experiment (in which $[K^+]_0$ was augmented from 5 to 10 mM) in terms of this diffusion model.

Fig. 1 shows the experimental values (o,●) and the theoretical time course of the $[K^+]$-sensitive electrode response. The continuous curves were calculated using the model to fit the experimental points. It may be seen that, with the tip of the microelectrode placed on the surface of the islet (0), the theoretical time course is similar to the time course of the $[K^+]$-sensitive microelectrode response (time to 63% of the maximum response, 5.1 s). To facilitate the comparison the response with the tip 28 μm within the islet tissue (●) and the pre-dicted time course are also shown in Fig. 1 (time to 63% of the re-sponse, 6.1 s). The predicted response was calculated adjusting the apparent diffusion coefficient for K^+ representing the diffusion coefficient in the intercellular spaces of the islet. The apparent diffusion coefficient from the fit is 0.9×10^{-5} cm^2/s.

An alternative method to estimate the apparent diffusion coeffi-cient follows from its definition,[10]

$$D_{K,I} = \Delta\, r^2 / \Delta t \quad (2)$$

where Δr is the radial distance from the tip of the electrode to the surface and Δt is the difference in time to achieve 63% of the re-sponse with the tip inside the islet minus the time taken when the tip is placed at the surface. For the experiment illustrated in Fig. 1, this difference of 1 s, must represent the time taken for K^+ to dif-fuse into the islet through the intercellular clefts. As Δr is 28 μm the apparent diffusion coefficient $D_{K,I}$ is calculated as 0.78×10^{-5} cm^2/s. For another experiment not shown, Δr was 56 μm and Δt was 4.2 s. The apparent diffusion coefficient $D_{K,I}$ is calculated as 0.75×10^{-5} cm^2/s.

Oscillatory Changes in the Intercellular Concentration of Potassium and Calcium in the Presence of Glucose

In the presence of glucose, cyclic changes in β-cell membrane

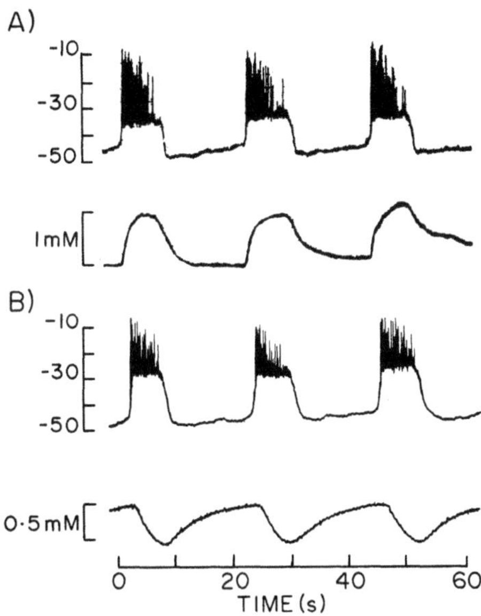

Fig. 2. Simultaneous recordings of the intracellular potential and
the cation-sensitive microelectrode potential. (A) Upper record
represents the membrane potential record from a cell about 20 μm from
the surface of the islet. V_K represents the [K$^+$]-sensitive microelectrode potential record made with the tip at a depth of about 65
μm. The microelectrode tips were separated by about 110 μm. Islet
perifused with Krebs solution plus 11 mM glucose at 37 °C. (B) Upper
record represents the membrane potential record from a β-cell in
another islet. V_{Ca} represents the Ca^{2+}-sensitive electrode responses.
Islet in the presence of 11 mM glucose at 37 °C.

K$^+$-permeability occur.[7] According to Atwater[2] the time spent in the
active phase depends on the Ca^{2+} entry and accumulation during the
active phase of the bursts and the stimulation of the [Ca^{2+}]$_I$-acti-
vated K$^+$-permeability (which causes membrane hyperpolarization and
termination of the burst). The oscillatory activation of voltage-
gated Ca-channels and K-channels during the action potentials occurring over the active phase of the bursts causes a net influx of Ca^{2+}
and a net efflux of K$^+$. Given the small volume of the extracellular
space and restricted intercellular diffusion, small depletions and
accumulations of these ions can be expected to occur.

Fig. 2 depicts the results of two typical experiments designed to
measure the changes in $[K^+]_I$ (Fig. 2A) or $[Ca^{2+}]_I$ (Fig. 2B) induced by
11 mM glucose. As shown in the upper part of Fig. 2A, the cell responded to 11 mM glucose with the characteristic burst pattern of
electrical activity. The $[K^+]$-sensitive microelectrode response, V_K
exhibits cyclic changes in phase with the slow waves of the burst
pattern.

Oscillations in V_K, similar to those shown in Fig. 2 were recorded from 25 islets and in all instances V_m and V_K always had the
same periodicity. The oscillations persisted at 27°C but ceased when
glucose was removed from the perifusion medium (see Fig. 4).

The time relationship between the electrical activity and $[Ca^{2+}]_I$
may be compared in Fig. 2B which shows two segments of the simultaneous records of V_m (upper part) and V_{Ca} (lower part) from another islet
in the presence of 11 mM glucose. It may be seen that the signal
(equivalent to a small concentration decrease of about 0.5 mM in the
region of 1.6 mM Ca^{2+}) reaches its minimum value at the end of the
active phase of the burst. The return to basal V_{Ca} is slower than the
abrupt return of V_m to the silent phase potential.

It is clear from Fig. 2B that the time course of the record of
the $[Ca^{2+}]$-sensitive microelectrode response is almost the mirror
image of the response of the $[K^+]$-sensitive microelectrode (Fig. 2A),
both oscillations being in phase with the bursts of electrical activity evoked by 11 mM glucose. Furthermore, the size of the concentration changes shows the expected balance of charge, the net Ca^{2+} loss
from the intercellular space amounts to about 0.5 mM and the corresponding accumulation of K^+ amounts to about 1 mM.

Effects of External Calcium on Glucose-Induced Oscillatory Changes in $[K^+]$

The burst pattern of glucose-induced electrical activity is
affected by $[Ca^{2+}]_o$. The duration of the silent phase is increased
and that of the active phase reduced. The size of the membrane potential change in the transition from silent to active phase is also
augmented.[5]

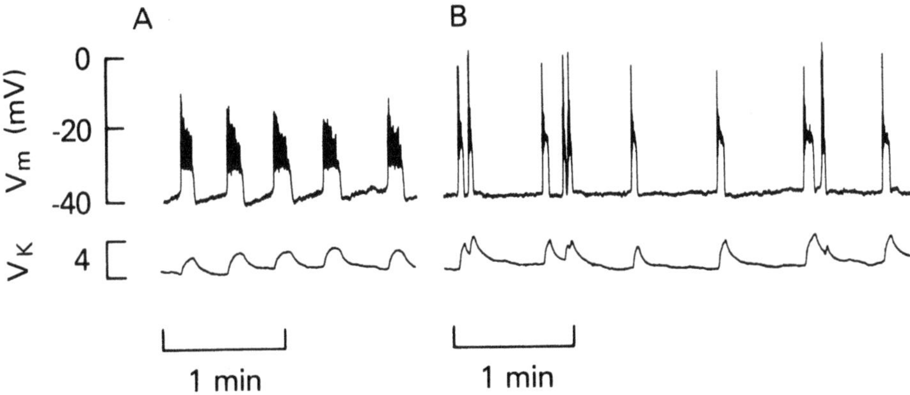

Fig. 3. Simultaneous recordings of V_m and V_K in the presence of
normal and high Ca^{2+}. External Ca^{2+} was increased from 2.56 for part
A to 7.5 mM 2 min before the start of the record shown in part B. The
analysis of the time course of V_K after bursts in part B taking the
initial phase of each burst as time zero, revealed that the curves
represent a single exponential decay.

Fig. 3 illustrates the records of V_m (upper part) and V_K (lower
part) made in the presence of 11 mM glucose Krebs solution with 2.6
(Fig. 3A) and 7.5 mM $CaCl_2$ (Fig. 3B) at 27°C. It may be seen that the
oscillatory changes in V_K are in phase with the bursts of electrical
activity. Furthermore, the rising phase of V_K, which commences with
the spike activity, is steeper in the presence of high Ca^{2+} when the
spike frequency is increased. $[K^+]_I$ reaches a peak value at the end
of the active phase of the bursts and, then, it decays towards the
base line during the silent phase. Even though the oscillations shown
in Fig. 3 are in phase, the time course of the V_m and V_K is fundamentally different. While the return to the silent phase potential in
the V_m record takes less than 1 s, the average time for an e-fold
decrease in V_K towards the base line is 6 s. The true time constant
(time for an e-fold decrease in V_K) calculated by fitting an
exponentially decaying function to V_K gave a mean value of 6.4 s.

If one considers that the brief burst of spikes represents a
sudden K^+ loading of the intercellular space, it is possible to estimate the apparent diffusion coefficient from the time course of the V_K
signal during the silent phase. Assuming that this time constant

represents the time constant of K^+-unloading of the intercellular
space then, using the time constant and the penetration depth, $D_{K,I}$
($=\Delta r^2/\Delta t$) is calculated to be 0.7×10^{-5} cm^2/s for the experiment illustrated in Fig. 3B.

Effect of Glucose Removal on $[K^+]$ in the Intercellular Space

The amplitude of the K^+-sensitive electrode response depends on
the penetration depth and on the local volume at the tip of the electrode. Shown in Fig. 4 are the records of the responses of two K^+-
sensitive microelectrodes implanted in the same islet at different
depths, i.e., 40 (middle record) and 80 μm (lower record). In the
presence of 11 mM glucose, the two recordings are synchronous although
the amplitude of the response from the microelectrode impaled deeper
in the islet tissue was larger. The removal of glucose (at the time
indicated in the upper trace) caused a rapid cessation of the oscillations in $[K^+]_I$. In the absence of glucose the level of K^+ increased
somewhat over the lowest value during the oscillations and remained at
an intermediate value between the maximum and the minimum recorded
during the oscillations in the presence of 11 mM glucose. The return

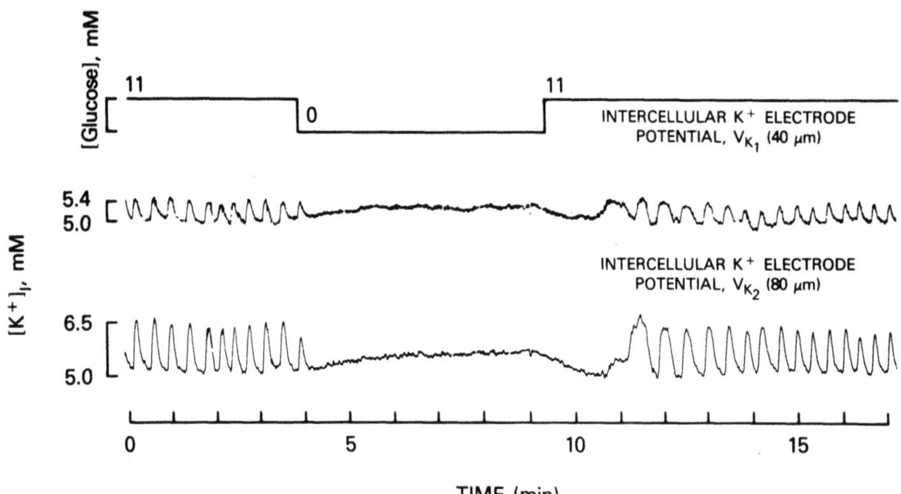

Fig. 4. Simultaneous recordings of the V_K responses from two
K^+-sensitive microelectrodes at two different depths. Mean value of
$\Delta[K^+]_I = 0.82 \pm 0.61$ (n = 13) and $\Delta[Ca^{2+}]_I = 0.58 \pm 0.1$ (n=5). Temperature 34°C.

40

to 11 mM glucose caused a rapid decrease and then a return to the
oscillatory pattern of changes in $[K^+]_I$ (n=5).

Effect of 16 mM Glucose Under Steady State Conditions

Fig. 5 shows two segments of the V_K record taken under conditions
in which the islet had been exposed for 30 min first to 11 mM glucose
(Fig. 5A) and then to 16 mM (Fig. 5B) while the K^+-sensitive microelectrode was kept in the same position. While the amplitude of the
oscillations remained almost constant (compare Fig. 5A and 5B), the
duration of each cyclic response increased from 13 (Fig. 5A) to 17 sec
(Fig. 5B). It is interesting to notice that in the presence of 16 mM
glucose, the decay in $[K^+]_I$ from the peak value (_ca._ 7 mM) to the
minimum level (5 mM) was clearly biphasic.

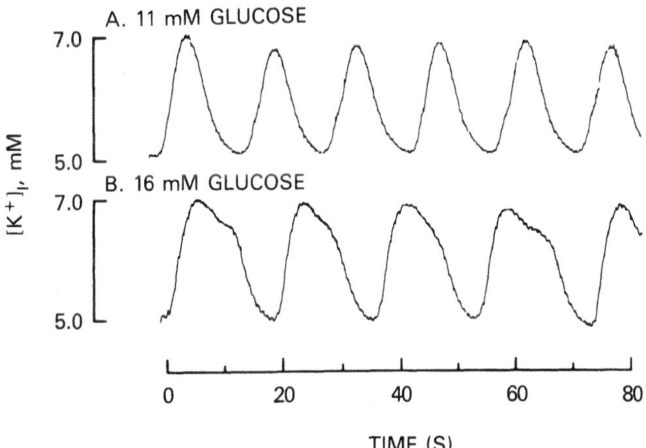

Fig. 5. Steady state oscillatory changes in K^+ in the intercellular
space in the presence of 11 (A) and 16 (B) mM glucose. Penetration
depth estimated as 87 μm. Temperature 36°C.

Compairson Between Spike Frequency and Potassium Concentration

The results presented so far show a close correlation between the
effects of glucose on electrical activity and the changes in $[K^+]_I$.

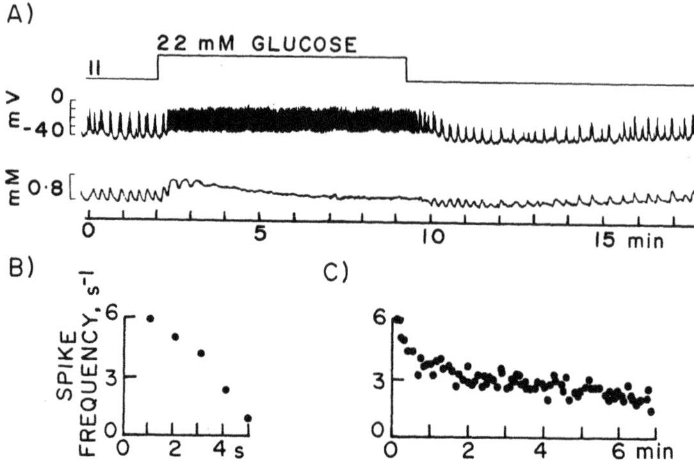

Fig. 6. Effect of a sudden change in glucose concentration from 11 to 22 mM on membrane potentials and on K^+ in the intercellular space. (A) K^+-sensitive microelectrode tip at a depth of about 53 μm. Mean amplitude of the $[K^+]_I$ oscillations 5.4 mM (11 mM glucose). (B) Spike frequency aling the active phase during the bursts recorded in the presence of 11 mM glucose. (C) Spike frequency along the continuous electrical activity evoked by 22 mM glucose. Temperature 34°C.

It is well documented that decreasing the concentration of glucose from 11 mM to zero induces a cessation of the electrical activity. Similarly, the oscillations in $[K^+]_I$ also ceased (Fig. 4). Augmenting the concentration of glucose from 11 to 16 mM is known to increase the duration of the bursts of electrical activity. Similarly, the duration of the oscillations in $[K^+]_I$ was increased (Fig. 5). In order to determine the relationship between the spike electrogenesis and $[K^+]_I$ we have calculated the spike frequency along the bursts at 11 mM and the spike frequency along the continuous electrical activity induced by 22 mM glucose and compared these data with the $[K^+]_I$ as measured by the K^+-sensitive microelectrode (Fig. 6). After switching to 22 mM glucose there is a rapid increase in $[K^+]_I$ to a value higher than the peak value recorded in the presence of 11 mM Glucose. In this experiment, a few small oscillations are apparent at the beginning of exposure to high glucose. Thereafter, $[K^+]_I$ slowly decreases towards a

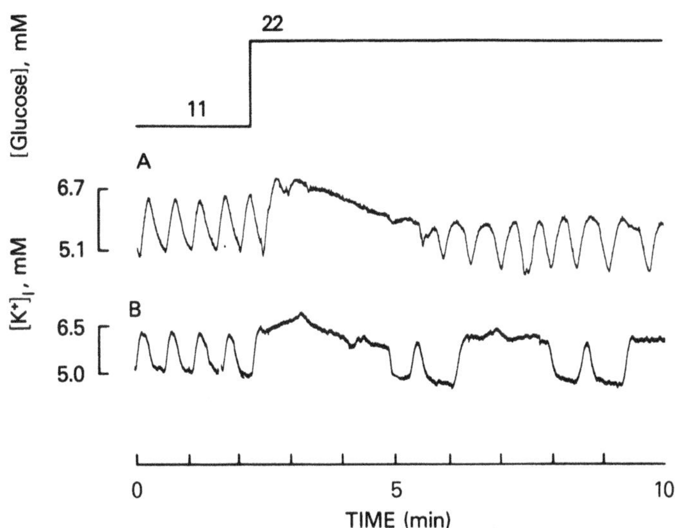

Fig. 7. Appearance of slow K$^+$ oscillatory changes in the presence of high glucose. (A) Mean amplitude of the oscillations in $[K^+]_I$ 6.2±0.08 mM (11 mM glucose) and 6.5 mM (22 mM glucose). Temperature 34°C.

value intermediate between the maximum and minimum value recorded during the oscillations in the presence of 11 mM glucose. The return to 11 mM glucose was accompanied by a return to the oscillatory pattern of changes in $[K^+]_I$. Fig. 6B shows that the spike frequency along an individual burst of activity (in the presence of 11 mM glucose) decreased from 6 to 1 spike/sec within about 5 sec. Fig. 6C shows that the spike frequency in the presence of 22 mM glucose decreased from 5 spikes/sec at the onset of continuous activity to 2 spikes/sec after several minutes.

In 50% of the islets used for measurement of V_K in the presence of 22 mM glucose, oscillations appeared in the V_K records after an initial phase of 2 to 5 min without oscillations in V_K. Two results of this type are shown in Fig. 7.

In the presence of 11 mM glucose both islets gave oscillatory responses. A sudden elevation of the glucose level to 22 mM induced a sustained high level of $[K^+]_I$. After about 3 min the oscillations were re-established. In some experiments, the oscillations were

regular, the shape of the V_K responses resembling those recorded in
the presence of 16 mM glucose (compare Fig. 7A with Fig. 7B). Elec-

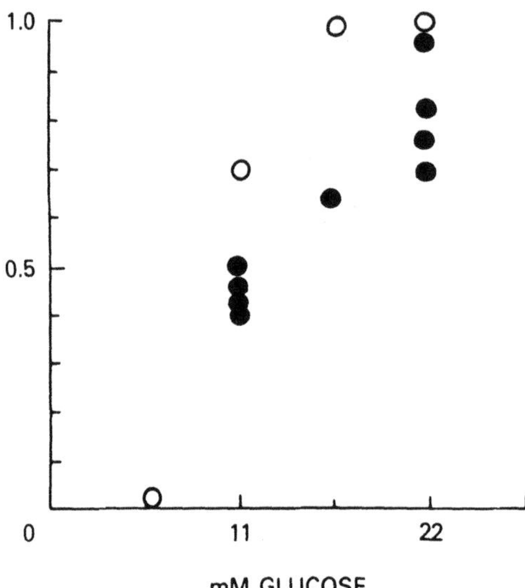

Fig. 8. Comparison of effects of glucose on the fraction of time in
the active phase (o) and the fraction of time in the high $[K^+]_I$ (●).
Data for the fraction of time in the active phase from Rosario et al.
(this book). $[K^+]_I$ as a function of glucose was measured in 5 experiments.

trical activity recorded in islet cells during the first 10 minutes of
exposure to 22 mM glucose typically exhibited long lasting bursts and
often showed irregular patterns.

Fig. 8 shows a comparison of the effects of glucose concentration
on the fraction of time in the active phase (open symbols) and the
fraction of time in high $[K^+]_I$ (filled symbols). The mid-point for
both curves is similar (just over 11 mM glucose), and the maximal
response is reached around 28 mM (see curve relating the fraction of
time in the active phase to glucose concentration).

DISCUSSION

Ion sensitive electrodes have been used to document the interrelationship between cell function and ion concentration in the restricted extracellular space surrounding most cells in in their normal
physiological habitat.[33,35] Transmembrane ion fluxes have also been

studied using ion-sensitive microelectrodes in the extracellular space
of ventricular muscle.[14]

In an earlier study, we reported the first use of cation-sensitive microelectrodes in islets of Langerhans and showed that $[K^+]_I$ was
oscillating in synchrony with the electrical burst pattern.[30] In that
study, we showed that the K^+ exchange between the intercellular space
in the islet and the external medium is diffusion limited. Because of
this restriction, the extra K^+ efflux, during each burst of glucose-
induced electrical activity, induces accumulation of K^+ in the intercellular space $[K^+]_I$. The changes in $[K^+]_I$ are oscillatory and synchronous with the burst pattern and the K^+ fluxes during electrical
activity could be estimated from the time course of the changes in
$[K^+]_I$. We proposed in that study that membrane parameters responsible
for the threshold of the response to glucose could be modulated by the
changes in $[K^+]_I$.[30]

In the present study we extend our observations with K^+-
sensititive electrodes to include measurements of $[K^+]_I$ at various
glucose concentrations and of the transient changes accompanying
sudden changes in glucose concentration. We have also used Ca^{2+}-
sensitive electrodes to show the expected concomitant oscillations in
$[Ca^{2+}]_I$ during the burst pattern. The main experimental finding is
that Ca^{2+} exchange between the intercellular space in the islet and
the external medium is also diffusion limited. Thus, the extra K^+
efflux and the extra Ca^{2+} influx, during each burst of electrical
activity, simultaneously induce Ca^{2+} depletion and K^+ accumulation in
the intercellular space, the overall charge transfer in and out of the
intercellular space being zero.

Apparent Diffusion Coefficient for K^+ Within the Islet Tissue

The apparent diffusion coefficient for K^+, D_I^K, could be estimated
from two different experimental protocols. Firstly, externally applied changes in $[K^+]_O$ which induced changes in $[K^+]_I$ within the islet
tissue. Solving the diffusion equation assuming free-diffusion of K^+
in the unstirred solution around the islet and an apparent diffusion
coefficient about half that of the K^+ in solution for K^+ diffusion
within the islet tissue could reproduce the measured changes in $[K^+]_I$
(see Fig. 1). Secondly, since the intercellular space is loaded with
K^+ by the cells during the active phase of the bursts, the time con-

stant for the return of $[K^+]_I$ to its level at the end of the silent phase of the bursts can be used to estimate D_I^K. This analysis gives a value for the diffusion coefficient of about half that in free solution.[30]

Oscillatory Changes in $[K^+]_I$ and $[Ca^{2+}]_I$

In the presence of glucose (between 11 and 16 mM), under steady state conditions, small oscillatory changes in both $[K^+]_I$ and $[Ca^{2+}]_I$ were detected. While $[K^+]_I$ increased, $[Ca^{2+}]_I$ decreased along the active phase of the bursts of electrical activity recorded intracellularly from a cell in the same islet. Maximum $[K^+]_I$ and minimum $[Ca^{2+}]_I$ were reached at the end of the active phase of the bursts. On average, the K^+ gain amounted to 1 mM and the Ca^{2+} loss to 0.5 mM. Furthermore, the amplitude of these oscillations did not change when increasing glucose from 11 to 16 mM. Membrane potential and spike frequency at the commencement of the bursts in 16.7 mM glucose are usually the same as at the beginning of the bursts in 11 mM glucose.[5]

During a sudden increase inglucose concentration, when the spike frequency remains high for more than a few seconds, the accumulation of K^+ is accentuated. Furthermore, the decay in spike frequency is closely correlated to the decay in $[K^+]_I$. Thus, the change in extracellular ion concentration reflects spike frequency of the surrounding cells. During a sudden decrease in glucose concentration, when the spike activity ceases, there is a transient depletion of K^+ for a few minutes, suggesting that active mechanisms may be involved and may be more slowly turned off than the electrical activity. The steady-state level of $[K^+]_I$ in the presence of 22 mM glucose or in the absence of glucose is about 30% higher than the minimum value recorded during the oscillations observed in 11 mM glucose.

Evaluation of K^+ and Ca^{2+} Fluxes

Under steady-state conditions, the concentration of K^+ in the islet intercellular space (now on referred to as v) is essentially determined by the relative size of four K^+ fluxes (per unit membrane area facing v): $_KJ_{c,v}$ from the β-cell to v, $_KJ_{v,c}$ from v into the β-cells, $_KJ_{o,v}$ from outside to v and, $_KJ_{v,o}$ from v to oustide the islet. It is possible to estimate $_KJ_{c,v}$ from available K^+-tracer flux data (Henquin, 1973) and from morphometric data.[13] In the absence of glucose, with a constant β-cell membrane potential the rate of

$^{42}K^+$-outflow is about 5×10^{-4} s^{-1} in rat islets[19] and 4×10^{-4} for 86Rb outflow in mouse islets.[12] The largest islet used in the present work had a volume of 2.3×10^{-5} cm^3. Assuming that 64% of this volume represents endocrine cells,[13] the β-cell islet volume is calculated as 1.5×10^{-5} cm^3. Taking the intracellular potassium concentration as 10^{-4} mole/cm^3,[7] the K^+ efflux per islet becomes 0.75 pmole/s (= volume of the β-cell x potassium concentration in the cells x rate of tracer K^+-outflow). Taking the β-cell radius as 6 μm, the β-cell membrane area is calculated as 4.5×10^{-6} cm^2 and thus, $_KJ_{c,v}$ is estimated as 10 pmole/cm^2 s.

Spike activity in β-cells is mainly due to the activation of voltage-gated Ca-channels and K-channels.[2,7,9,25] Depending on the frequency, a train of action potentials along the active phase of the bursts may induce K^+ accumulation and Ca^{2+} depletion in a restricted intercellular space.[17]

Assuming that in the presence of glucose $_KJ_{c,v}$ (through K-channels) is reduced,[1,9] the extra efflux from the cells during the action potentials (now on referred to as $_KJ^*_{c,v}$) would induce changes in $[K^+]_I$. One may calculate the rate of change of $[K^+]_I$ as

$$v/A \; d[K^+]_{I,t}/dt = {}_KJ^*_{c,v} \quad (3)$$

where v/A is the volume to surface ratio or effective width of the intercellular space for the islet tissue. The underlying assumption here is that the passive fluxes (presumably $_KJ_{c,v}$, $_KJ_{v,o}$ and $_KJ_{o,v}$) and the K^+ uptake through the Na^+/K^+ pump are constant in the presence of glucose.[23] From time records of V_K it is possible to calculate the extra K^+-efflux when dV_K/dt is zero (i.e., $d[K^+]_{I,t}/dt$ equal to zero). Taking the initial rate of increase in $[K^+]_I$ as 10^{-6} mole/cm^3 s (see Fig. 2A), for an effective width of the intercellular space of 4×10^{-6} cm, the extra K^+-efflux is calculated as 4 pmole/cm^2 s (= 10^{-6} mole/cm^3 s x 4×10^{-6} cm).

In rat islets, in presence of glucose (from 11 to 22 mM), the average $^{42}K^+$-efflux is about 4 pmole/cm^2 s.[19] The agreement between the tracer flux value and the value estimated from the K^+-sensitive electrode measurements, supports the hypothesis that, in the presence of glucose, there is a complete inhibition of efflux $_KJ_{c,v}$ presumably due to the inhibition of the glucose-blockable K^+-channel (see

Ashcroft <u>et al</u>., Chapt. 4 in this book) and that the increase in $[K^+]_I$ is caused by the activation of voltage-gated K-channels and/or $[Ca^{2+}]_I$-activated K-channels.[9]

As the initial spike frequency along the active phase of each burst ranges between 3 to 5 spkes/s,[3] the extra K^+-efflux is calculated as 0.8 to 1.3 pmole/cm^2 per spike. Thus, due to the balance of charge in the intercellular space v, the corresponding Ca^{2+} influx should be half as large, i.e., 0.4 to 0.65 pmole/cm^2 per action potential. Similarly, from above estimated K^+ efflux (10 pmole/cm^2 s) the associated Ca^{2+} influx should be about 5 pmole/cm^2 s. It is proposed that the spaces between β-cells constitute a restricted diffusion system where K^+ accumulation and Ca^{2+} depletion occur during the repetitive firing of action potentials as a consequence of the activation of voltage-gated potassium and calcium channels. It is also supposed that the ion content of the islet intercellular space reflects β-cell activity as a consequence of electrical coupling between β-cells.

ACKNOWLEDGEMENTS

The authors are pleased to thank Eduardo Rojas for his constant and enthusiastic support.

REFERENCES

1. F. M. Ashcroft, D. E. Harrison, and S. J. H. Ashcroft, Glucose induces closure of single potassium channels in isolated rat pancreatic β-cells, <u>Nature</u> 312:446 (1984).
2. I. Atwater, Control mechanisms for glucose induced changes in the membrane potential of mouse pancreatic β-cell, <u>Cienc. Biol.</u> (Portugal) 5:299 (1980).
3. I. Atwater and P. M. Beigelman, Dynamic characteristics of electrical activity in pancreatic β-cells. Effects of calcium and magnesium removal, <u>J. Physiol.</u> (Paris) 72:769 (1976).
4. I. Atwater, C. M. Dawson, B. Ribalet, and E. Rojas, Potassium permeability activated by intracellular calcium ion concentration in the pancreatic β-cell, <u>J. Physiol.</u> 288:575 (1979).
5. I. Atwater, C. M. Dawson, A. Scott, G. Eddlestone, and E. Rojas, The nature of the oscillatory behaviour in electrical activ-

ity from pancreatic β-cell, <u>J. Horm. Metab. Res.</u> (suppl) 10:100 (1980).

6. I. Atwater, A. Goncalves, A. Herchuelz, P. Lebrun, W. J. Malaisse, E. Rojas, and A. Scott, Cooling dissociates glucose-induced insulin release from electrical activity and cationic fluxes in pancreatic islets, <u>J. Physiol.</u> 348:615 (1984).

7. I. Atwater, B. Ribalet, and E. Rojas, Cyclic changes in potential and resistance of the β-cell membrane induced by glucose in islets of Langerhans from mouse, <u>J. Physiol.</u> 278:117 (1978).

8. I. Atwater, B. Ribalet, and E. Rojas, Mouse pancreatic β-cells: tetraethylammonium blockage of the potassium permeability increase induced by depolarization, <u>J. Physiol.</u> 288:561 (1979).

9. I. Atwater, L. Rosario, and E. Rojas, Properties of the Ca-activated K-channel in pancreatic β-cells, <u>Cell Calcium</u> 4:451 (1983).

10. J. Crank, The mathematics of diffusion, Clarendon Press, Oxford (1956).

11. D. L. Z. Cook and C. N. Hales, Intracellular ATP directly blocks K^+ channels in pancreatic β-cells, <u>Nature</u> 311:271 (1984).

12. C. M. Dawson, P. C. Croghan, I. Atwater, and E. Rojas, Estimation of potassium permeability in mouse islets of Langerhans, <u>Biomedical Research</u> 4:389 (1983).

13. P. M. Dean, Ultrastructural morphometry of the pancreatic β-cell, <u>Diabetologia</u> 9:115 (1973).

14. K. P. Dresdner and R. P. Kline, Extracellular calcium ion depletion in frog cardiac ventricular muscle, <u>Biophys. J.</u> 48:33 (1985).

15. G. T. Eddlestone, A. Goncalves, J. A. Bangham, and E. Rojas, Electrical coupling between β-cells in islets of Langerhans from mouse, <u>J. Mem. Biol.</u> 77:1 (1984).

16. R. Ferrer, B. Soria, C. M. Dawson, I. Atwater, and E. Rojas, Effects of Zn^{2+} on glucose-induced electrical activity and insulin release from mouse pancreatic islets, <u>Amer. J. Physiol.</u> 246:C520 (1984).

17. B. Frankenhaeuser and A. L. Hodgkin, The after-effects of impulses in the giant nerve fibres of Loligo, <u>J. Physiol.</u> 131:341 (1956).

18. D. E. Goldman, Potential, impedance and rectification in membranes, <u>J. Gen. Physiol.</u> 27:37 (1943).

19. J. C. Henquin, D-glucose inhibits potassium efflux from
    pancreatic islet cells, _Nature_ 271:271 (1978).

20. A. L. Hodgkin and B. Katz, The effect of sodium ions on the
    electrical activity of the giant axon of the squid, _J.
    Physiol._ 108:37 (1949).

21. K. H. Kilb and R. Stampfli, A new stopcock for pharmacological
    purposes, _Naunyn-Schmiedeberg's Archive Pharmacology_ 285:293
    (1974).

22. E. Kohen, C. Kohen, B. Thorell, D. H. Mintz, and A. Rabinovitch,
    Intercellular communication in pancreatic islet monolayer
    cultures: a microfluorometric study, _Science_ 204:862
    (1979).

23. S. Kawazu, A. C. Boschero, C. Delcroix and W. J. Malaisse, The
    stimulus-secretion coupling of glucose-induced insulin
    release. XXVII Effect of glucose on Na^+ fluxes in isolated
    islets, _Pflugers Arch._ 375:197 (1978).

24. W. Loewenstein, Junctional intercellular communication: the
    cell-to-cell membrane channel, _Physiological Reviews_ 61:829
    (1981).

25. E. K. Matthews and Y. Sakamoto, Electrical characteristics of
    pancreatic islet cells, _J. Physiol._ 246:421 (1975).

26. P. Meda, I. Atwater, A. Goncalves, A. Bangham, L. Orci, and E.
    Rojas, The topography of electrical synchrony among β-cells
    in the mouse islet of Langerhans, _Quart. J. Exp. Physiol._
    69:719 (1984).

27. P. Meda, A. Perrelet, and L. Orci, Increase of gap junctions
    between pancreatic β-cells during stimulation of insulin
    secretion, _J. Cell Biol._ 82:441 (1979).

28. P. Meda, J. F. Denef, A. Perrelet and L. Orci, Nonrandom distri-
    bution of gap junctions between pancreatic β-cells, _Am. J.
    Physiol._ 238:C114 (1980).

29. L. Orci, R. H. Unger, and A. E. Renold, Structural coupling
    between pancreatic islet cells, _Experimentia_ 29:1015
    (1973).

30. E. Perez-Armendariz, I. Atwater, and E. Rojas, Glucose-induced
    oscillatory changes in extracellular ionized potassium
    concentration in mouse islets of Langerhans, _Biophys. J._
    48:741 (1985).

31. B. Ribalet and P. Beigelman, Calcium action potentials and potas-
    sium permeability activation in pancreatic β-cells, _Am. J.
    Physiol._ 239:C124 (1980).

32. L. M. Rosario and E. Rojas, Potassium channel selectivity in
        mouse pancreatic β-cells, <u>Am. J. Physiol</u>. 250:C90 (1986).

33. P. A. Rutecki, F. J. Lebeda, and D. Johnston, Epileptiform activ-
        ity induced by changes in extracellular potassium in
        hippocampus, <u>J. Neurophysiol</u>. 54:1363 (1985).

34. L. S. Satin and D. L. Cook, Voltage-gated Ca current in
        pancreatic islet β-cells, <u>Pflugers Arch</u>. 404:385 (1985).

35. E. Sykoba, Extracellular K^+ accumulation in the central nervous
        system, <u>Prog. Biophys. Mol. Biol</u>. 42:135 (1983).

A POTASSIUM CHANNEL MODULATED BY GLUCOSE METABOLISM IN RAT PANCREATIC
β-CELLS

F.M. Ashcroft, D.E. Harrison and
S.J.H. Ashcroft

University Laboratory of Physiology
Parks Rd., Oxford OX1 3PT, England

 Nuffield Department of Clinical Biochemistry
John Radcliffe Hospital
Headington, Oxford OX3 9DU, England

It is well established that glucose-stimulated insulin release from
the pancreatic β-cell is associated with the initiation of electrical
activity (see recent review by Henquin & Meissner[10]). Microelectrode
recordings have shown that in the absence of glucose the β-cell is
electrically silent. Increasing plasma glucose depolarises the membrane by
an amount that is dependent on the glucose concentration and electrical
activity is initiated when this depolarisation exceeds a threshold level.
There is considerable evidence that the initial slow depolarisation induced
by glucose results from a decrease in the resting potassium permeability of
the membrane[5,8,14] as a consequence of glucose metabolism[9].

This paper is concerned with the initial decrease in membrane
potassium permeability and consequent depolarisation induced by glucose.
Our results suggest that both of these result from the closing of a
particular class of potassium channel in the β-cell membrane. For
convenience we have referred elsewhere to this glucose-sensitive channel as
the G-channel[2]. Our experiments are also consistent with the idea that
glucose-induced channel inhibition is mediated by an increase in
intracellular ATP[3].

METHODS

Experiments were carried out on single pancreatic β-cells isolated
from rat islets of Langerhans[12]. We have shown elsewhere that this
preparation secretes insulin in response to glucose[2]. The cells were
maintained in RPMI 1640 tissue culture medium supplemented with penicillin
(100 U/ml), streptomycin (0.1 mg/ml), fetal calf serum (10%) and glucose
(11 mM), and were used within 1 to 6 days after plating onto glass
coverslips (see Ashcroft et al.[2] for further details). An equilibration
period of 30-60 min in glucose-free solution was allowed before starting
experiments. Single channel recordings were made using standard patch
clamp methods[7]. All records are displayed using the normal sign
conventions; i.e. potentials relative to 0 mV external and inward currents
across the membrane as downward deflections of the current trace.

The pipette was filled with a solution (I), containing (mM): 140 KCl,
5 $CaCl_2$, 5 $MgSO_4$, 5 NaHEPES (pH 7.4). For those experiments in which it
was important that the metabolism of the cell was retained intact, we used
cell-attached membrane patches and the bath contained a standard external
solution (II) consisting of (mM): 5 KCl, 135 NaCl, 5 $CaCl_2$, 5 $MgSO_4$, 5
NaHEPES (pH 7.4). In other experiments we used isolated inside-out
membrane patches and an intracellular solution (III) containing (mM): 119
KCl, 11 KEGTA, 10 KHEPES (pH 7.1) 5 $MgCL_2$, 1 $CaCl_2$ in the bath. All
experiments were carried out at room temperature (20-25°C).

RESULTS

We used cell-attached patches to investigate the effect of glucose on
the resting potassium permeability of the β-cell. Under physiological
conditions potassium currents are likely to be very small at the resting
potential because it is almost the same as the potassium equilibrium
potential. Therefore the pipette was filled with a solution containing a
high concentration of potassium (solution I) to increase the amplitude of
K-channel currents and enable them to be recorded at the resting potential
of the cell.

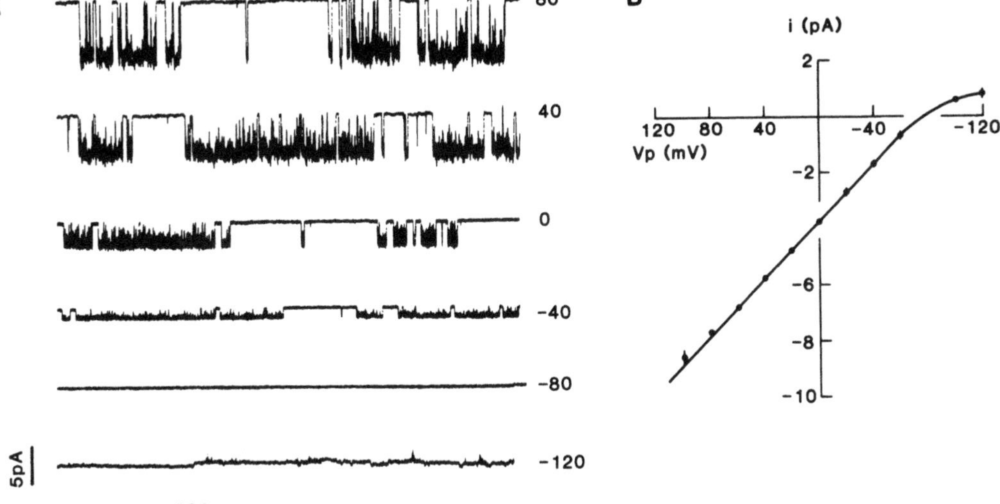

Fig. 1.   A. Single channel currents recorded at different pipette
          potentials (indicated to the right of each trace) from a
          cell-attached patch in glucose-free solution. Pipette, solution I.
          B. Mean relation between single channel current and pipette
          potential (n = 16).   From Ashcroft et al.[2].

Fig. 1A shows single channel currents recorded at different pipette
potentials from a cell-attached patch on a β-cell exposed to glucose-free
solution.   The mean current-voltage relation for this channel is plotted on
the right.   Single channel currents decreased as the membrane was
depolarised and reversed at a mean pipette potential of -76 mV.   In a
separate series of experiments we measured the mean zero current potential
under whole-cell voltage clamp to be -68 mV.   Using this value for the
resting potential, the membrane potential at which the single channel
current reverses (+8 mV) lies close to the estimated potassium equilibrium
potential under our experimental conditions (+9 mV) and suggests the
G-channel is principally permeable to potassium.   Changing the potassium
concentration in the pipette solution shifted the reversal potential in
accordance with the Nernst equation for a K-selective channel[2], providing
support for this idea.

The single channel conductance, measured from the slope of the line
between +80 and -60 mV had a mean value of 51 pS.   The open probability of
the channel varied greatly between cells but was independent of membrane

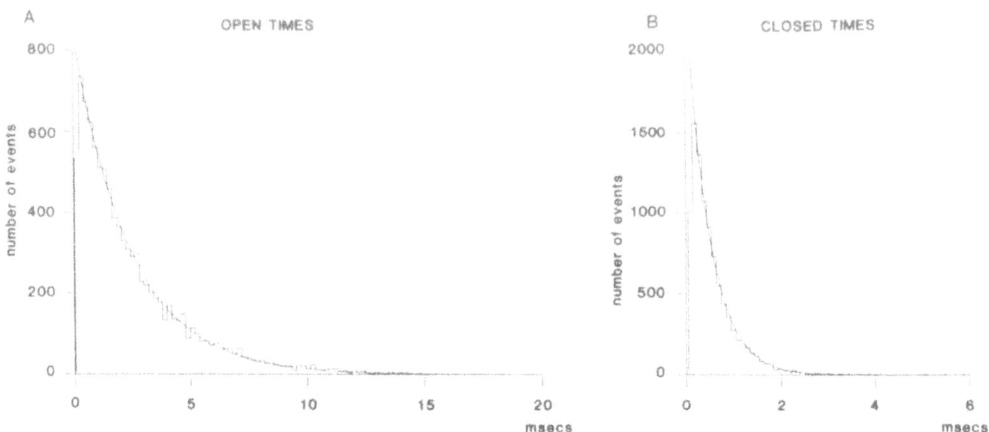

Fig. 2. Analysis of burst kinetics. Frequency histograms of open times
(A) and closed times (B). The line is drawn to a single
exponential with a time. constant of 2.4 msec (A, open time) or of
0.48 msec (B, closed time). Frequency histograms were constructed
from samples of approximately 2 min duration.

potential (between +80 and -60 mV). The variability in the open
probability probably reflects differences in the metabolic states of the
β-cells (see later). It is clear from Fig. 1A that the channel kinetics
are complex since the channel shows bursts of activity interrupted by long
closed periods. This implies that the channel can enter at least two
closed states, a short closed state represented by the brief closings found
within the burst, and a much longer closed state that determines the
interburst interval. We have confined our analysis of open and closed
times to the burst kinetics. This is shown in Fig. 2. At the resting
potential the open time frequency histogram could be fitted by a single
exponential with a time constant of 2.4 msec. The time constant of the
fast component of the closed times (i.e. that within the burst) was 0.48
msec.

The channel illustrated in Fig. 1 was the one we observed most
frequently at the resting potential (85% of patches). We also identified
two other channels in cell-attached patches[2], but the open state probability and conductance of these channels suggest they do not contribute
significantly to the resting potassium permeability of the β-cell in
glucose-free solutions. Our results suggest that this is principally
determined by the G-channel.

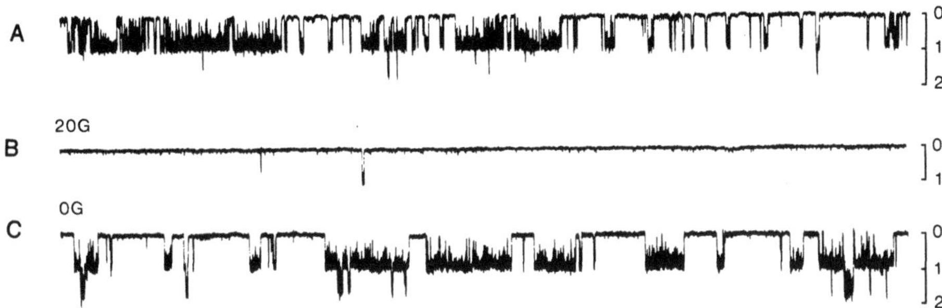

Fig. 3.   Effect of 20 mM glucose on G-channel activity. Records were
obtained successively from a cell-attached patch at a pipette
potential of 0 mV (i.e. at the resting potential). A.
Glucose-free solution. B. 20 mM glucose solution, 7 mins.
C. Glucose-free solution, 12 mins. The scale to the right of each
trace indicates the current level when 0, 1 or 2 channels were
open.

The addition of glucose to the bath resulted in the rapid and
reversible inhibition of G-channel activity. Fig. 3 shows that channel
inhibition is almost complete in 20 mM glucose solution; only occasional,
very brief, openings are found. In glucose concentrations above 10 mM
channel inhibition is accompanied by the appearance of action potentials[2].
Action potential activity recorded from single cells was invariably
irregular and did not show the steady bursting pattern characteristic of
microelectrode recordings from intact islets[5].

The most marked effect of glucose on channel activity is to decrease
the duration of the bursts and cause the channel to remain in the long-
lasting closed state for very much longer. The open state probability of
the channel is consequently significantly decreased (Fig. 5A). The
reduction in open probability is glucose-dependent, the maximal effect
taking place between 0 and 3 mM glucose and inhibition being almost
complete by 7 mM glucose. This relationship between glucose concentration
and channel open probability is in agreement with the inhibitory effect of
glucose on membrane K-permeability[8], and provides support for the idea that
the initial depolarisation induced by glucose results from closure of the
G-channel. Analysis of the burst kinetics suggests that glucose does not
affect the fast component of closed times, but there is a progressive
decrease in the open time associated with increasing glucose concentration.
The open time decreases by around 50% on increasing glucose from 0 to 10
mM.

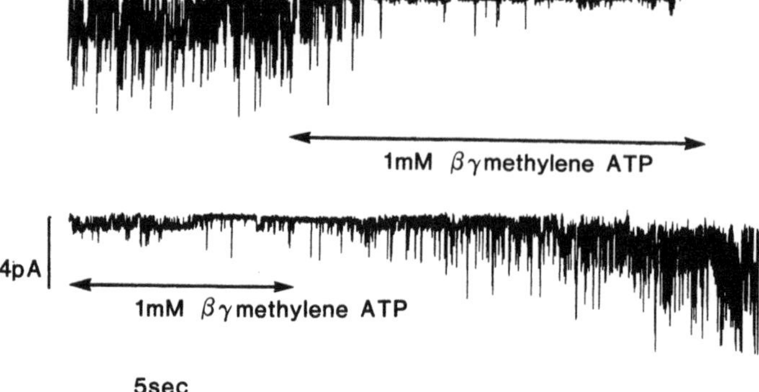

Fig. 4. Effect of 1 mM βγmethylene ATP on G-channel activity in an inside-out patch at a pipette potential of 40 mV (i.e. membrane potential = -40 mV). Pipette, solution I; Bath, solution III. The bars indicate the time during which the patch was exposed to βγ methylene ATP.

It is well established that both glucose-induced electrical activity and insulin secretion require the metabolism of glucose[1,9]. As expected, G-channel inhibition also requires glucose metabolism since the addition of 20 mM mannoheptulose to the bath solution restored channel activity previously abolished by glucose[2]. Furthermore, glucose applied directly to the intracellular membrane surface (in inside-out patches) was without effect on channel open probability.

To identify the product of glucose metabolism responsible for channel inhibition we tested the effects of possible intracellular regulators of G-channel activity in inside-out membrane patches. Cook and Hales[6] recently reported a K-channel in neonatal rat β-cells which is blocked by ATP, and which appears similar to the G-channel in its conductance and kinetic properties. Furthermore in intact islets of Langerhans intracellular ATP is known to increase when glucose is metabolised[4]. We therefore tested the effect of ATP and its non-hydrolysable analogue βγ methylene ATP on G-channel activity in inside-out patches.

The addition of 1 mM βγ methylene ATP applied to the bath solution resulted in the rapid and reversible inhibition of G-channel activity (Fig. 4). This patch contained at least 2 glucose-sensitive channels and another type of channel of lower conductance. 1 mM βγ methylene ATP applied to the bath solution (i.e., the intracellular membrane surface) selectively inhibited the G-channel. ATP itself had a similar effect. βγ methylene

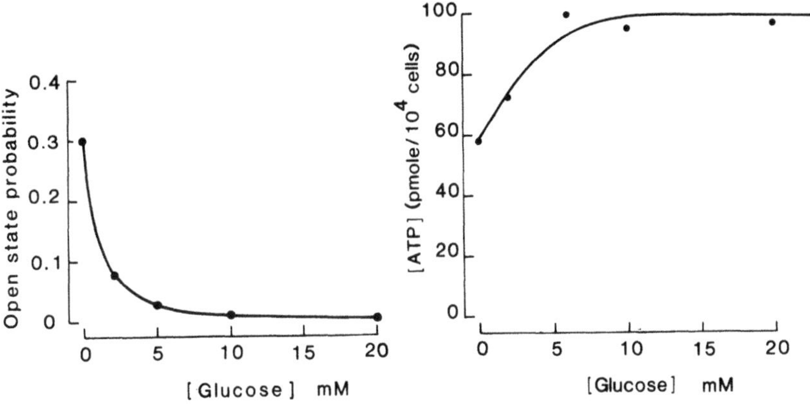

Fig. 5.   Relationship between glucose concentration and open state
probability. Open probability was measured for a single cell from
recordings of at least 2 min duration between 5-10 min after
introduction of the test solution. B. Relationship between
glucose concentration and intracellular ATP. ATP was measured in
dispersed cells at 37°C using a luciferin-luciferase assay[4] about
10 min after exposure to the test solution. An equilibration
period of 60 min in glucose-free solution was allowed before
introduction of test solutions.

ATP is a non-hydrolysable analogue of ATP and its ability to inhibit
channel activity indicates that phosphorylation is not required for channel
inhibition. The inhibitory effect of ATP suggests that the G-channel may
be identical with the ATP-sensitive potassium channel found by Cook and
Hales[6] in neonatal rat beta cells and with the larger of the ATP-sensitive
channels described by Petersen[13]. The results further suggest that ATP may
act as a link between glucose metabolism and G-channel inhibition in the
intact cell[3].

In order to identify ATP as the physiological channel modulator
mediating glucose-induced inhibition we have to show: 1) that the effect
of ATP on channel kinetics resembles that found for glucose, 2) that there
is a correlation between the effect of glucose concentration on the channel
open probability and its effect on the intracellular ATP concentration, and
3) the ATP concentration in the cell can account for the level of channel
activity in cell-attached patches.

We have not yet quantified the effect of ATP on channel kinetics but
our preliminary data suggests that, like glucose, ATP causes the channel to
enter a long-lasting closed state. This also appears to be the case for
the ATP-sensitive channels in both cardiac[16,11] and skeletal muscle[15].

Fig. 5 compares the effect of glucose concentration on the channel open
probability (A) with its effect on the intracellular ATP concentration of
suspensions of dispersed cells (B). Raising glucose increases intracellular ATP and decreases channel open probability. In both cases the greatest
effect is found between 0 and 7 mM glucose and reaches a limiting value at
higher concentrations. The data presented in Fig. 5B was obtained at 37°C;
a smaller increase in ATP is found when glucose is increased at 22°C. The
correlation between the effect of glucose on ATP and on the open
probability supports the idea that ATP may modulate channel activity in the
intact cell. The large variability in open state probability might also be
explained if it were dependent on the metabolic state of the cell.

Metabolic inhibitors such as rotenone (100 ng/ml) which lower ATP
levels, induce massive activation of the channel in cell-attached patches
on cells exposed to both glucose and glucose-free solutions[3]. This is also
consistent with modulation of channel open probability by ATP in intact
cells. Indeed the channel appears to be substantially inhibited even in
glucose-free solutions, since the open probability increases on patch
excision and when ATP production is inhibited in glucose-free solutions[3].

However, there appears to be a discrepancy between the ATP sensitivity
of the channel found in the intact cell and that seen in the isolated
patch. Thus 1 mM ATP is sufficient to inhibit activity completely in the
isolated patch experiments – and indeed Cook and Hales[6] report a Ki as low
as 15 μM – whereas the intracellular ATP of intact β-cells is around 5 mM
in glucose-free solutions and significant channel activity is found in
cell-attached patches.

We can suggest several explanations for this discrepancy. First, ATP
may be compartmentalised within the beta-cell – within the secretory
granules, for example – although it is generally considered unlikely that
ATP is compartmentalised to this extent. Alternatively, ATP sensitivity in
the isolated patch may be altered due to the loss of an additional intracellular modulator. Finally, it may be that the β-cell contains very many
G-channels and that only a very small percentage of these are active even
in glucose-free solution.

In summary, therefore, glucose metabolism causes a dose-dependent
inhibition of G-channel activity. Since the G-channel is open at rest its
closure can account for the decrease in potassium permeability and initial
membrane depolarisation induced by glucose. Our results are also
consistent with the idea that glucose metabolism inhibits G-channel
activity by raising intracellular ATP.

ACKNOWLEDGEMENTS

We thank Drs. N.B. Standen and P.R. Stanfield for the use of computing
facilities and Mr. A.C. Ashcroft for technical assistance. We also thank
the M.R.C. and the B.D.A. for support. DEH is a Lawrence Fellow of the
BDA.

REFERENCES

1. S.J.H. Ashcroft, Glucoreceptor mechanisms and the control of insulin
        release and biosynthesis, Diabetologia 18:5 (1980).
2. F.M. Ashcroft, D.E. Harrison, and S.J.H. Ashcroft, Glucose induces
        closure of single potassium channels in isolated rat pancreatic
        beta-cells, Nature 312:446 (1984).
3. F.M. Ashcroft, D.E. Harrison, and S.J.H. Ashcroft, The
        glucose-sensitive potassium channel in rat pancreatic beta-cells is
        inhibited by intracellular ATP, J. Physiol. 369:101P(1985).
4. S.J.H. Ashcroft, L.C.C. Weerasinghe, and P.J. Randle, Interrelation-
        ship of islet metabolism, adenosine triphosphate content and
        insulin release, Biochem J. 132:223 (1973).
5. I. Atwater, B. Ribalet, and E. Rojas, Cyclic changes in potential and
        resistance of the β-cell membrane induced by glucose in islets of
        Langerhans from mouse, J. Physiol. 278:117 (1978).
6. D. Cook and N. Hales, Intracellular ATP directly blocks K^+ channels in
        pancreatic β-cells, Nature 311:271 (1984).
7. O.P. Hamil, A. Marty, E. Neher, B. Sakmann, and F. Sigworth, Improved
        patch clamp techniques for high resolution current recordings from
        cells and cell-free membrane patches, Pflügers Arch. 391:85 (1981).

8. J.C. Henquin, D-glucose inhibits potassium efflux from pancreatic
    islet cells, _Nature_ 271:271 (1978).

9. J.C. Henquin, Metabolic control of the potassium permeability in
    pancreatic islet cells, _Biochem. J._ 186:541 (1980).

10. J.C. Henquin and H.P. Meissner, Significance of ionic fluxes and
    changes in membrane potential for stimulus-secretion coupling in
    pancreatic beta-cells, _Experientia_ 40:1024 (1984).

11. M. Kakei, A. Noma, and T. Shibasaki, Properties of adenosine-
    triphosphate-regulated potassium channels in guinea-pig ventricular
    cells, _J. Physiol._ 363:441 (1985).

12. A. Lernmark, The preparation of, and studies on, free cell suspensions
    from mouse pancreatic islets, _Diabetologia_ 10:445 (1974).

13. O.H. Petersen, this book.

14. J. Sehlin and I.-B. Taljedal, Transport of rubidium and sodium in
    pancreatic islets, _Nature_ 253:635 (1975).

15. A.E. Spruce, N.B. Standen, and P.R. Stanfield, Voltage-dependent
    ATP-sensitive potassium channels of skeletal muscle membrane,
    Nature 316:736 (1985).

16. G. Trube and J. Hescheler, Inward-rectifying channels in isolated
    patches of the heart cell-membrane: ATP-dependence and comparison
    with cell-attached patches, _Pflügers Arch._ 410:178 (1984).

GLUCOSE SUPPRESSES ATP-INHIBITED K-CHANNELS in PANCREATIC β-CELLS

D.L. Cook*, C. N. Hales[+] and L.S. Satin*

*Departments of Physiology and Biophysics
University of Washington, and
Department of Medicine,
The Seattle VA Medical Center
Seattle, WA 98108

[+]Department of Clinical Biochemistry
Cambridge University
Cambridge, England

Glucose-induced insulin release involves a cascade of events in the pancreatic β-cell: the metabolism of glucose enhances the net uptake of calcium by the cell, mobilizes calcium from stores within the cell and thus triggers the exocytosis of insulin[15]. Calcium uptake occurs when glucose depolarizes the β-cell membrane to trigger periodic bursts of calcium action potentials[2,3,10,13]. The mechanisms which couple glucose metabolism to the depolarization have eluded workers since the late 1960's. Depolarization is accompanied by increased cell input impedance, decreased dependence of membrane potential on external K^+[2], and decreased efflux of $^{42}K^+$ and $^{86}Rb^+$[8] all of which depend on glucose metabolism[9,12]. These findings have suggested the existence of membrane K-channels which are open in low glucose and are closed by a process linked to glucose metabolism. As possibilities, it has been suggested that Ca-activated K-channels are closed by a glucose-induced fall of intracellular free Ca^{2+}[2], or that K-channels are closed by metabolic proton production and intracellular acidification[14].

METHODS

These possibilities were examined by studying K-permeable membrane
channels using patch-clamp methods[7] in monolayer cultures of neonatal rat
β-cells identified by immunohistochemistry. A Dagan 8900 patch-clamp
system was used with fire-polished soda glass pipettes filled with a
solution containing (in mM): 140 KCl, 10 HEPES, 2 $MgCl_2$, 0.1 EGTA with
free Ca^{2+} set to 3 μM and pH = 7.2. Excised, inside-out patches were
exposed to an identical solution having different Ca^{2+} and pH levels, or
having various added compounds. For cell-attached recordings, cells were
perifused with oxygenated saline containing (in mM): 140 NaCl, 5 KCl, 2
$MgCl_2$, 2 $CaCl_2$ and 10 HEPES with pH = 7.2.

RESULTS AND DISCUSSION

In support of the first proposal, we found[4] that Ca-activated
K-channels do exist in β-cells and are activated by increasing
intracellular $[Ca^{2+}]$ and membrane depolarization as expected from results
in other systems. In support of the second proposal, we found that
activation of these channels by Ca^{2+} is inhibited by intracellular
acidification. While acidification is effective in the physiologic pH
range (i.e. 6.8 to 7.6) at physiologic membrane potentials, Ca-activation
generally requires more than 1 μM $[Ca^{2+}]$[4,6]. The results of recent dye
studies have shown, however, that in low glucose, intracellular $[Ca^{2+}]$ is
in the 0.1 μM range and actually increases with glucose stimulation[17],
while dye studies of intracellular pH[11] show that intracellular pH may
actually increase with glucose stimulation. Thus, these results would
predict that β-cell Ca-activated K-channels are not significantly active in
low glucose and that glucose-dependent pH_i changes are unlikely to account
for deactivation of these channels.

As an alternative, we have also recently described a novel β-cell
K-channel[5] which coexists in about equal numbers with the Ca-activated
K-channel, but whose activation is unaffected by membrane potential (-100
to +100 mV), intracellular $[Ca^{2+}]$ (10^{-9} to 10^{-3} M) or pH (6.8 to 7.6). The
channel has a conductance of 54 pS (versus 244 pS for the Ca-activated
K-channel), is K-selective and is tonically active in excised patches. Its
most interesting feature is that the channel is rapidly, completely and
reversibly blocked by the application of 1 mM ATP to the inner membrane
surface. Lower ATP doses block channel activity in a dose-dependent manner

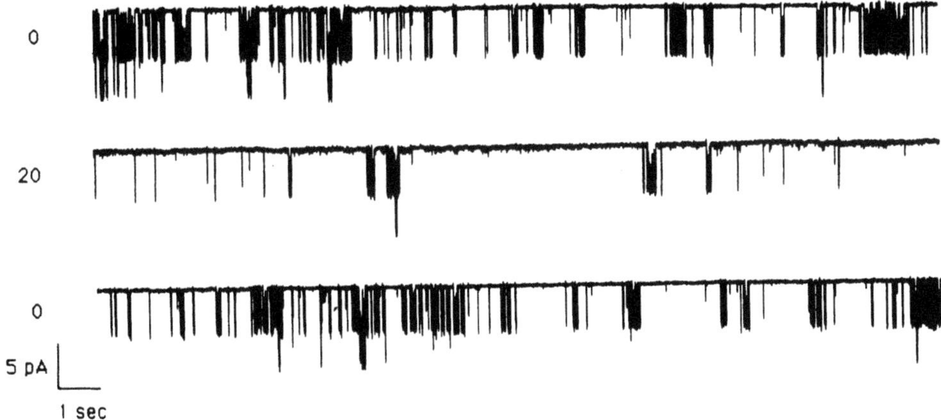

Fig. 1.   K-currents recorded in a voltage-clamped patch of cultured
neonatal β-cell membrane in cell-attached mode. Top: Steady
state channel activation when the cell is perifused with
oxygenated saline with 0 glucose. Middle: Steady state activity
seven minutes following addition of 20 mM glucose. Bottom:
Activity five minutes after return to zero glucose.

with half-maximal inhibition at 0.015 mM, and a Hill coefficient of 1.18
suggesting single site regulation. The effect is specific for ATP since 1
mM ADP has less effect than 0.1 mM ATP while 1 mM each of AMP, GTP, UTP or
cAMP have no effect. ATP hydrolysis appears not to be required since 1 mM
each of Mg-free ATP and AMP-PNP, a slowly hydrolyzable analogue of ATP, are
as rapidly effective as 1 mM ATP in blocking channel activity.

The existence of this channel, the first K-channel to be described
whose activity is regulated by a metabolic product of glucose, suggested
immediately that ATP may be the "second messenger" responsible for this
effect. The problem, however, is that the channel, as studied in excised
patches, is almost completely blocked by 0.1 mM ATP[5] and would be expected
to be totally blocked by the generally accepted level of free ATP in cells
(i.e. 2-5 mM). One would predict that only under the most dire of
metabolic conditions would ATP levels be low enough to allow the channel to
open. It is possible, however, that in intact cells the channel may be
less sensitive to ATP. This is supported by studies of a similar
ATP-inhibited K-channel in cardiac cells where the K_i for ATP in excised
patches was 0.1 mM while the K_i for the channel, _in situ_ and still exposed
to the cytoplasm, was 0.5 mM[16]. Although there is no supporting data, it

may also be that Ca-activated K-channels are more Ca-sensitive in intact cells than in excised patches. Because of these possibilities, it is important to determine which, if either, of these channels are active in intact β-cells in low glucose and, which, if either, is turned off by glucose stimulation.

We and others[1] have monitored channel activity in membrane patches still attached to perifused β-cells (cell-attached mode) as bath glucose level was increased from zero to 10 mM and then returned to zero. In zero glucose, the only active channel is insensitive to membrane voltage and has a conductance which matches that of the ATP-inhibited K-channel. These channels are reversibly, but only partially, suppressed by 10 mM glucose (Fig. 1) and appear to account for virtually all of the K-permeable channel activity. We also observe very occasionally a larger channel (conductance, 120 pS) and a smaller channel (10 pS) both of which are voltage-dependent but appear to contribute little to the total membrane K-conductance (data not shown). From this we conclude that, in neonatal (this study) and adult[1] rat β-cells it is the ATP-inhibited K-channel which accounts for the resting K-permeability in low glucose and which is suppressed by increasing glucose metabolism. This suggests the possibility that ATP is the second messenger coupling glucose metabolism to the suppression of membrane K-permeability. Indeed, since the channels are active even in high glucose (Fig. 1; ref. 1), then the channel's ATP sensitivity, <u>in situ</u>, must lie within the range of ATP levels which exist near the β-cell membrane.

ACKNOWLEDGEMENTS

The authors thank Dr. Wilfred Y. Fujimoto for cell cultures and Dr. Denis Baskin for assistance in the immunohistochemical identification of the cells. This work was supported by the US Veterans Administration and NIH Research Grant AM29816. C.N.H. was supported by a Visiting Scientist Award from the American Heart Association of Washington. L.S.S. was supported by the Neurobiology Training Grant in the Department of Physiology and Biophysics, University of Washington.

REFERENCES

1. F.M. Ashcroft, D.E. Harrison, and S.J.H. Ashcroft, Glucose induces closure of single potassium channels in isolated rat pancreatic β-cells, <u>Nature</u> 312:446, (1984).

2. I. Atwater, B. Ribalet, and E. Rojas, Cyclic changes in potential and
    resistance of the β-cell membrane induced by glucose in islets of
    Langerhans from mouse, J. Physiol. 278:117 (1978).

3. D.L. Cook, Electrical pacemaker mechanisms of pancreatic islet cells,
    Fed. Proc. 43:2368 (1984).

4. D.L. Cook, M. Ikeuchi, and W.Y. Fujimoto, Lowering of pH_i inhibits
    Ca^{2+}-activated K^+ channels in pancreatic β-cells, Nature 311:269
    (1984).

5. D.L. Cook and C.N. Hales, Intracellular ATP directly blocks K^+
    channels in pancreatic β-cells, Nature 311:271 (1984).

6. I. Findlay, M.J. Dunne, and O.H. Petersen, High-conductance K^+ channel
    in pancreatic islet cells can be activated and inactivated by
    internal calcium, J. Memb. Biol. 83:169 (1985).

7. O.P. Hamill, et al., Improved patch-clamp techniques for
    high-resolution current recordings from cells and cell-free
    membrane patches, Pflügers Arch. 391:85 (1981).

8. J.-C. Henquin, D-glucose inhibits potassium efflux from pancreatic
    islet cells, Nature 271:271 (1978).

9. J.-C. Henquin, Metabolic control of potassium permeability in
    pancreatic islet cells, Biochem. J. 186:541 (1980).

10. J.-C. Henquin and H.P. Meissner, Significance of ionic fluxes and
    changes of membrane potential for stimulus-secretion coupling in
    pancreatic β-cells, Experientia 40:1043 (1984).

11. P. Lindstrom and J. Sehlin, Effect of glucose on the intracellular pH
    of pancreatic islet cells, Biochem. J. 218:887 (1984).

12. W.J. Malaisse, et al., Insulin release: the fuel hypothesis,
    Metabolism 28:373 (1978).

13. E.K. Matthews, Electrophysiology of pancreatic islet β-cells, in: "The
    Electrophysiology of the Secretory Cell", Poisner and Trifaro,
    eds., Elsevier Science Publishers (1985).

14. C.S. Pace, J.T. Tarvin, and J.S. Smith, Stimulus-secretion coupling in
    β-cells: modulation by pH, Am. J. Physiol. 244:E3 (1983).

15. M. Prentki and C.B. Wollheim, Cytosolic free Ca^{2+} in insulin secreting
    cells and its regulation by isolated organelles, Experientia
    40:1052 (1984).

16. M. Soejima and A. Noma, Mode of regulation of the ACh-sensitive
    K-channel by the muscarinic receptor in rabbit atrial cells,
    Pflügers Arch. 400:424 (1984).

17. C.B. Wollheim and T. Pozzan, Correlation between cytosolic free Ca^{2+}
    and insulin release in an insulin-secreting cell line, J. Biol.
    Chem. 259:2262 (1984).

K-CHANNELS IN AN INSULIN-SECRETING CELL LINE: EFFECTS OF ATP AND
SULPHONYLUREAS

M.L.J. Ashford[*], N.C. Sturgess[*+], D.L. Cook[†] and C.N. Hales[+]

[*]Department of Pharmacology
[+]Department of Clinical Biochemistry
University of Cambridge
Cambridge, England

[†]Departments of Physiology and Biophysics, and Medicine
University of Washington and
Seattle VA Medical Center
Seattle, WA 98108

The recent application of the patch-clamp technique to the study of
single channels present in the β-cell plasma membrane has led to the
discovery of a channel the activity of which is inhibited by the
application of ATP to the membrane cytoplasmic surface[7]. It has been
suggested that this channel may be responsible for glucose-induced β-cell
depolarisation and that ATP may be the long-sought connection between
glucose metabolism and the excitation of insulin secretion[7,8]. Cell lines
are proving to be very convenient and reproducible systems with which to
study metabolic processes and their regulation and we have found them
particularly suitable for patch clamp procedures. We have therefore
investigated a new insulin-secreting cell line[5] using the patch-clamp
technique in order to study potassium channels and their role in the
control of insulin secretion.

METHODS

The rat cell line (CRI-G1) selected for these studies was derived from
a transplantable islet cell tumor[6] and contained a relatively high insulin
content[5]. Cells were maintained in Dulbecco's Modified Eagles medium.
Patch-clamp studies, using both inside-out and outside-out membrane

patches, were carried out 4-8 days after plating and by standard methods[9].
Single channel currents were recorded through a Dagan 8900 patch clamp
amplifier and records were stored on an FM tape recorder (Racal Store
4D/4DS) with a bandwidth of DC to 2.5 kHz. The solution in the patch
pipette for inside-out studies contained (in mM): 140 KCl; 1.0 $MgCl_2$; 1.0
$CaCl_2$ and 10 HEPES. The pH was adjusted to 7.2 with KOH. The bath
solution contained (in mM): 140 KCl; 1.0 $MgCl_2$; 1.0 EGTA; 10 HEPES and
$CaCl_2$ was added to give a calculated free Ca^{2+} concentration of 10^{-6} M at
pH 7.2. Outside-out patches were studied using pipettes filled with (in
mM): 140 KCl; 1.0 EGTA and 10 HEPES (pH 7.2). For these experiments the
bath contained the same buffer with the addition of (in mM): 1.0 $MgCl_2$ and
0.9 $CaCl_2$. The solution to which the patch was exposed was changed either
by changing the bath solution or by local superfusion of the pipette tip[7].

RESULTS

Three different classes of K-channels were observed in inside-out
patches. They could be distinguished by their conductance and sensitivity
to calcium, membrane voltage and ATP. Two of these K-channels were
infrequently observed, a 7 pS channel which was insensitive to all three
conditions and a 200 pS channel which was calcium and voltage sensitive.
By far the most frequently observed channel (86% patches, n = 117) was one
which was reversibly inhibited by the application of 1mM ATP to the
cytoplasmic surface of the patch but not when ATP was applied to the
external surface (Fig. 1a). This channel had a conductance of 55.1 ± 0.35
pS (mean ± SEM n = 27) but in 16% of the patches what appeared to be a
subconductance state (27.8 ± 0.52 pS, mean ± SEM n = 17) was observed. It
was also sensitive to ATP. Detailed studies of the 55 pS channel showed it
to be indistinguishable from that previously observed in neonatal rat
β-cells with respect to the conductance, concentration of ATP producing
half-maximal inhibition and the Hill coefficient of the ATP effect[15].

In a series of studies of outside-out patches, the effects of the
sulphonylureas tolbutamide and glibenclamide have been investigated[16].
Tolbutamide (1 mM) applied to the external surface of patches produced a
50% reduction in the number of open single channels (n = 8). This effect
was only partially reversible upon removal of the drug over a time period
of 10 min. This effect of tolbutamide was concentration dependent
producing reductions in the average current to 65% and 35% at
concentrations of 0.22 mM and 2.0 mM, respectively. Glibenclamide (10 µM)
produced an effect comparable to that of tolbutamide 1 mM (Fig. 1b).

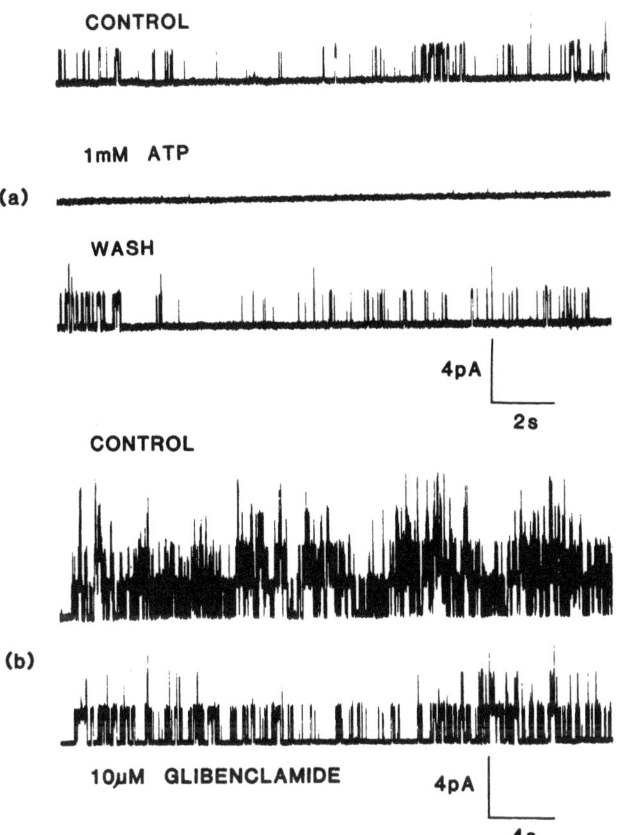

Fig. 1. (a) Effect of 1 mM ATP on single K-channel currents when applied
to the cytoplasmic side (bath applied) of an inside-out patch
excised from cell-line CRI-G1. Currents were recorded under
symmetrical 140 mM KCl conditions, and the membrane potential was
held at +50 mV. Single channel openings are upward deflections.
Top: control single channel activity. Middle: 2 seconds after
the addition of 1 mM ATP to the bathing solution. Bottom: 5
seconds after wash-out of ATP. (b) Effect of 10 μM Glibenclamide
on single K-channel activity recorded from an outside-out patch
excised from CRI-G1. The membrane potential was held at +50 mV,
and currents were recorded under symmetrical 140 mM KCl. Single
channel openings are upward. Top: at least 4 channels are active
in the patch under control conditions. Bottom: 20 seconds after
the addition of 10 μM Glibenclamide to the bathing solution:
single K-channel activity is reduced to a maximum of 3 channels
opening simultaneously.

The cell line used in these studies has been found to be sensitive to the stimulation of insulin secretion by leucine, leucine and theophylline, forskolin, raised extracellular K^+ concentration and tolbutamide[5] (and unpublished observations by C.A. Carrington and C.N. Hales). It is not sensitive to stimulation by glucose over the range of glucose concentrations 2.8-20 mM. It therefore provides a convenient and reproducible model with which to study the mechanisms of action of some insulin secretagogues. The finding that an ATP-sensitive K-channel is by far the most common K-channel observed strengthens our belief that this channel is of great importance in the determination of the membrane potential in insulin-secreting cells. It is also likely that modulation of the K^+ permeability of this channel plays a major role in the regulation of secretion. Hence the finding that sulphonylureas are capable of acting on the channel in isolated outside-out patches of membrane is of great interest[16]. Other studies have indicated that sulphonylureas probably exert their effects by an action on the plasma membrane and without entering the cell[12]. They decrease potassium permeability[10] and depolarise β-cells, actions which then lead to an increased uptake of calcium and insulin secretion. Thus the finding that sulphonylureas act on an ATP-sensitive potassium channel in the plasma membrane is entirely consistent with what is known of their actions to increase insulin secretion. Our experiments do not allow us to conclude whether the site of sulphonylurea action is on a polypeptide which is an integral part of the channel itself, or a membrane protein which is capable of closely associating with it. A related but probably different ATP-sensitive K^- channel has been described in the heart[14] and we have discussed elsewhere what the consequences might be if this channel was also susceptible to a similar effect of tolbutamide[16].

The ATP-sensitive K-channel of the present cell line and of the neonatal β-cell exhibits a degree of ATP sensitivity which at first sight might seem to exclude physiological changes in ATP as a means of its control. However it is clear from studies of β-cells using cell-attached membrane patches that a channel with a very similar conductance is active under conditions in which it would be expected that the ATP content of the cell would be more than sufficient to close the channel completely[1,8]. Such a finding could be explained either by compartmentalisation of ATP within the β-cell and/or by the activities of other factors modulating ATP sensitivity. Studies of the effects of ATP on ischaemic contracture and membrane permeability in skeletal muscle and the heart provide a precedent for the former explanation. In skeletal muscle, ATP concentrations as low

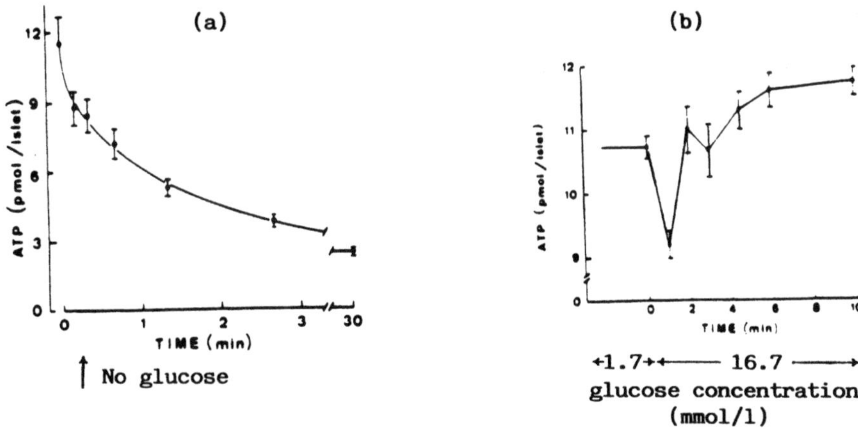
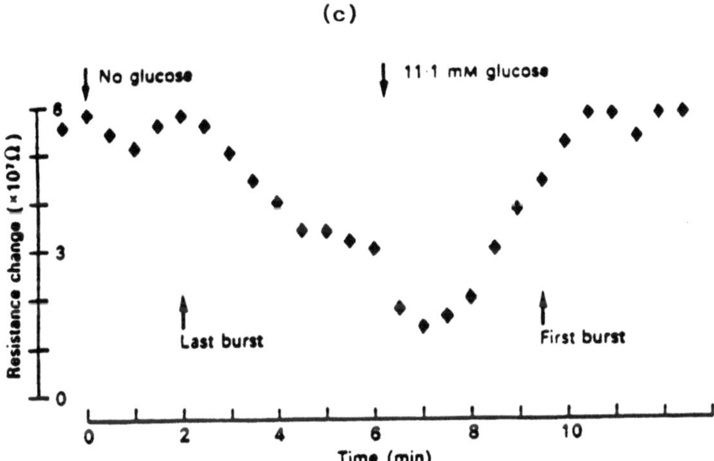

Fig. 2.   Illustration of the close temporal correlation between changes in islet ATP content (a+b) and changes in β-cell membrane potassium permeability as measured by the changes in membrane resistance (c) when the extracellular glucose concentration is suddenly decreased or increased. The acute reduction of membrane resistance was produced by the sudden removal of glucose (at arrow), the acute reduction of islet ATP was produced by the sudden removal of glucose plus the addition of rotenone (2.0 μM), antimycin A (10.0 μM) and KCN (1.0 mM) at time zero. The acute changes in membrane resistance upon the sudden reintroduction of 11.1 mM glucose are shown in (c) and the acute changes in islet ATP when the extracellular glucose was suddenly increased at time zero from 1.7 mM to 16.7 mM in (b). (a) and (b) are reproduced from ref 13 and (c) from ref.3 with the kind permission of the journals concerned.

as 10 μM are required for calcium-insensitive rigor complex formation and
it has been pointed out that there is a 240-fold discrepancy between this
concentration and the measured tissue ATP content[4]. Furthermore ATP
produced by glycolysis has been found to be much more effective than
mitochondrial ATP in preserving heart membrane functions[11]. It is
conceivable therefore that the β-cell cytosolic ATP concentration is not
reflected by the whole tissue ATP content and that this compartment of ATP
could have profound effects on the properties of the plasma membrane. Thus
a failure of islet ATP content to parallel insulin secretion under all
conditions is not good evidence that ATP does not mediate the effects of
glucose metabolism on insulin secretion.

Examination of the published data on the correlation of islet ATP
content with the rate of insulin secretion does in fact show, in many
instances, a remarkably good correlation. The effect of glucose to
increase islet ATP and of the effects of metabolic inhibitors to decrease
ATP and insulin secretion are entirely compatible with the postulated role
of ATP as a mediator[2]. Whilst complex mixtures of substrates, stimuli and
metabolic inhibitors give a less clear correlation, this could be due to a
protocol which included stimulated cAMP production and did not discriminate
between manipulations of cytosolic and mitochondrial ATP. Perhaps the most
striking example of correlations between islet ATP content and potassium
permeability are provided by two independent studies carried out for other
purposes but which involved the sudden removal or addition of glucose to
incubation media[13,3]. Figure 2 reproduces these findings in a composite
form. It illustrates how the time course of the fall in islet ATP closely
parallels the increase in potassium permeability as measured by β-cell
membrane resistance. Conversely the sudden reintroduction of a high
glucose concentration leads to a curious biphasic curve of ATP content and
potassium permeability. Again the parallels between the two are quite
striking.

Thus while it is clear that further experiments and techniques need to
be established to look at this question in a much more direct way, there is
already in the literature considerable evidence to support the proposal
that changes in cytosolic ATP are a major factor in the mechanism by which
glucose regulates β-cell membrane potential. The fact that the K-channel
upon which ATP acts is also sensitive to sulphonylureas acting on an
extracellular site is an additional indication that this channel plays a
key role in the regulation of insulin secretion.

ACKNOWLEDGEMENTS

We thank Dr. Christine Carrington for the culture of CRI-G1 cells,
Mrs. C.N. Lee Creswell for secretarial assistance and Hoechst U.K. Ltd. for
the supply of tolbutamide and glibenclamide. This work was supported by
the British Diabetic Association, Medical Research Council, Wellcome Trust,
US Veterans Administration and NIH Research Grant AM 29816.

REFERENCES

1. F.M. Ashcroft, D.E. Harrison, and S.J.H. Ashcroft, Glucose induces
       closure of single potassium channels in isolated rat pancreatic
       cells, Nature 312:446 (1984).
2. S.J.H. Ashcroft, L.C.C Weerasinghe, and P.J. Randle, Interrelationship
       of islet metabolism, adenosine triphosphate content and insulin
       release, Biochem. J. 132:223 (1973).
3. I. Atwater, B. Ribalet, and E. Rojas, Cyclic changes in potential and
       resistance of the β-cell membrane induced by glucose in islets of
       Langerhans from mouse, J. Physiol. 278:117 (1978).
4. O.L. Bricknell, P.S. Daries, and L.H. Opie, A relationship between
       adenosine triphosphate, glycolysis and ischaemic contracture in the
       isolated rat heart, J. Molec. and Cell Cardiol. 13:941 (1981).
5. C.A. Carrington, E.D. Rubery, E.C. Pearson, and C.N. Hales, Five new
       insulin-producing cell lines with differing secretory properties,
       J. Endocr. 109:in press(1986).
6. W.L. Chick, S. Warren, R.N. Chute, A.A. Like, V. Lauris and K.C.
       Kitchen, A transplantable insulinoma in the rat, Proc. Natl. Acad.
       Sci. USA 74:628 (1977).
7. D.L. Cook and C.N. Hales, Intracellular ATP directly blocks K^+
       channels in pancreatic β-cells, Nature 311:271 (1984).
8. D.L. Cook, C.N. Hales, and L.S. Satin, Glucose suppresses
       ATP-inhibited K-channels in pancreatic β-cells. This volume (1986).
9. O.P. Hamill, A. Marty, E. Neher, B. Sakmann and F.J. Sigworth,
       Improved patch-clamp techniques for high resolution current
       recording from cells and cell-free membrane patches, Pflügers Arch.
       391:85 (1981).
10. J.-C. Henquin, Tolbutamide stimulation and inhibition of insulin
       release: studies of the underlying ionic mechanisms in isolated
       rat islets, Diabetologia 18:151 (1980).

11. T.J.C. Higgins and P.J. Bailey, The effects of cyanide and iodoacetate intoxication and ischaemia on enzyme release from the perfused rat heart, <u>Biochim. Biophys. Acta</u> 762:67 (1983).

12. H.E. Lebovitz, Oral hypoglycaemic agents, in: "The Diabetes Annual", K.G.M.M. Alberti and L.P. Krall, eds., Elsevier, 1:93 (1985).

13. W.J. Malaisse, J.C. Hutton, S. Kawazu, A. Herchuelz, I. Valverde, and A. Sener, The stimulus-secretion coupling of glucose-induced insulin release. XXXV, The links between metabolic and cationic events, <u>Diabetologia</u> 16:331 (1979).

14. A. Noma, ATP-regulated K^+ channels in cardiac muscle, <u>Nature</u> 305:147 (1983).

15. N.C. Sturgess, M.L.J. Ashford, C.A. Carrington, and C.N. Hales Single Channel Recordings of Potassium Currents in an Insulin Secreting Cell Line, J. Endocr. 109: in press (1986).

16. N.C. Sturgess, M.L.J. Ashford, D.L. Cook, and C.N. Hales, The sulphonylurea receptor may be an ATP-sensitive potassium channel, <u>Lancet</u> ii:474 (1985).

INHIBITION OF K-CHANNELS IN INSULIN SECRETING CELLS

O. H. Petersen, I. Findlay, and M. J. Dunne

M. R. C. Secretory Control Research Group
The Physiological Laboratory
University of Liverpool
P.O. Box 147
Brownlow Hill
Liverpool, L69 3BX, U.K.

INTRODUCTION

The major physiological stimulus for insulin secretion from
pancreatic β-cells is an elevation of the plasma glucose concentration. The initial effect is to evoke membrane depolarization[4] followed by a cyclical pattern of electrical activity consisting of slow
waves of depolarization with action potential-like spikes.[9,12] Tracer
flux studies have indicated that glucose metabolism is associated with
a decrease in membrane K^+ permeability[17] and Ashcroft, Harrison and
Ashcroft[2] have demonstrated glucose-induced closure of single K-
channels in isolated rat pancreatic β-cells (this book).

RESULTS AND DISCUSSION

Fig. 1 shows an example of secretagogue-evoked inhibition of
single K-channels in an insulin-secreting cell. This experiment was
carried out on the rat insulinoma cell line m5F which does not respond
to glucose stimulation, but does secrete insulin when challenged with
glyceraldehyde.[13] Before stimulation repeated openings of K-channels
were observed (Fig. 1a) whereas about 30s after start of stimulation
with 10 mM D-glyceraldehyde channel openings were rare, the single-
channel current amplitude was markedly reduced and action potentials
were observed (Fig. 1b). The effect of glyceraldehyde stimulation was

Fig. 1. Rat insulinoma cell line m5F. Single-channel current recording obtained in cell-attached configuration. The recording pipette
was filled with K^+-rich intracellular solution whereas the bath contained the Na^+-rich extracellular solution. a-d are consecutive
traces from the same membrane patch. (a) control, (b) 30s after start
of stimulation with 10 mM, D-glyceraldehyde (an action potential is
seen), (c) 1.5 min after start of glyceraldehyde stimulation, (d) 1.5
min after return to control solution. c represents current level when
all channels are closed and o the current with one channel open. The
pipette voltage was kept constant at 0 mV.

partially transient since 1.5 min after start of stimulation the
channel open-state probability had increased somewhat and action
potential firing had ceased (Fig. 1c). 1.5 min after return to the
control solution the single-channel current amplitude was back to full
size and the open-state probability was as high as in the first control period (Fig. 1d). The shape of the single-channel current-voltage relationship before and after stimulation was similar but shifted
along the x-axis corresponding to a depolarization of about 40 mV
(from about -60 mV to -20 mV). The decrease in single-channel current
amplitude during glyceraldehyde stimulation (Fig. 1) is therefore
simply due to the membrane depolarization (inward currents through the
K-channel were recorded since the patch pipette was filled with
K^+-rich 'intracellular' solution whereas the bath was filled with
Na^+-rich extracellular solution).

The channels shown in Fig. 1 are the dominant ones in intact
resting pancreatic islet cells and have the same conductance properties as the ATP-sensitive K-channels studied in excised patches.[2,7]
Rorsman and Trube[14] have compared the kinetics of the glucose-sensitive resting K-channels in intact β-cells with the ATP-sensitive
K-channel in excised inside-out membrane patches from the same cells
and concluded that the channels are identical. Evidence that the

ATP-sensitive K- channels dominate the overall K^+ conductance comes
from whole-cell recording experiments in which a very high input
resistance was found if the medium dialysing the cell interior contained 3 mM ATP. The absence of ATP evoked a large additional K^+
conductance.[14] It has been known for some time that quinine is able
markedly to depolarize the pancreactic β-cell membrane[9] and since it
is now clear that quinine in low concentrations (about 100 μM) is a
relatively specific inhibitor of the ATP-sensitive K-channel[8] this
result is consistent with the view that the resting potential is
mainly due to the presence of ATP-sensitive K-channels.

In cardiac cells the ATP-sensitive K-channel is not operational
under resting conditions. This is consistent with the knowledge that
the intracellular ATP concentration is about 3-4 mM and that in excised inside-out or open cell-attached membrane patches ATP in a
concentration of 2 mM totally inhibits channel opening.[10] With regard
to pancreatic islet cells even lower ATP concentrations have been
reported to abolish K-channel openings in excised membrane patches[3]
yet these channels are nevertheless operational in the intact cells.
It turns out that immediately after excision of a membrane patch from
an islet cell into the inside-out configuration there is a dramatic
increase in the number of active channels from about 1-2 to 10-20.[6,7]
This is consistent with the hypothesis that these are ATP-sensitive
channels since excision exposes the membrane inside to the ATP-free
bathing solution. Subsequent ATP addition to the bath closes all the
channels. Similar results can be obtained by opening up intact cells
from which single-channel recording is being made in the cell-attached
configuration by saponin or digitonin treatment. The typical pattern
of activity seen in intact resting cells (Fig. 1) must therefore be
interpreted as a state in which the vast majority of the many ATP-
sensitive K-channels present are closed due to the action of intracellular ATP, i.e., the open-state probability of the channels is very
low. In excised patches there is a marked run-down of ATP-sensitive
K-channels[6,7] and therefore many single-channel recordings obtained
from excised patches after extensive run-down of channel activity look
not dissimilar to those obtained in the cell-attached configuration
from intact resting cells. This has caused considerable confusion as
it has been suggested that the degree of channel activation in the
intact cell is far too high in relation to the known ATP-sensitivity
in excised patches and the known ATP concentration in intact resting

cells.[1,14] The discrepancy is, however, much smaller than generally
assumed since the channel open-state probability is in fact extremely
low in the intact cell and it is only because of the high channel
density that openings can be observed.[7]

It is known that the intracellular ATP concentration increases
after glucose stimulation[1,11] and it is therefore attractive to suggest that the ATP generated during glycolysis (Fig. 2) is responsible
for the closure of K-channels. As pointed out by Dean, Matthews and
Sakamoto[5] one of the crucial metabolic steps in the glycolytic pathway
appears to be the one catalyzed by phosphoglycerate kinase (PGK) which
is an enzyme bound to the plasma membrane (Fig. 2). ATP generated in
this reaction step could therefore act locally to decrease the open-
state probability of the resting K-channel and thereby depolarize the
cell.

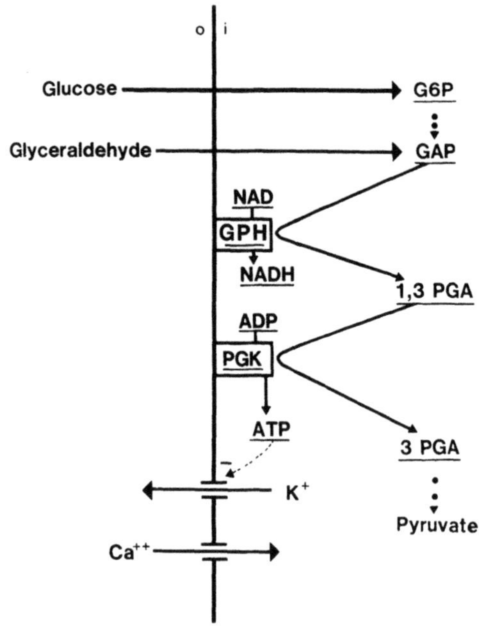

Fig. 2. Schematic diagramme indicating the mechanism by which glucose
or glyceraldehyde closes K-channels in insulin-secreting cells. Abbreviations: G6P, glucose-6-phosphate; GAP, glyceraldehyde phosphate;
GPH, glyceraldehyde phosphate dehydrogenase; 1,3-PGA, 1,3-diphospho-
glycerate; 3-PGA, 3-phosphoglycerate. The closure of K-channels
evokes depolarization and this activates the voltage-gated Ca-channels.

The importance of secretagogue-evoked inhibition of K-channels in islet cells lies in the resulting depolarization. The insulin-secreting cells possess Ca-channels activated by membrane depolarization. Voltage-activated Ca^{2+} currents have recently been studied with the help of patch-clamp whole-cell current recording in mouse pancreatic β-cells[15,16] and the rat insulinoma cell line m5F.[6] The voltage-activated Ca^{2+} current in the insulinoma cell line has properties adequately explaining the action potential[6] and therefore the increase in $[Ca^{2+}]_i$ observed after glyceraldehyde stimulation.[19] The stimulant-evoked increase in $[Ca^{2+}]_i$ is an important signal for insulin secretion by exocytosis[18] although the mechanisms by which Ca^{2+} acts are still obscure.

REFERENCES

1. F. M. Ashcroft, S. J. H. Ashcroft, and D. E. Harrison, The glucose-sensitive potassium channel in rat pancreatic beta-cells is inhibited by intracellular ATP, J. Physiol. (Lond.) 369:101P (1985).

2. F. M. Ashcroft, D. E. Harrison, and S. J. Ashcroft, Glucose induces closure of single potassium channels in isolated rat pancreatic beta-cells, Nature 312:446 (1984).

3. D. L. Cook and C. N. Hales, Intracellular ATP directly blocks K^+ channels in pancreatic β-cells, Nature 311:271 (1984).

4. P. M. Dean and E. K. Matthews, Electrical activity in pancreatic islet cells, Nature 219:389 (1968).

5. P. M. Dean, E. K. Matthews, and Y. Sakamoto, Pancreatic islet cells: effects of mono-saccharides, glycolytic intermediates and metabolic inhibitors on membrane potential and electrical activity. J. Physiol. 246:459 (1985).

6. I. Findlay and M. J. Dunne, Voltage-activated Ca^{2+} currents in insulin-secreting cells, FEBS Lett. 189:281 (1985).

7. I. Findlay, M. J. Dunne, and O. H. Petersen, ATP-sensitive inward rectifier and voltage- and calcium-activated K^+ channels in cultured pancreatic islet cells, J. Membr. Biol. 88:165 (1985).

8. I. Findlay, M. J. Dunne, S. Ullrich, C. B. Wollheim, and O. H. Petersen, Quinine inhibits Ca^{2+}-independent K^+ channels whereas tetraethylammonium inhibits Ca^{2+}-activated K^+ channels in insulin-secreting cells, FEBS Lett. 185:4 (1985).

9.   J. C. Henquin and H. P. Meissner, Significance of ionic fluxes
     and changes in membrane potential for stimulus-secretion
     coupling in pancreatic β-cells, Experientia, 40:1043
     (1984).

10. M. Kakei, A. Noma, and T. Shibasaki, Properties of adenosine-
     triphosphate-regulated potassium channels in quinea-pig
     ventricular cells, J. Physiol. (Lond.) 363:441 (1985).

11. W. J. Malaisse and A. Herchuelz, Nutritional regulation of K^+
     conductance: an unsettled aspect of pancreatic β cell physi-
     ology, in: "Biochemical Actions of Hormones, Vol. IX," G.
     Litwack, ed., Academic Press, N.Y., pp. 69-92 (1982).

12. O. H. Petersen, "Electrophysiology of gland cells," Academic
     Press, London (1980).

13. G. A. Praz, P. A. Halban, C. B. Wollheim, B. Blondel, A. J.
     Strauss, and A. E. Renold, Regulation of immunoreactive-
     insulin release from a rat cell line (RINm5F), Biochem. J.
     210:345 (1983).

14. P. Rorsman and G. Trube, Glucose-dependent K^+ channels in
     pancreatic β-cells are regulated by intracellular ATP,
     Pflugers Arch. 405:305 (1985).

15. P. Rorsman and G. Trube, Calcium and delayed potassium currents
     in mouse pancreatic β-cells under voltage-clamp conditions,
     J. Physiol. (Lond.), (in press 1986).

16. L. S. Satin and D. L. Cook, Voltage-gated Ca^{2+} current in
     pancreatic β-cells, Pflugers Arch. 404:385 (1985).

17. J. Sehlin and I. -B. Taljedal, Glucose-induced decrease in Rb^+
     permeability in pancreatic β-cells, Nature 253:635 (1975).

18. C. B. Wollheim and G. W. G. Sharp, Regulation of insulin release
     by calcium, Physiol. Rev. 61: 914 (1981).

19. C. B. Wollheim, S. Ullrich, and T. Pozzan, Glyceraldehyde, but
     not cyclic AMP-stimulated insulin release is preceded by a
     rise in cytosolic free Ca^{2+}, FEBS Lett. 177:17 (1984).

PHARMACOLOGICAL CONTROL OF ^{86}Rb EFFLUX FROM MOUSE PANCREATIC ISLETS

J.C. Henquin, M.G. Garrino, M. Nenquin
G. Paolisso and M. Hermans

Unite de Diabetologie et Nutrition
University of Louvain
Faculty of Medicine
UCL 54.74, B1200 Brussels
Belgium

Much attention has been paid to the K^+ permeability of the plasma membrane in pancreatic β-cells because its control by physiological or pharmacological agents is an important step in the sequence of events leading to insulin release[20]. The existence of different K-channels in β-cells, first suggested on a pharmacological basis[3,4,22], has been demonstrated recently by the patch clamp technique[2,8,11]. Elucidation of their respective physiological functions would be greatly facilitated by the availability of relatively specific inhibitors devoid of untoward effects in intact cells.

In the present study, we have investigated the effects of several established or putative blockers of K-channels on ^{86}Rb efflux from mouse islet cells. Experiments using islets loaded with both [42]K and ^{86}Rb have shown that ^{86}Rb is a suitable tracer for K efflux in this tissue[17].

MATERIALS AND METHODS

All experiments were performed with islets isolated after collagenase digestion of the pancreas of fed female NMRI mice. After loading with ^{86}Rb for 90 min in the presence of 10 mM glucose, the islets were placed in a perifusion system to monitor the efflux of the tracer[16]. In certain

experiments a portion of the effluent fractions was drawn for measurement
of insulin release.

Theophylline and tetraethylammonium chloride were from Merck A.G.,
quinine hydrochloride from Sigma Chemical Co., sparteine sulfate from
Janssen Chimica, 4-aminopyridine and 3,4-diaminopyridine from Aldrich
Europe, apamin and capsaicin from Serva, tubocurarine from Wellcome Ltd,
and gallamine triethiodide was provided by Rhone-Poulenc.

RESULTS AND DISCUSSION

Comparison of the Effects of Tetraethylammonium and Quinine

Tetraethylammonium (TEA) and quinine inhibit ^{86}Rb efflux from islet
cells[5,15,16], but differently affect the membrane potential of
β-cells[3,4,22,28]. Certain authors have considered TEA and quinine as
specific blockers of voltage-dependent and Ca-activated K-channels,
respectively[3,4]. Others have shown, however, that TEA can also inhibit
Ca-activated K^+ permeability[16,28] and raised doubts about a specific action
of quinine on the Ca-activated K^+ permeability[18]. Very recently, a patch
clamp study, performed with insulin-secreting tumor cells, suggested that
quinine is a rather specific blocker of the inward rectifier K-channels,
whereas TEA is a better blocker of the voltage- and Ca-activated
K-channels[12]. We have thus reevaluated the effects of both agents on ^{86}Rb
efflux from mouse islet cells.

The experiments have been carried out in the presence of 7 mM glucose,
a concentration that causes an almost maximal inhibition of ^{86}Rb efflux
from islet cells and depolarizes the β-cell membrane to the threshold at
which electrical activity starts[26]. One may thus expect that under these
conditions the inward rectifier K-channels are essentially closed, since
the depolarizing effect of glucose is thought to result from their
inhibition[2,8]. Moreover, Ca-sensitive K-channels should be activated only
weakly since electrical activity is present in only 25% of β-cells and
remains small in the presence of 7 mM glucose[21].

Addition of 10 mM theophylline to a medium containing 7 mM glucose
induced a marked acceleration of ^{86}Rb efflux from islet cells (Fig. 1) and

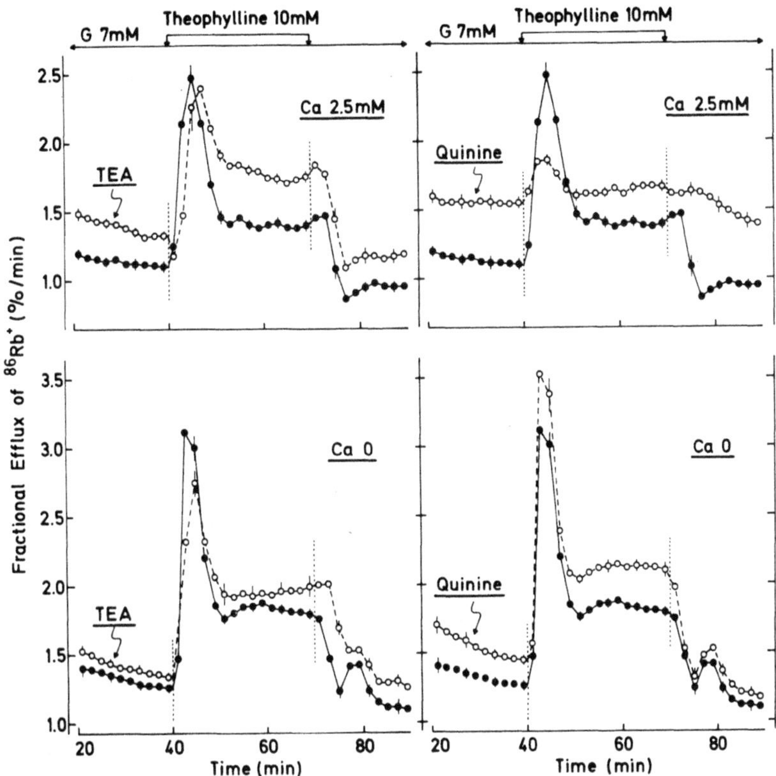

Fig. 1. Effects of tetraethylammonium (TEA) and quinine on theophylline-
induced acceleration of ^{86}Rb efflux from mouse islets perifused
with a medium containing 7 mM glucose (G). TEA (20 mM) and
quinine (50 μM) were added to the test solutions (o) for the whole
experiments. Values are means ± SEM for 4-6 experiments.

hyperpolarized the β-cell membrane[21]. This hyperpolarization can thus be
ascribed to an increase in K$^+$ permeability. It is unlikely that
theophylline itself increases K$^+$ permeability since it decreases ^{86}Rb
efflux and depolarizes the β-cell membrane in the presence of 3 mM
glucose[21]. It is also unlikely that the effect is mediated by cAMP since
dibutyryl-cAMP and forskolin accelerate ^{86}Rb efflux only slightly and
depolarize the membrane under similar conditions[20,21]. It is more likely
that a mobilization of cellular Ca by the methylxanthine underlies this
increase in K$^+$ permeability. Thus, both ^{86}Rb efflux (Fig. 1) and ^{45}Ca
efflux (not shown) were accelerated by theophylline in the presence or
absence of extracellular Ca.

Figure 1 shows that TEA did not prevent the acceleration of ^{86}Rb
efflux by theophylline, whereas quinine largely suppressed it in the

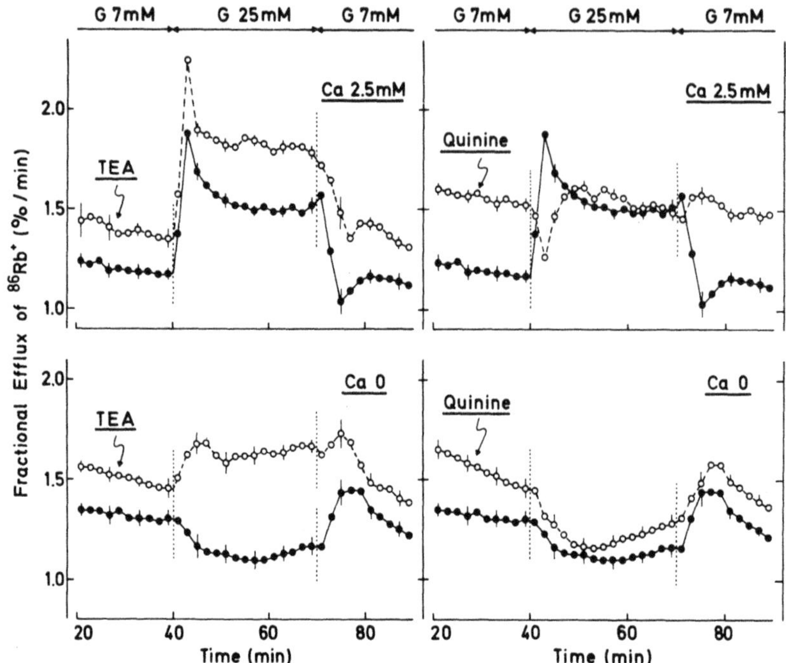

Fig. 2. Effects of tetraethylammonium (TEA) and quinine on the changes in ^{86}Rb efflux from mouse islets induced by a rise in the glucose (G) concentration from 7 to 25 mM. TEA (20 mM) and quinine (50 μM) were added to test solutions (o) for the whole experiments. Values are means ± SEM for 5-6 experiments.

presence of extracellular Ca. Surprisingly, however, quinine was ineffective in a Ca-free medium.

Raising the concentration of glucose from 7 to 25 mM is known to depolarize the β-cell membrane to the plateau level and to induce a continuous electrical activity[26]. The acceleration of ^{86}Rb efflux that occurs under these conditions (Fig. 2) may be due to the activation of Ca-sensitive K-channels by the incoming Ca and/or the greater depolarization of the membrane. As previously reported[25], quinine but not TEA inhibited this acceleration. In a Ca-free medium, the rise in glucose concentration slightly decreased ^{86}Rb efflux (Fig. 2). This effect was not altered by quinine. In the presence of TEA, however, a paradoxical increase was observed.

Interpretation of these experiments of ^{86}Rb efflux must be cautious because it is not exactly known how the membrane potential of β-cells changes under all experimental conditions. It is clear, however, that quinine, but not TEA, practically abolished ^{86}Rb efflux expected to occur, at least partially, through Ca-sensitive K-channels. These data, therefore, do not support the recent suggestion that TEA, but not quinine, blocks these channels[12]. It is possible that the discrepancy is due to abnormalities of K-channels in the tumor cells used for patch clamp experiments. Thus, the inward rectifier K-channel was only sensitive to quinine in these cells[12], whereas it was sensitive to both quinine and TEA in neonatal rat β-cells[8].

In summary, the available evidence does not permit one to conclude that TEA or quinine is a specific blocker of one type of K-channel in β-cells of normal mice.

<u>Effects of Other Putative Inhibitors of K Channels</u>

<u>Aminopyridines</u> are potent K-channel blockers in various tissues[14]. This property conceivably could explain the potentiation of glucose-induced insulin release by <u>4-aminopyridine</u> (4-AP) in mice deprived of their sympatho-adrenal system[1]. Figure 3 illustrates the complex, multiphasic, but reproducible, changes in ^{86}Rb efflux and insulin release brought about by 4-AP in mouse islets. In the presence of 3 mM glucose, 1 mM 4-AP accelerated ^{86}Rb efflux, whereas 10 mM caused an initial decrease, a secondary increase and finally a sustained inhibition. Removal of 4-AP was followed by an increase in ^{86}Rb efflux. Basal insulin release was not affected by 4-AP. In the presence of 10 mM glucose, 1 mM 4-AP caused first a slight decrease and secondarily a slight acceleration of the efflux rate (Fig. 3). Raising the concentration of 4-AP to 10 mM produced a large transient efflux of ^{86}Rb after an initial decrease. Removal of 4-AP essentially decreased the efflux rate. The changes in insulin release were fairly parallel to the changes in ^{86}Rb efflux. When the same experiments were repeated in the absence of Ca, a marked and sustained increase in ^{86}Rb efflux was already produced by 1 mM 4-AP, whereas insulin release was unaffected. A related compound, <u>3,4 diaminopyridine</u> which is more potent than 4-AP in other tissues[14], had no significant effect on ^{86}Rb efflux or insulin release from mouse islets perifused at 3 or 10 mM glucose (Table 1).

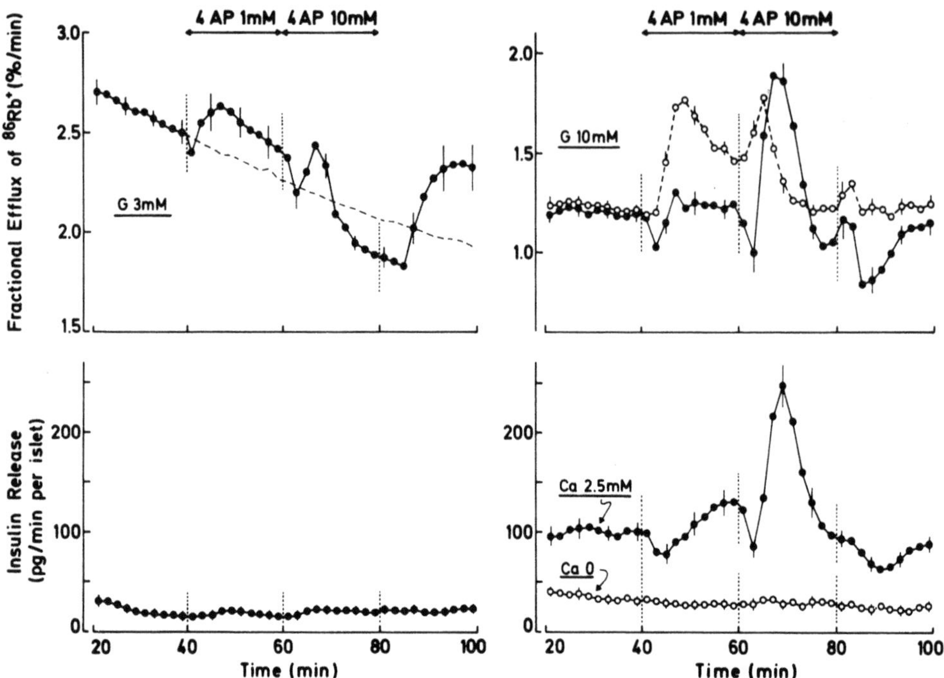

Fig. 3. Effects of 4-aminopyridine (4-AP) on ^{86}Rb efflux and insulin
release from mouse islets perifused with a medium containing 3 mM
glucose (G) (left panels) or 10 mM glucose (right panels), with Ca
(•) or without Ca (o). Values are means ± SEM for 4-6
experiments.

4-AP (10 mM) was found to block K-channel activity in inside-out
patches of pancreatic β-cells[8], and the batch of 4-AP used in the present
studies effectively blocked K^+ currents in the frog node of Ranvier (T.D.
Plant, personal communication). On the other hand, no clear-cut inhibition
of K^+ permeability by 4-AP could be demonstrated in intact islets, from
which the drug rather accelerated ^{86}Rb efflux, even in the absence of Ca.
The complex effects of 4-AP in islet cells could be due to effects
unrelated to K-channels and/or to opposite effects on different K-channels
as in certain neurons[14]. It is therefore unlikely that aminopyridines will
prove very helpful in further studies of stimulus-secretion coupling in
β-cells.

Sparteine is an alkaloid known to inhibit K^+ conductance in nerve
cells, which may be either sensitive[29] or resistant[23] to the action of
aminopyridines. In the presence of 3 mM glucose, sparteine decreased the

Table 1

Test substance			^{86}Rb efflux	Insulin release
			(% of Controls)	
Glucose 3 mM				
+ 3,4-diaminopyridine	50	µM	100 ± 1	102 ± 2
	500	µM	97 ± 2	104 ± 5
+ sparteine	200	µM	60 ± 2^c	130 ± 7^a
+ capsaicin	10	µM	102 ± 3	90 ± 11
	100	µM	75 ± 1^c	90 ± 9
+ apamin	0.01	µM	97 ± 3	101 ± 16
	0.1	µM	95 ± 2	116 ± 10
+ gallamine	50	µM	103 ± 1	104 ± 12
	500	µM	98 ± 1	110 ± 11
+ tubocurarine	50	µM	98 ± 3	107 ± 13
Glucose 10 mM				
+ 3,4-diaminopyridine	50	µM	102 ± 3	100 ± 5
	500	µM	101 ± 2	97 ± 4
+ sparteine	200	µM	118 ± 2^b	495 ± 23^c
+ capsaicin	10	µM	100 ± 4	115 ± 12
	100	µM	113 ± 4^a	80 ± 11
+ apamin	0.01	µM	102 ± 2	97 ± 16
	0.1	µM	102 ± 2	94 ± 11
+ gallamine	50	µM	102 ± 3	100 ± 9
	500	µM	104 ± 3	101 ± 8
+ tubocurarine	50	µM	101 ± 5	119 ± 14

After loading with ^{86}Rb$^+$, the islets were perifused in the presence of the indicated concentrations of glucose. Test substances were added to the perifusion medium after 40 min. When a substance was studied at two concentrations, these were tested in the same experiment, during two successive periods of 20 min. For each individual experiment, the average fractional rate of ^{86}Rb$^+$ efflux and the average rate of insulin release were calculated for the period 16-20 min of stimulation with the different test substances. These values are expressed as percentages of the control values, measured during the same period, in the absence of test substance. Results are means ± SEM for 4-5 experiments. [a] $p < 0.05$; [b] $p < 0.01$; [c] $p < 0.001$ vs controls.

rate of ^{86}Rb efflux from islet cells (Table 1), depolarized the β-cell
membrane, induced an electrical activity similar to that triggered by
glucose or tolbutamide[27] and increased insulin release (Table 1). In the
presence of 10 mM glucose, it slightly increased ^{86}Rb efflux, markedly
potentiated insulin release (Table 1) and increased electrical activity in
β-cells[27]. If the medium contained glucose but no calcium, sparteine
inhibited ^{86}Rb efflux but did not affect insulin release. Sparteine also
reversed the inhibition of insulin release by diazoxide, a substance known
to increase K^+ permeability in β-cells[19]. The effects of this alkaloid on
β-cells are thus very similar to those of hypoglycaemic sulfonylureas[19].

Capsaicin, a pungent substance found in red peppers, is a neurotoxin
best known for its ability to release substance P from sensory neurons[13].
It has been found to block one component of the fast K^+ current in the frog
node of Ranvier[10].

In the presence of 3 mM glucose, 100 μM capsaicin slightly decreased
[86]Rb efflux without changing basal insulin release (Table 1). A higher
concentration (250 μM) markedly inhibited ^{86}Rb efflux but only caused a
small and sluggish increase in release (not shown). In the presence of 10
mM glucose, 100 μM capsaicin slightly increased the efflux rate but tended
to inhibit insulin release (Table 1). 250 μM capsaicin had no clear effect
on ^{86}Rb efflux but markedly inhibited release (not shown). Certain
K-channels in islet cells seem thus to be sensitive to capsaicin. However,
the high concentrations required (ten-fold higher than in nerves) may alter
cellular metabolism[7], an effect that could contribute to the paradoxical
inhibition of insulin release in the presence of 10 mM glucose.

Apamin, a polypeptide extracted from the bee venom, is a very potent
blocker of Ca-activated K-channels in certain but not all tissues[9]. Table
1 confirms that apamin is without effect on insulin release[6] and [86]Rb
efflux[24] from islet cells. Gallamine, which consists of three TEA moieties
attached to a benzene ring, and tubocurarine are two neuromuscular blocking
agents known to decrease Ca-induced K loss from guinea-pig hepatocytes[9].
Both were ineffective in islet cells (Table 1). It is thus clear that
apamin, gallamine and tubocurarine the most potent blockers of Ca-sensitive
K-channels in hepatocytes are inactive in β-cells as in erythrocytes[9].

ACKNOWLEDGEMENTS

These studies were supported by grant 3.4552.81 from the FRSM,
Brussels. J.C.H. is "Maitre de Recherches" and M.H. is "Aspirant" of the
FNRS, Brussels. G.P. is a Research Fellow on leave from the First Medical
School of the University of Naples.

REFERENCES

1. B. Ahren, S. Leander, and I. Lundquist, Effects of 4-aminopyridine on
       insulin secretion and plasma glucose levels in intact and
       adrenalectomized-chemically sympathectomized mice, Europ. J.
       Pharmacol. 74:221 (1981).
2. F.M. Ashcroft, D.E. Harrison, and S.J.H. Ashcroft, Glucose induces
       closure of single potassium channels in isolated rat pancreatic
       β-cells, Nature 312:446 (1984).
3. I. Atwater, C.M. Dawson, B. Ribalet, and E. Rojas, Potassium
       permeability activated by intracellular calcium ion concentration
       in the pancreatic β-cell, J. Physiol. 288:575 (1979)
4. I. Atwater, B. Ribalet, and E. Rojas, Mouse pancreatic β-cells:
       tetraethylammonium blockage of the potassium permeability increase
       induced by depolarization, J. Physiol. 288:561 (1979).
5. A. Carpinelli and W.J. Malaisse, Regulation of ^{86}Rb outflow from
       pancreatic islets: Reciprocal changes in the response to glucose,
       tetraethylammonium and quinine, Mol. Cell. Endocrinol. 17:103
       (1980).
6. J. Chapal and M.M. Loubatieres-Mariani, Attempt to antagonize the
       stimulatory effect of ATP on insulin secretion, Europ. J.
       Pharmacol. 74:127 (1981).
7. P. Chudapongse and W. Janthasoot, Mechanism of the inhibitory action
       of capsaicin on energy metabolism by rat liver mitochondria,
       Biochem. Pharmacol. 30:735 (1981).
8. D.L. Cook and C.N. Hales, Intracellular ATP directly blocks K^+
       channels in pancreatic β-cells, Nature 311:271 (1984).
9. N.S. Cook and D.G. Haylett, Effects of apamin, quinine and
       neuromuscular blockers on calcium-activated potassium channels in
       quinea-pig hepatocytes, J. Physiol. 358:373 (1985).

10. J.M. Dubois, Capsaicin blocks one class of K^+ channels in the frog node of Ranvier, _Brain Res._ 245:372 (1982).

11. I. Findlay, M.J. Dunne, and O.H. Petersen, High conductance K^+ channel in pancreatic islet cells can be activated and inactivated by internal calcium, _J. Membr. Biol._ 83:169 (1985).

12. I. Findlay, M.J. Dunne, S. Ullrich, C.B. Wollheim, and O.H. Petersen, Quinine inhibits Ca^{2+}-independent K^+ channels, whereas tetraethylammonium inhibits Ca^{2+}-activated K^+ channels in insulin secreting cells, _FEBS Lett._ 185:4 (1985).

13. M. Fitzgerald, Capsaicin and sensory neurons, _Pain_ 15:109 (1983).

14. W.E. Glover, The aminopyridines, _Gen. Pharmacol._ 13:259 (1982).

15. J.C. Henquin, Tetraethylammonium potentiation of insulin release and inhibition of rubidium efflux in pancreatic islets, _Biochem. Biophys. Res. Commun._ 77:551 (1977).

16. J.C. Henquin, Opposite effects of intracellular calcium and glucose on the potassium permeability of pancreatic islet cells, _Nature_ 180:66 (1979).

17. J.C. Henquin, The potassium permeability of pancreatic islet cells: mechanisms of control and influence on insulin release, _Horm. Metab. Res. Suppl._ 10:66 (1980).

18. J.C. Henquin, Quinine and the stimulus-secretion coupling in pancreatic β-cells: glucose-like effects on potassium permeability and insulin release, _Endocrinology_ 110:1325 (1982).

19. J.C. Henquin and H.P. Meissner, Opposite effects of tolbutamide and diazoxide on ^{86}Rb fluxes and membrane potential in pancreatic B-cells, _Biochem. Pharmacol._ 31:1407 (1982).

20. J.C. Henquin and H.P. Meissner, Significance of ionic fluxes and changes in membrane potential for stimulus-secretion coupling in pancreatic B-cells, _Experientia_ 40:1043 (1984).

21. J.C. Henquin and H.P. Meissner, Effects of theophylline and dibutyryl cyclic adenosine monophosphate on the membrane potential of mouse B-cells, _J. Physiol._ 351:595 (1984).

22. J.C. Henquin, H.P. Meissner, and M. Preissler, 9-aminoacridine- and tetraethylammonium-induced reduction of the potassium permeability in pancreatic B-cells. Effects on insulin release and electrical properties, _Biochim. Biophys. Acta_ 587:579 (1979).

23. A.L. Kleinhaus, Segregation of leech neurons by the effect of sparteine on action potential duration, _J. Physiol._ 299:309 (1980).

24. P. Lebrun, I. Atwater, M. Claret, W.J. Malaisse, and A. Herchuelz, Resistance to apamin of the Ca^{2+}-activated K^+ permeability in pancreatic B-cells, <u>FEBS</u> <u>Lett.</u> 161:41 (1983).

25. P. Lebrun, W.J. Malaisse, and A. Herchuelz, Paradoxical activation by glucose of quinine-sensitive potassium channels in the pancreatic B-cell, <u>Biochem.</u> <u>Biophys.</u> <u>Res.</u> <u>Commun.</u> 107:350 (1982).

26. H.P. Meissner, Electrical characteristics of the beta cells in pancreatic islets, <u>J.</u> <u>Physiol.</u> (<u>Paris</u>) 72:757 (1976).

27. G. Paolisso, M. Nenquin, W. Schmeer, F. Mathot, H.P. Meissner, and J.C. Henquin, Sparteine increases insulin release by decreasing the K^+ permeability of the B-cell membrane, <u>Biochem.</u> <u>Pharmacol.</u> 34:2355 (1985).

28. B. Ribalet and P.M. Beigelman, Calcium action potentials and potassium permeability activation in pancreatic β-cells, <u>Am.</u> <u>J.</u> <u>Physiol.</u> 239:C124 (1980).

29. C.L. Schauf, C.A. Colton, J.S. Colton, and F.A. Davis, Aminopyridines and sparteine as inhibitors of membrane K^+ conductance, <u>J.</u> <u>Pharmacol.</u> <u>Exp.</u> <u>Ther.</u> 197:414 (1976).

ELECTROPHYSIOLOGICAL MEASUREMENTS SHOW MARKED DIFFERENCES IN THE PROPERTIES
OF THE PANCREATIC β-CELL K-CHANNELS FROM ALBINO MICE AND A STRAIN OF ob/ob
(OBESE) MICE

L. M. Rosario

Laboratory of Cell Biology and Genetics
NIDDK, National Institutes of Health
Building 8, Room 403
Bethesda, MD 20892

Several mouse mutants exhibiting obesity or diabetes-like syndromes
(e.g. yellow obese, obese, diabetes, KK, New Zealand obese) have been used
as animal models of diabetes mellitus[11,12,28]. They all have
hyperglycemia, hyperinsulinemia, and obesity as common metabolic
characteristics[28]. Among these mouse mutants, particular attention has
been paid to the homozygous diabetes mouse carrying the diabetes gene
(commonly known as the db/db mouse), and the homozygous obese mouse
carrying the obese gene (the ob/ob mouse). Besides the metabolic disorders
and the insulin receptor defects characteristic of the diabetes-like
syndromes, little is known about the possible malfunction of the pancreatic
β-cell. In this context, the earlier membrane potential measurements
carried out by Meissner and Schmidt in pancreatic β-cells from C57BL/KsJ
db/db-mouse islets[33] assume a special significance. These authors have
shown that the db/db β-cells exhibit a pattern of bursting or continuous
electrical activity which is virtually insensitive to the concentration of
the natural secretagogue D-glucose, thus suggesting an impairment of the
db/db β-cell K^+-permeability[9]. In pancreatic β-cells from normal mouse
islets, K^+-permeability is now known to play a key role in both the
generation and regulation of the glucose-induced bursts of electrical
activity[3-5,7]. The glucose-induced electrical activity (calcium spikes),

in turn, is closely correlated with insulin release under normal
physiological conditions[43], and is commonly regarded as an early event in
stimulus-secretion coupling[27]. These considerations have encouraged us to
undertake a detailed investigation of the membrane potential
characteristics of β-cells from ob/ob mice, with a special emphasis on the
characterization of K^+-permeability. Indeed, the data demonstrate that
pancreatic β-cells from a strain of ob/ob mice (Norwich colony) display a
profound impairment of their K^+-permeability system[37-39]. Thus, when
compared to β-cells from albino mouse islets, the ob/ob β-cells respond in
a completely different way to agents known to inhibit, either directly or
indirectly, K-channels[37,38]. Recently, the extraordinary development of
techniques suitable for recording ionic channel currents from either single
small cells or patches of cell membranes, has made possible the
characterization of single K-channels in pancreatic β-cells from various
sources[1,15,16,20]. Therefore, the main purpose of this paper is to focus
on the implications of the membrane potential data obtained from the ob/ob
pancreatic β-cells, with a special reference to the recent advances in the
understanding of the single K-channel events. It should be made clear that
these measurements have been made using a particular strain of ob/ob mice
(Norwich colony). As the expression of the obese gene depends on the
background genome carrying the gene[14,22], it would be inappropriate to
generalize the conclusions to other strains of ob/ob mice.

RESULTS AND DISCUSSION

<u>ob/ob and Normal β-cells: Different Patterns of Electrical Response to
D-glucose</u>

D-glucose is a well known inhibitor of K^+-permeability in normal mouse
and rat islets[3,10,18,25]. In normal β-cells, glucose induces bursts of
electrical activity at a threshold concentration of around 7 mM, the cells
remaining silent and hyperpolarized below threshold. In contrast, the
ob/ob β-cells often exhibit spiking or bursting electrical activity at
sub-threshold glucose concentrations[38] (Fig. 1). Although the excitability
pattern of ob/ob β-cells is variable from islet to islet electrical
activity has been recorded in the absence of glucose[38], a situation which
contrasts sharply with normal β-cells. In the absence of the secretagogue,
the ob/ob β-cells are, on average, 14 mV more depolarized than normal
β-cells[38]. The amplitude of this depolarization is in good agreement with

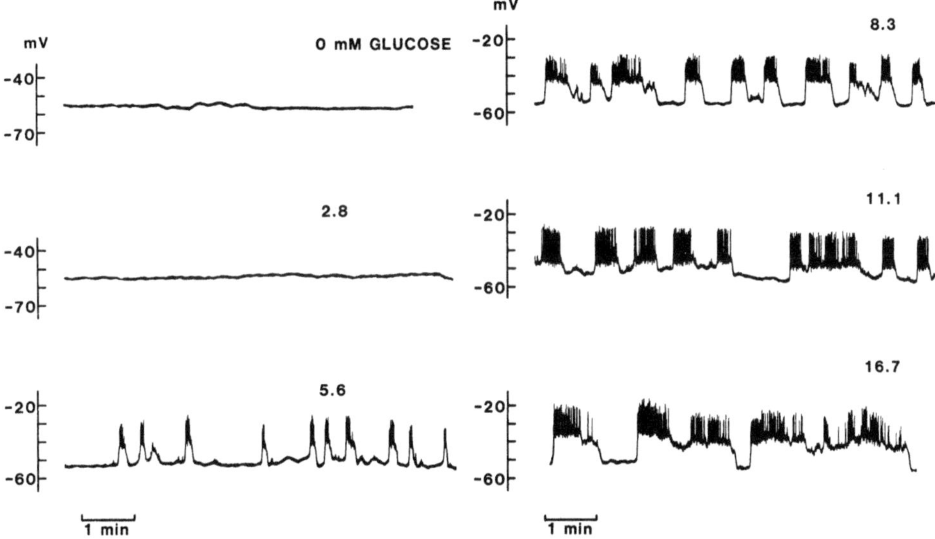

Fig. 1. Membrane potential responses to increasing glucose concentrations
in the ob/ob β-cell. The membrane potential was recorded with
high-resistance microelectrodes as previously described[38]. The
ob/ob islet was sequentially exposed to increasing glucose
concentrations. Each record shown was obtained towards the end of
a 15 min exposure to a given glucose level (same cell throughout).
Medium composition (mM): 120 NaCl, 25 $NaHCO_3$, 5 KCl, 2.6 $CaCl_2$,
1.1 $MgCl_2$ (pH 7.4).

the selectivity ratios P_{Na}/P_K calculated for both types of cells: 0.046
(normal)[40] and 0.11 (ob/ob)[39] (L.M. Rosario and E. Rojas, manuscript in

preparation). As was the case for islets from C57BL/KsJ db/db mice, a

decreased basal K^+-permeability has been found for islets from the Norwich

colony of ob/ob mice[42]. Thus, it is possible that the size of the stimulus

(glucose concentration) required for further membrane depolarization,

activation of voltage-sensitive Ca-channels and consequent triggering of

bursting electrical activity, is smaller in ob/ob than in normal β-cells.

This would provide a satisfactory explanation for the appearance of bursts

of electrical activity at sub-threshold glucose concentrations.

At high glucose concentrations, normal β-cells display a pattern of

continuous electrical activity. However, ob/ob β-cells often exhibit an

irregular pattern of bursting at high glucose concentrations[38] (Fig. 1),

and this difference is even more accentuated in β-cells from the C57BL/6J

strain of ob/ob mice (Rosario, manuscript in preparation), thus

demonstrating that such a feature is not peculiar to the strain of mice

used in the present studies (Norwich strain).

It has been previously reported that the curve relating spike frequency to glucose concentration in ob/ob β-cells is shifted towards lower glucose concentrations (half-maximal stimulation was obtained at 6.9 mM glucose as opposed to 10.2 mM for normal β-cells)[38]. In the concentration range 10-20 mM, this curve is much less steep for ob/ob β-cells than for normal β-cells, becoming virtually flat at around 12 mM. It is interesting to notice that the mean blood glucose levels reported for fed ob/ob mice fall in the range 10-15 mM[23,42]. As spike frequency may be considered a measure of Ca^{2+}-influx through voltage-sensitive Ca-channels, a well known requirement for glucose-induced insulin secretion, the electrical data suggest that the ob/ob mice would exhibit an impaired plasma insulin response to an intraperitoneal glucose challenge. This has been observed in ob/ob mice from the Aston (U.K.) strain[23]. Hence the data emphasize the physiological relevance of the K-channels that underline glucose-induced electrical activity.

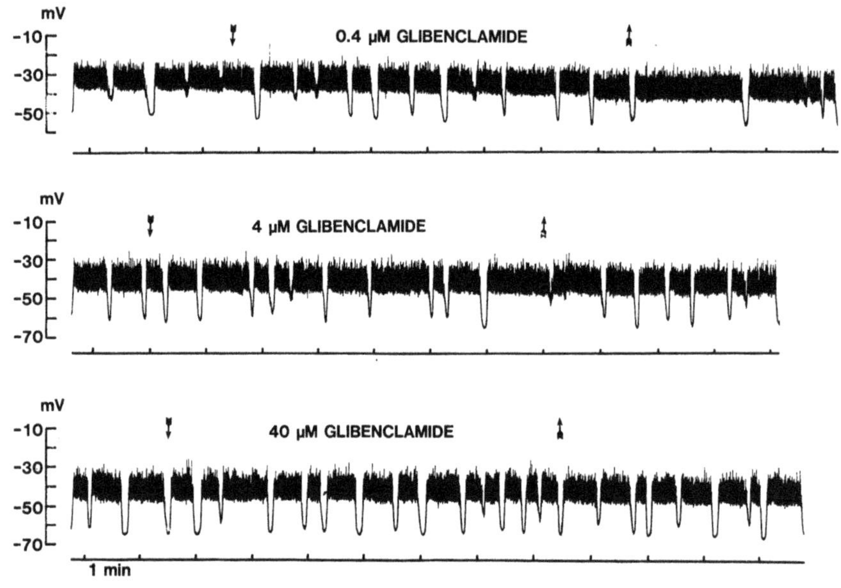

Fig. 2.   Effects of increasing concentrations of glibenclamide on glucose-induced electrical activity in the ob/ob β-cell. Consecutive segments of the membrane potential record from the same ob/ob β-cell (11 mM glucose throughout). Glibenclamide was added to and removed from the medium as indicated by the arrows.

<u>Effects of Sulphonylureas on ob/ob β-cell electrical activity:</u>
<u>Insensitivity to Glibenclamide and Reduced Sensitivity to Tolbutamide</u>

The sulphonylurea glibenclamide has long been known to depolarize the normal β-cell membrane and induce continuous electrical activity[31,32]. These effects have been attributed to inhibition of K^+-permeability[19]. However in ob/ob β-cells, glibenclamide (0.4-40 μM) failed to suppress the silent phases of glucose-induced bursts of electrical activity[38] (Fig. 2). These silent phases are thought to reflect activity of the Ca^{2+}-activated K-channel[4]. Furthermore, this lack of effect of glibenclamide has been observed in ob/ob β-cells basically unresponsive to glucose in the medium-high concentration range (Fig. 3). Recently, preliminary patch-clamp studies have indicated that glibenclamide blocks single K-channels in normal adult mouse β-cells in culture[6]. Hence, the present data suggest that: i) the ob/ob β-cell membrane either lacks the K-channel receptor for glibenclamide or contains a pharmacologically distinct K-channel as far as binding to sulphonylurea is concerned; and, ii) in normal β-cells, this K-channel may play a significant role on the modulation of the bursts of electrical activity in the medium-high glucose concentration range. Additionally, in the light of the observation that glibenclamide potentiates glucose-induced insulin release in islets from the Umeå (Sweden) colony of ob/ob mice[24], it is tempting to speculate that the glibenclamide-stimulated Ca influx is not the only aspect of sulphonylurea action as far as insulin release is concerned. However the possibility cannot be ruled out that β-cells from different colonies of ob/ob mice have different K-channel characteristics. Interestingly, it has been observed

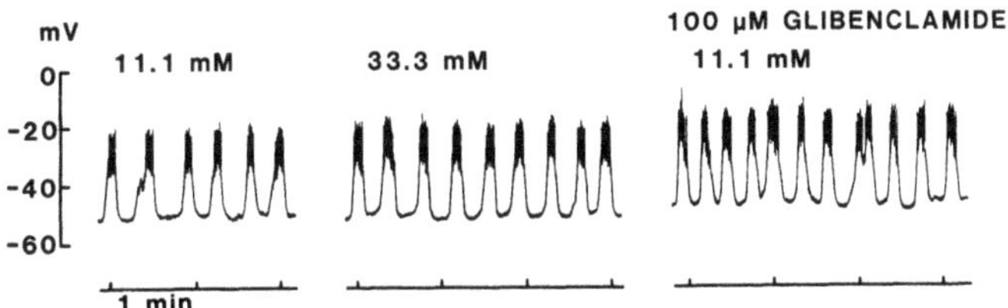

Fig. 3.   Comparison of the effects of a high glucose concentration and
glibenclamide on the bursting electrical activity of an ob/ob
β-cell. Left record: islet in 11.1 mM glucose; Middle record:
towards the end (8 min) of an exposure to 33.3 mM glucose; Right
record: towards the end (10 min) of an exposure to 100 μM
glibenclamide in the presence of 11.1 mM glucose. Same cell
throughout.

that intraperitoneal glibenclamide failed to alter the plasma insulin and
glucose concentration of fed ob/ob mice from the Aston (U.K.) Colony[8], thus
reinforcing the similarity between this strain and the Norwich Strain of
ob/ob mice.

Another sulphonylurea, tolbutamide, has also been observed to affect
differentially glucose-induced electrical activity in β-cells from normal
and ob/ob mice (Fig. 4). Thus, tolbutamide (70 μM) rapidly induced
continuous electrical activity in β-cells from normal mice[26] (Fig. 4A),
whereas in ob/ob β-cells the stimulatory effect of the sulphonylurea was

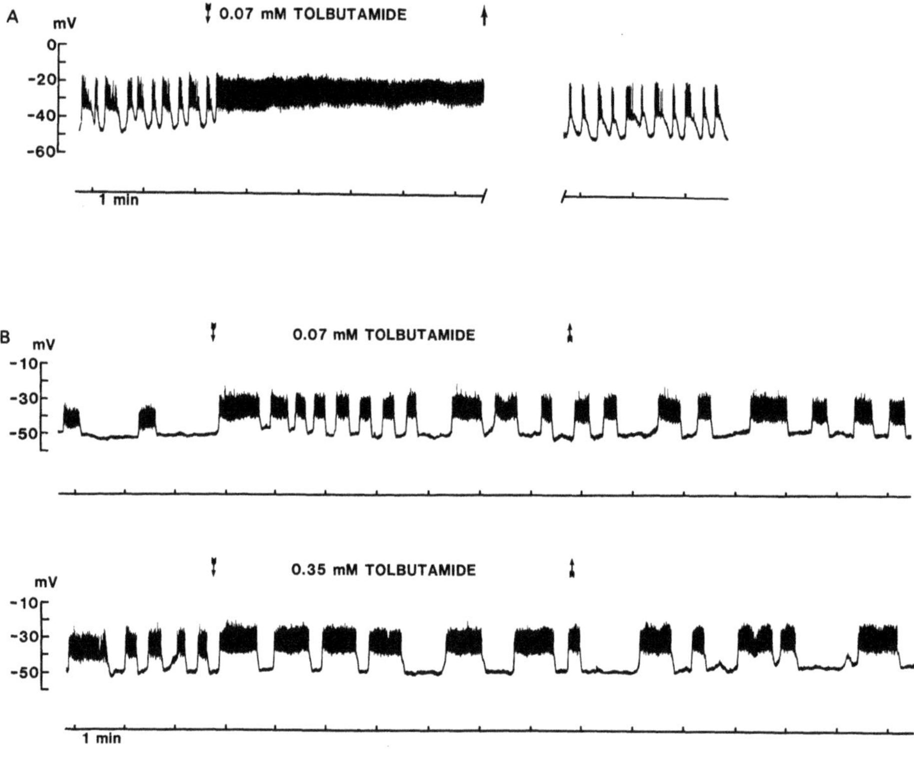

Fig. 4. Comparison of the effects of tolbutamide on glucose-induced
electrical activity in the normal and the ob/ob β-cell. A: an
albino mouse islet was exposed to 70 μM tolbutamide for about 5
min as indicated by the arrows. Seven min later the electrical
activity was as shown on the right side (11 mM glucose
throughout). B: an ob/ob islet was exposed to 70 μM and 350 μM
tolbutamide as indicated by the arrows. The lower record is the
continuation of the record shown in the upper part (11 mM glucose
throughout).

100

less remarkable (Fig. 4B; at 70 µM tolbutamide, an average 70% increase in the fraction of time in the active phase was measured in two experiments). It has been proposed that tolbutamide inhibits K^+-permeability in the normal β-cell[26]. Thus, taken together with the lack of effects of glibenclamide, the tolbutamide effects reported here suggest that these sulphonylureas, at least in the ob/ob β-cells, inhibit two distinct K-channels and/or bind to different sites on the same channel.

Effects of K-Channel Blockers on ob/ob β-cell electrical activity: Reduced Sensitivity to Quinine and Increased Sensitivity to Apamin

In the presence of glucose, quinine, a well known inhibitor of Ca^{2+}-activated K^+-permeability in red cells[30,34], is known to depolarize the normal β-cell membrane and induce continuous electrical activity[3]. On the other hand, in β-cells from ob/ob mice, quinine depolarizes the silent phase of the bursts of electrical activity in a dose-dependent fashion, but fails to suppress the bursting electrical activity even at very high quinine concentrations[38] (Fig. 5). According to the model of Atwater et al.[4] and to the theoretical modelling of the bursts of electrical activity carried out by Chay and Keizer[13], the slow oscillations of β-cell membrane potential may be explained by a feedback regulatory cycle involving $[Ca^{2+}]_i$ and activation/inactivation of Ca^{2+}-dependent K-channels. Thus, it seems plausible that the lack of steady-state effects of low concentrations of quinine (e.g. 100 µM) on glucose-induced electrical activity reflects an alteration in Ca^{2+}-activated K^+-permeability, as previously suggested[38]. This hypothesis is further supported by the observation that another

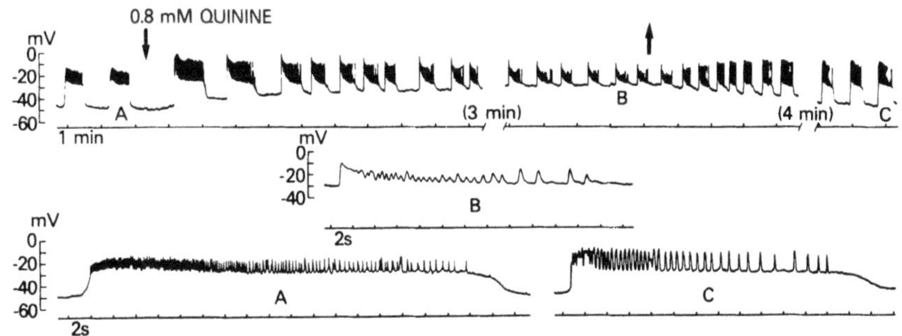

Fig. 5. Effects of a high concentration of quinine on glucose-induced electrical activity in the ob/ob β-cell. Quinine (0.8 mM) was added to and removed from the medium as indicated by the arrows (11 mM glucose and 5 mM $CaCl_2$ throughout). A, B and C: bursts from the upper record on an expanded time base. Notice the decrease in both the depolarization and repolarization rates of the spikes upon addition of quinine (bursts A and B).

Ca^{2+}-activated K-channel blocker, apamin[35], increases the duration of the bursts of electrical activity in ob/ob β-cells[37], but has no effect on electrical activity in normal β-cells[29].

CONCLUSIONS

ob/ob and normal pancreatic β-cells display striking differences in the membrane potential responses to several agents known to inhibit K^+-permeability in the normal β-cell and/or other cell types. These agents include quinine, apamin, the sulphonylureas glibenclamide and tolbutamide, and D-glucose. In summary, i) glibenclamide and quinine (50-200 μM) fail to affect the glucose-induced burst pattern of electrical activity in ob/ob β-cells, whereas in normal β-cells both agents induce continuous spiking activity; ii) apamin increases the duration of the bursts in ob/ob β-cells, but not in normal β-cells; and iii) the dose-response curve of electrical activity versus glucose concentration for ob/ob β-cells is shifted towards lower glucose concentrations compared to that of normal β-cells. These observations reveal profound pharmacological differences, and possibly functional differences as well, between the K-channels operating in the normal and ob/ob β-cell membranes.

It is becoming increasingly clear that at least two large conductance K-channels operate in pancreatic β-cells, namely a Ca^{2+}-activated K-channel[16,20] and an ATP-blockable K^+ channel[15]. The former channel has a very large conductance (about 250 pS), is voltage-sensitive[16,20] and may be blocked by lowering pH on the cytoplasmic side of the membrane[16,41]. Although in a tumor cell line this K-channel has been claimed to be non-blockable by quinine (100 μM)[21], there are indications that, at least in neonatal rat β-cells in culture, the channel is blocked by quinine[16]. The ATP-blockable channel has a conductance of about 50 pS, is voltage-insensitive, both Ca^{2+}- and pH-insensitive[15], seems to be blockable by glucose metabolism[1,36], and has been claimed to be blockable by quinine in a tumor β-cell line[21]. Due in part to the diversity of the patch-clamp techniques and the β-cell preparations used so far, the relative contributions of the Ca^{2+}-activated and of the ATP-blockable K-channels to the resting membrane potential in the absence of glucose, to the

glucose-induced depolarization and, finally, to the glucose-induced slow
oscillations of membrane potential, are still unclear.

It has been shown that glibenclamide blocks a 50 pS conductance,
voltage-insensitive K-channel in normal adult mouse pancreatic β-cells in
culture[6]. This channel resembles the ATP-blockable K-channel mentioned
above. As yet, there is no direct information on the specificity of
glibenclamide as a K-channel blocker. However, there are indications that
glibenclamide may also block the Ca^{2+}-activated K-channel. Thus, glibenclamide has been shown to impair the hyperpolarizing effect of a Ca-load
caused by high K^+ or by a mitochondrial uncoupler (CCCP) known to increase
$[Ca^{2+}]_i$[17,19]. Therefore, the lack of effect of glibenclamide on the ob/ob
β-cell bursting electrical activity, when the ATP blockable K-channel is
already totally inhibited by glucose[2], suggests that the Ca^{2+}-activated K-
channel in the ob/ob β-cell membrane is glibenclamide-insensitive. It is
also possible that this K-channel is quinine-insensitive and blockable by
apamin, as previously suggested[37]. If this was the case, the ob/ob and
normal β-cell K-channel networks would differ primarily in the pharmacological characteristics of the Ca^{2+}-activated K-channel.

Whatever the ultimate differences in the K-channel characteristics of
both types of cells may be, it seems clear that the ob/ob β-cell membrane
is provided with a modified K^+-permeability. Because the heterozygous lean
mice (litter mates) carrying the ob gene respond to quinine in the usual
way (L.M. Rosario, unpublished observations), this modified K^+-permeability
is highly unlikely to be a property of the particular genetic background of
the ob/ob mice used in these studies. However the possibility cannot be
ruled out that K-channels in pancreatic β-cells from other strains of
ob/ob mice might have different properties, as it is well known that the
expression of the ob gene depends on the background genome carrying the
gene[14,22]. It is therefore important that detailed membrane potential and
patch-clamp studies are carried out in pancreatic β-cells from other
strains of ob/ob mice. Although the possible link between the ob/ob β-cell
ionic channel impairment and the ob/ob diabetes-like syndrome is unknown,
the ob/ob mouse proves to be an interesting and potentially rich animal
model for the study of the β-cell K-channels.

ACKNOWLEDGEMENTS

The author thanks Drs. E. Rojas and I. Atwater for many discussions
and suggestions. Thanks are also due to Drs. P. Flatt, D.L. Coleman and
R.M. Santos for helpful comments and to Dr. K. Brocklehurst for careful
revision of the manuscript.

REFERENCES

1. F.M. Ashcroft, D.E. Harrison, and S.J.H. Ashcroft, Glucose induces
   closure of single potassium channels in isolated rat pancreatic
   β-cells, Nature 312:446 (1984).

2. F.M. Ashcroft, D.E. Harrison, and S.J.M. Ashcroft, A potassium channel
   modulated by glucose metabolism in rat pancreatic β-cells (this
   book).

3. I. Atwater, C.M. Dawson, B. Ribalet, and E. Rojas, Potassium
   permeability activated by intracellular calcium ion concentration
   in the pancreatic β-cell, J. Physiol. 288:575 (1979).

4. I. Atwater, C.M. Dawson, A. Scott, G. Eddlestone, and E. Rojas, The
   nature of the oscillatory behaviour in electrical activity from
   pancreatic β-cell, Horm. Metab. Res. (Suppl.) 10:100 (1980).

5. I. Atwater, B. Ribalet, and E. Rojas, Cyclic changes in potential and
   resistance of the β-cell membrane induced by glucose in islets of
   Langerhans from mouse, J. Physiol. 278:117 (1978).

6. I. Atwater, P. Rorsman, B.A. Suarez-Isla, and E. Rojas, Single K^+
   channels blocked by glibenclamide in normal adult mouse pancreatic
   β-cells in culture, Diabetes 34:94A (1985).

7. I. Atwater, L.M. Rosario, and E. Rojas, Properties of the Ca-activated
   K-channel in pancreatic β-cells, Cell Calcium 4:451 (1983).

8. C.J. Bailey and P.R. Flatt, Characterization of the insulin secretory
   defect in fed adult Aston ob/ob mice: lack of effect of
   glibenclamide, Bioch. Soc. Trans. 10:28 (1982).

9. O. Berglund, J. Sehlin, and I.-B Taljedal, Defective ion transport in
   diabetic mouse islet cells, Acta Physiol. Scand 106:489 (1979).

10. A.C. Boschero and W.J. Malaisse, Stimulus-secretion coupling of
    glucose-induced insulin release, XXIX. Regulation of Rb^+ efflux
    from perifused islets, Am. J. Physiol. 236:E139 (1979).

11. G.A. Bray and D.A. York, Hypothalamic and genetic obesity in experimental animals: an anatomic and endocrine hypothesis, Physiol. Rev. 59:719 (1979).

12. H.M. Charlton, Mouse mutants as models in endocrine research, Quart. J. Exp. Physiol. 69:655 (1984).

13. T.R. Chay and J. Keizer, Minimal model for membrane oscillations in the pancreatic β-cell, Biophys. J. 42:181 (1983).

14. D.L. Coleman and K.P. Hummel, The influence of genetic background on the expression of the obese (ob) gene in the mouse, Diabetologia 9:287 (1973).

15. D.L. Cook and C.N. Hales, Intracellular ATP directly blocks K^+ channels in pancreatic β-cells, Nature 311:271 (1984).

16. D.L. Cook, M. Ikeuchi, and W.Y. Fujimoto, Lowering of pH_i inhibits Ca^{2+}-activated K^+ channels in pancreatic β-cells, Nature 311:269 (1984).

17. C.M. Dawson, I. Atwater, and E. Rojas, The response of pancreatic β-cell membrane potential to potassium-induced calcium influx in the presence of glucose, Quart. J. Exp. Physiol. 69:819 (1984).

18. C.M. Dawson, P.C. Croghan, I. Atwater, and E. Rojas, Estimation of potassium permeability in mouse islets of Langerhans, Biom. Res. 4:389 (1983).

19. R. Ferrer, I. Atwater, E.M. Omer, A.A. Goncalves, P.C. Croghan, and E. Rojas, Electrophysiological evidence for the inhibition of potassium permeability in pancreatic β-cells by glibenclamide, Quart. J. Exp. Physiol. 69:831 (1984).

20. I. Findlay, M.J. Dunne, and O.H. Petersen, High-conductance K^+ channel in pancreatic islet cells can be activated and inactivated by internal calcium, J. Membrane Biol. 83:169 (1985).

21. I. Findlay, M.J. Dunne, S. Ullrich, C.B. Wollheim, and O.H. Petersen, Quinine inhibits Ca^{2+}-independent K^+ channels whereas tetraethylammonium inhibits Ca^{2+}-activated K^+ channels in insulin-secreting cells, FEBS Lett. 185:4 (1985).

22. P.R. Flatt and T.W. Atkins, Influence of genetic background and age on the expression of the obese hyperglycaemic syndrome in Aston ob/ob mice, Int. J. Obesity 6:11 (1982).

23. P.R. Flatt and C.J. Bailey, Development of glucose intolerance and impaired plasma insulin response to glucose on obese hyperglycaemic (ob/ob) mice, Horm. Metab. Res. 13:556 (1981).

24. B. Hellman, A. Lernmark, J. Sehlin, M. Soderberg, and I.-B. Taljedal, On the possible role of thiol groups in the insulin-releasing action of mercurials, organic disulfides, alkylating agents and sulfonylureas, _Endocrinology_ 99:1398 (1976).

25. J.C. Henquin, D-glucose inhibits potassium efflux from pancreatic islet cells, _Nature_ 271:271 (1978).

26. J.C. Henquin and H.P. Meissner, Opposite effects of tolbutamide and diazoxide on $^{86}Rb^+$ fluxes and membrane potential in pancreatic β-cells, _Biochem. Pharmacol._ 31:1407 (1982).

27. J.C. Henquin and H.P. Meissner, Significance of ionic fluxes and changes in membrane potential for stimulus-secretion coupling in pancreatic β-cells, _Experientia_ 40:1043 (1984).

28. L. Herberg and D.L. Coleman, Laboratory animals exhibiting obesity and diabetes syndromes, _Metabolism_ 26:59 (1977).

29. P. Lebrun, I. Atwater, M. Claret, W.J. Malaisse, and A. Herchuelz, Resistance to Apamin of the Ca^{2+} activated K^+-permeability in Pancreatic β-cells, _FEBS Lett._ 161:41 (1983).

30. V.L. Lew and H.G. Ferreira, _in_: "Membrane Transport in Red Cells", J.C. Ellory and V.L. Lew, eds, Academic Press (1977).

31. E.K. Matthews, P.M. Dean, and Y. Sakamoto, _in_: "Pharmacology and the Future of Man", Proceedings of the 5th International Congress on Pharmacology, San Francisco 1972, Vol. 3 (1973).

32. H.P. Meissner and I.J. Atwater, The kinetics of electrical activity of β-cells in response to a 'square wave' stimulation with glucose or glibenclamide, _Horm. Metab. Res._ 8:11 (1976).

33. H.P. Meissner and H. Schmidt, The electrical activity of pancreatic β-cells of diabetic mice, _FEBS Lett._ 67:371 (1976).

34. E. Reichstein and R. Rothstein, Effects of quinine on Ca^{2+}-induced K^+-efflux from human red blood cells, _J. Membrane Biol._ 59:371 (1981).

35. G. Romey, M. Hugues, H. Schmid-Automarchi, and M. Lazdunski, Apamin: A Specific Toxin to Study a Clan of Ca^{2+}-dependent K^+ Channels, _J. Physiol._ (Paris) 79:259 (1984).

36. P. Rorsman and G. Trube, Glucose dependent K^+-channels in pancreatic β-cells are regulated by intracellular ATP, _Pflüg. Arch._ 405:305 (1985).

37. L.M. Rosario, Differential effects of the K^+ channel blockers apamin and quinine on glucose-induced electrical activity in pancreatic β-cells from a strain of ob/ob (obese) mice, _FEBS Lett._ 188:302 (1985).

38. L.M. Rosario, I. Atwater, and E. Rojas, Membrane potential
    measurements in islets of Langerhans from ob/ob obese mice suggest
    an alteration in $[Ca^{2+}]_i$-activated K^+-permeability, Quart. J. Exp.
    Physiol. 70:137 (1985).

39. L.M. Rosario, I. Atwater and E. Rojas, Electrophysiological
    measurements show marked differences in the properties of the
    K^+-channels from ob/ob and albino mouse pancreatic β-cells, Diab.
    Res. Clin. Pract. (Suppl.) 1:S478 (1985).

40. L.M. Rosario and E. Rojas, Potassium channel selectivity in mouse
    pancreatic β-cells, Am. J. Physiol. 250:C90 (1986).

41. L.M. Rosario and E. Rojas, Modulation of K^+-conductance by
    intracellular pH in pancreatic β-cells, FEBS Lett. 200:203-209
    (1986).

42. A. Scott, C.M. Dawson, and A.A. Goncalves, Comparison of glucose-
    induced changes in electrical activity, insulin release, lactate
    output and potassium permeability between normal and ob/ob mouse
    islets: effects of cooling, J. Endocrinol. 107:265 (1985).

43. A.M. Scott, I. Atwater, and E. Rojas, A method for the simultaneous
    measurement of insulin release and β-cell membrane potential in
    single mouse islets of Langerhans, Diabetologia 21:470 (1981).

SINGLE K-CHANNEL ACTIVITY IN FISH ISLET CELLS

R.M. Santos, H. Finol* and E. Rojas

Laboratory of Cell Biology and Genetics
NIDDK, National Institutes of Health
Bethesda, MD 20892

*Escuela de Biologia
Universidad Central de Venezuela
Apartado 21201
Caracas, Venezuela

INTRODUCTION

The development of the patch-clamp technique, nearly ten years
ago, allowed electrical measurements to be made under voltage-clamp
conditions on small cells unsuitable for two microeletrodes voltage-
clamp techniques (see[14]).

Following the original observations of Marty and Neher,[21] several
K$^+$ channels have been characterized in pancreatic islet cells by
single-channel patch-clamp recording techniques.[1, 6, 7, 12, 13] These
studies have been performed using cultured islet cells from various
mammalian sources, either on intact cells or on excised patches of
membranes. To improve the chances of getting a stable, high-resistance seal between the islet cell membrane and the glass micropipette
used for single-channel recording, the experiments have been carried
out at room temperature. However, it is well known that both glucose
metabolism and insulin release are inhibited by lowering the temperature below 37°C,[2,20] thus raising some questions as to the significance of the electrophysiological data for _in vivo_ conditions. In an
attempt to circumvent this problem, we have characterized the K$^+$
channels from the principal islet of the rainbow trout _Salmo
gairdneri_.

Immunocytochemical identification of well defined cellular domains,[23, 24] together with the demonstration of cell-to-cell electrical coupling,[9, 10] appears to support the idea that islet function is
modulated by the coordinated action of groups of cells within the
islet[22] (see also [3]). Hence, it was considered important to keep
the islet preparation as intact as possible and single K^+ channel
currents have been recorded from clusters of cells using the cell
attached configuration. In addition, an electron microscopy study was
carried out to characterize the different cell types in the principal
islet.

METHODS

Patch-Clamp Experiments

The fish (rainbow trout _Salmo gairdneri_ 50-100 g) was killed by a
blow on the head. The abdomen was exposed by a ventral incision and
the principal islet removed by microdissection under microscope. The
membrane surrounding the islet was cut open and a small portion from
the central part was transferred to the experimental chamber, where it
was held by a suction pipette. The tissue was then exposed for 50 min
to a modified Krebs solution containing collagenase (1 mg/ml, Sigma
type V) at room temperature.

Gigaohm resistance seals were produced as described before.[14]
The output of the patch-clamp amplifier (10^9 ohm feedback resistance)
was low-pass filtered using a 8-pole Bessel filter set at 300 Hz and
recorded on FM tape (Racal Store 4, Racal Recorders Ltd., Southampton,
U.K.).

The ionic composition of the modified Krebs solution in the bath
was as follows (mM): 120 NaCl, 2 KCl, 2.5 $CaCl_2$, 1 $MgCl_2$, 5 $NaHCO_3$,
20 Na-Hepes, 2.8 glucose (pH 7.4). The patch-clamp pipette was filled
with a solution containing (mM): 120 KCl and 20 Na-Hepes (pH 7.4).
The experiments were performed in a temperature-controlled room
(16°C).

Islet Treatment for Electron Microscopy

Immediately after ventral incision and just before the veins
collapsed, glutaraldehyde was applied on the region of the principal

islet. The islet was then microdissected and cut in two portions.
Both halves were fixed during 30 min in 3.6% glutaraldehyde solution,
rinsed for 5 minutes and exposed to 1% osmium tetroxide solution
during 60 min. After a 10 min wash, the tissue was incubated for 10
min in 1% uranile acetate. Specimens were then dehydrated in graded
ethanol solutions, transferred through three fresh solutions of
propylene oxide and embedded in resin. Ultrathin sections were observed under a transmission electron microscope (Hitachi, model H-500,
Japan).

RESULTS

Patch-Clamp Experiments in Fish Islet Cells

Typically, seal resistances in the range 1 to 5 gigaohms were
obtained on the cell attached patch-clamp configuration. After a seal
was achieved at a 0 mV pipette potential, inward current jumps of
various durations were immediately observed (Fig. 1A). The amplitude
of the current noise recorded during the conducting state was larger
than in the non-conducting state. Also, brief closures were often
observed during a longer opening, resembling the flickering described
before.[8] These fast returns to the non-conducting state were sometimes incomplete due probably to the low pass filter used.

At 0 mV pipette potential, the channels have been observed to
spend most of the time in the closed configuration (see Fig. 2).
Thus, assuming a two-state (open and closed) model, the probability of
N channels existing in the patch of membrane is given by:

$$P_{obs} = \exp \{-n_o (N-1) m_o/(N\ m_s)\}$$

where n_o is the number of events observed, m_o the mean open time and
m_s the mean closed time [see 5]. For the experiment depicted in Fig.
1, 390 events were measured. The probability that two channels in the
patch (N=2) generated the fluctuations recorded is very small
(<0.004), suggesting that the patch-clamp records reflect the activity
of one single channel.

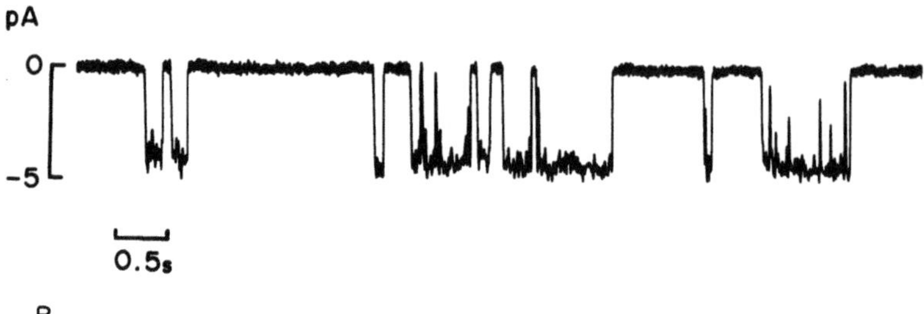

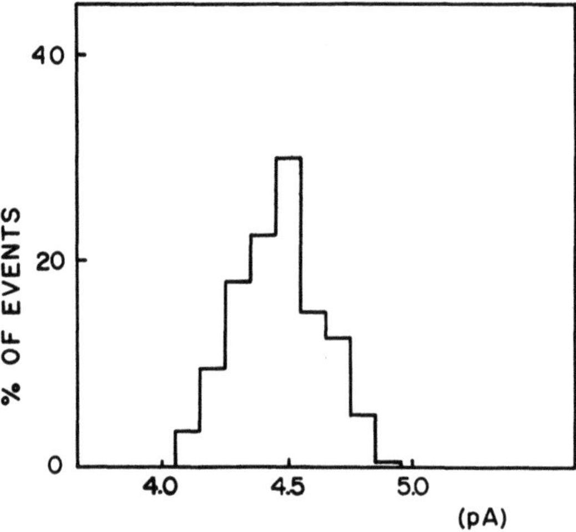

Fig. 1. Single-channel current recording (A) and frequency-amplitude histogram (B). The cell-attached recording mode was used. Pipette contained 120 mM KCl, 20 mM Na-Hepes (pH 7.4). High Na-Krebs was used in the bath. Pipette potential equal to 0 mV. No. of events in the histogram: 148.

Fig. 1B shows the frequency-amplitude histogram. It may be seen that the amplitude of the events appeared to be normally distributed around a mean value of -4.5 pA.

Fig. 2 depicts current records obtained at different pipette potentials (Vp). As the pipette potential was made more positive, the amplitude of the events increased continuously. This relationship is shown on Fig. 3 for two similar experiments. It may be seen that the amplitude of the events appear to decrease linearly with membrane potential. The mean single channel conductance and reversal pipette

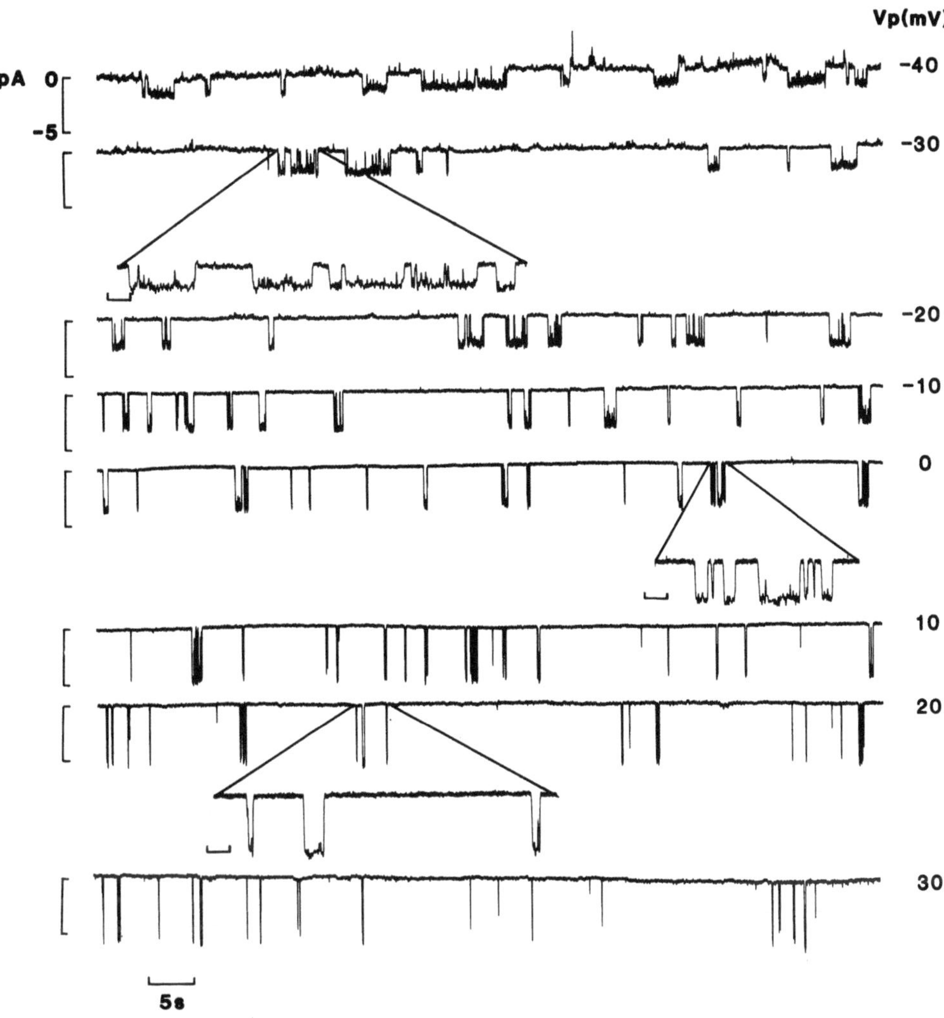

Fig. 2. Single-channel current recordings at different pipette potentials. Pipette potential is indicated on the right side of the corresponding current record. Calibration for expanded time-base records
shown at -30, 0 and 20 mV pipette potentials: 0.5 s.

potential, calculated from the slopes of the straight lines fitting
the current-voltage relationships and from their intercepts on the Vp
axis, are 74 pS and -58 mV respectively. Thus, assuming that only a
negligible K^+ gradient exists across the islet cell membrane (120 mM
K^+ in the patch-clamp pipette filling solution; $[K^+]_i$ = 110 mM), the
resting membrane potential can be taken, in a first approximation, as
-58 mV.

The channel events seemed to appear in groups, separated by long

113

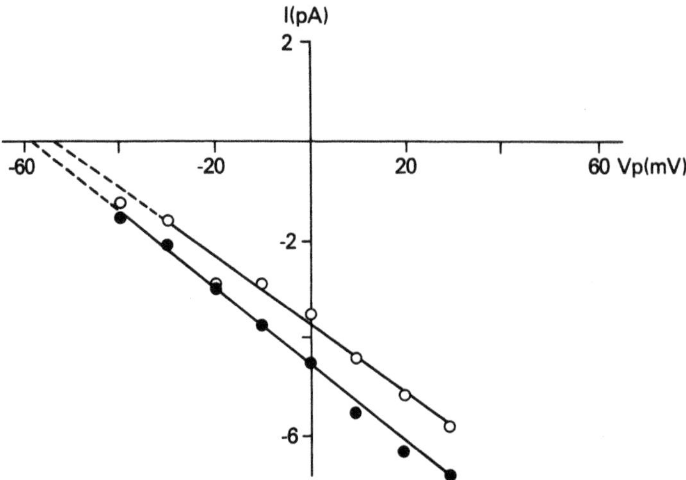

Fig. 3. Current-voltage relationships. Open and filled symbols correspond to two different experiments. Voltage values correspond to pipette holding potentials. Straight lines were fitted by eye.

periods in the non-conducting state (Fig. 2). During the periods of activity, the state of the channel fluctuated rapidly between the open and closed configuration (Fig. 2, amplified records).

The frequency distribution of the closed times at 0 mV pipette potential (Fig. 4) shows that the decay cannot be fitted by a single

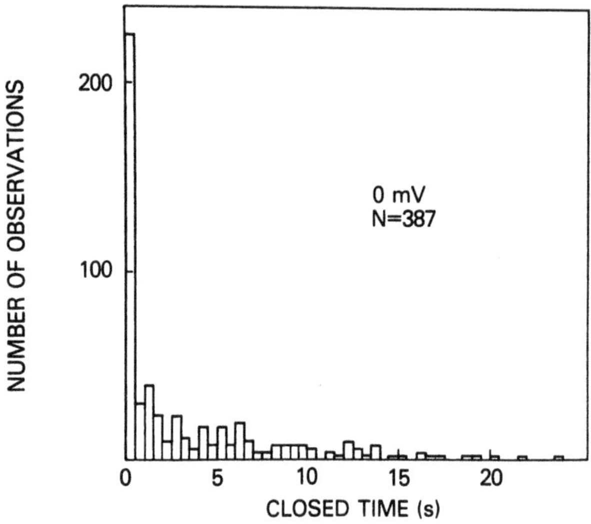

Fig. 4. Closed time histogram. The measurements were done at 0 mV pipette potential. N = number of observations.

exponential. This result suggests that a single first-order process
cannot be used to describe the channel closing kinetics.

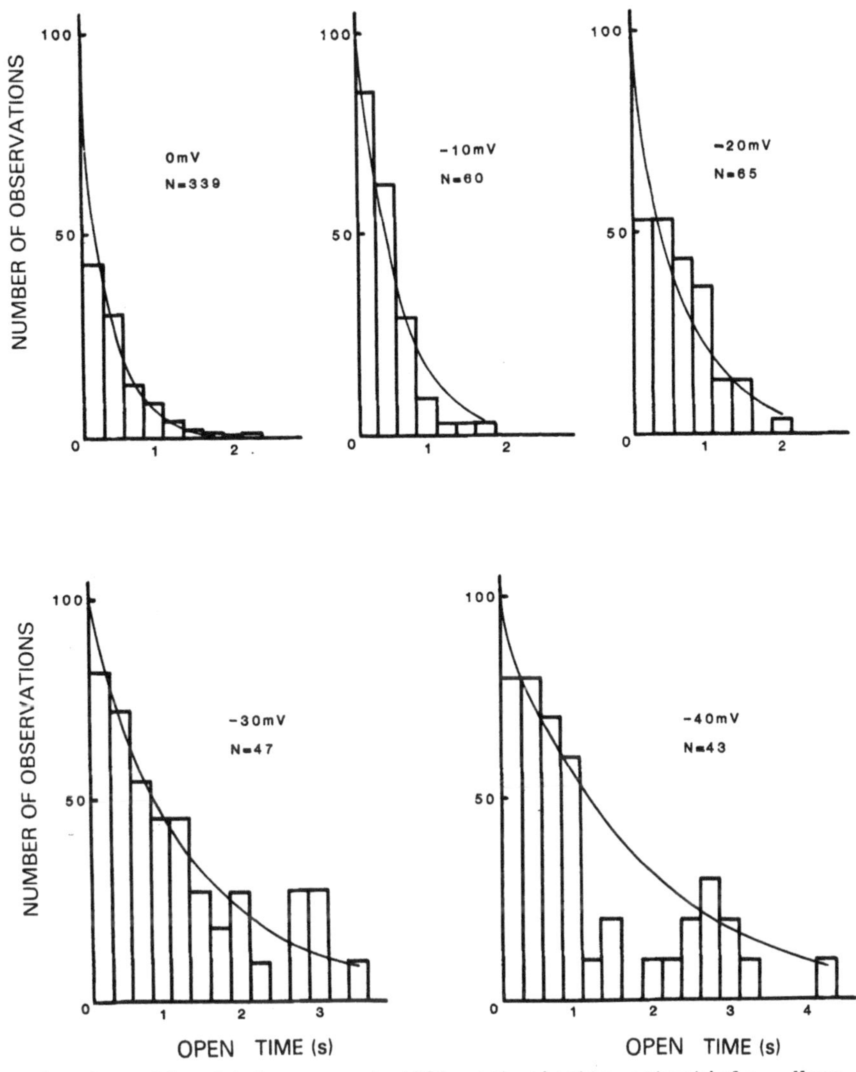

Fig. 5. Open-time histograms at different pipette potentials. Normalized histograms (i.e., the origin of the fitted exponential curve
taken as 100). Curves fitted by linear regression of the logarithm of
the data. Pipette potentials are given next to the corresponding
histogram. N = number of observations.

Also apparent from the records in Fig. 2 is the inverse voltage
dependence of the time spent in the conducting state. The frequency
distributions of the open time at different pipette potentials could
be fitted by single exponentials (Fig. 5). The channel open mean
lifetime became shorter as the pipette potential was made more posi -

tive, an approximate 5-fold decrease in the half-life being obtained
for a 40 mV change in V_p. Not only the channel half-life is the
voltage-dependent but also the overall time spent in the open state
(Fig. 6). It may be seen that the fraction of time the channel spent
in the open state decreased monotonically with pipette potential.

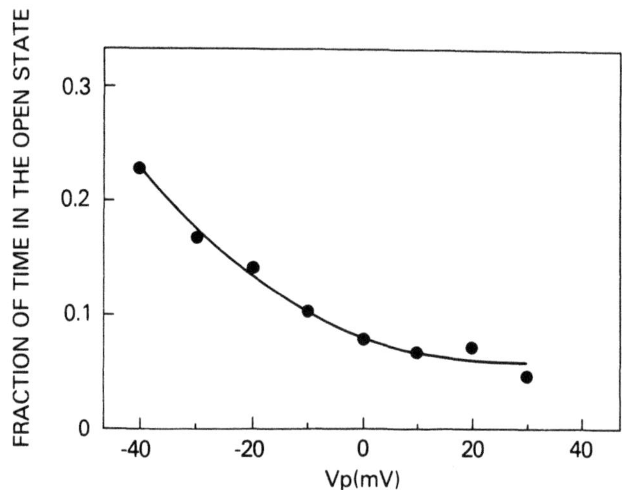

Fig. 6. Voltage dependence of the fraction of time in the open state.
Vp corresponds to pipette holding potentials. Curve fitted by eye.

Ultrastructural characterization of the principal islet

The principal islet in the rainbow trout _Salmo gairdneri_ is a
giant, highly vascularized oval body (c.a. 2 mm) located next to the
gallbladder. The endocrine cells, found in the center of the organ
surrounded by exocrine tissue, are concentrated in two bulk regions as
follows: β and δ cells in the central part of the islet; α and
pancreatic polypeptide (PP)-cells in the periphery.

β-cells, found mainly around the capillaries, are the predominant
cell type in the central region (Fig. 7B). β-cells are characterized
by a round, sometimes elongated nucleus with an heterogeneous
granulation. The endoplasmic reticulum appears scattered throughout
the cytosol and also, although less frequently, around the nucleus.
Round as well as elongated mitochondria are often seen. β-cell
secretory granules (larger diameter 150-450 nm) are abundant and

116

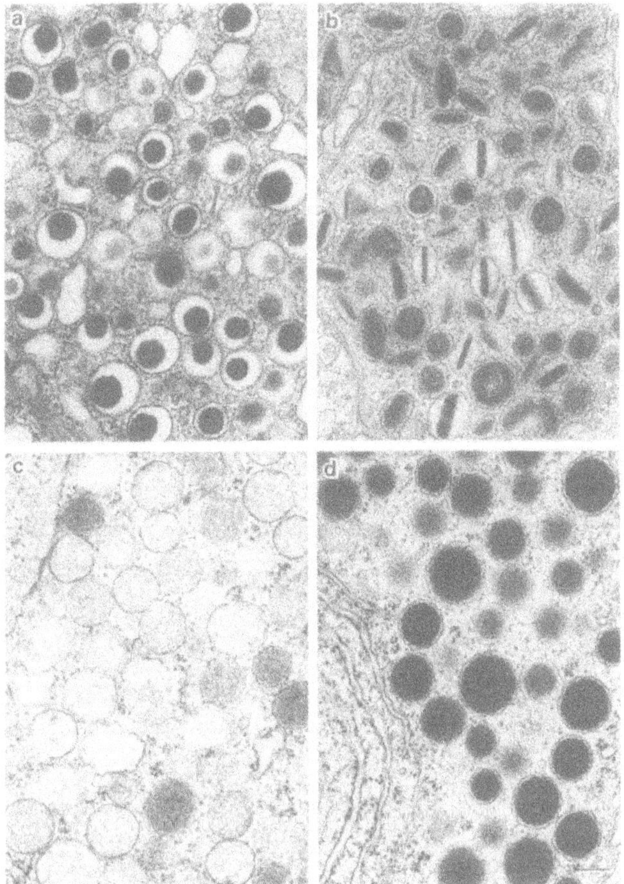

Fig. 7. Endocrine cell types in the Principal islet of rainbow trout
Salmo gairdneri. a) α-cell; b) β-cell; c) δ-cell; d) PP-cell. Calibration bar: 200nm.

distributed throughout the cytoplasm. Two different granular content
arrangements, coexisting in the cell, are apparent in the β-granules,
namely i) a configuration characterized by a round-structured electron
dense substance, separated from the granule membrane by a narrow clear
halo; and, ii) a rod-like electron dense substance with a large clear
halo (see Fig. 7B).

δ-cells (Fig. 7C) are often found in close association with
β-cells, frequently surrounding groups of these cells. The δ-cell
nucleus is round-shaped as in the β-cell, but less dense. A few oval
or elongated mitochondria are often seen in the cytosol, the endoplasmic reticulum being found mainly around the nucleus. The δ-cell
granules (220-430 nm) are characterized by a low electron density
content of rough, even floccular appearance, no gap between the matrix
and the granule membrane being apparent.

α-cells (Fig. 7A) are the predominant cell type in the peripheral
region. In these cells, the endoplasmic reticulum is situated mainly
around the nucleus, which in general is irregularly shaped. The
α-cell granules (150-450 nm) are round and contain an electron dense
core, which is located asymetrically, and separated from the granule
membrane by a very large and clear halo. The core of the α-cell
granule appears to have an hexagonal shape.

PP-cells (Fig. 7D) are found as the smallest proportion of the
endocrine cells in the periphery of the islet. PP-cells are characterized by an irregular nucleus with large lobes and an endoplasmic
reticulum usually forming a large whorl. The PP-cell granules
(180-470 nm) exhibit a highly electron dense core, separated from the
granule membrane by a very narrow clear halo.

Scattered throughout the principal islet, a poorly granulated
cell type (usually denominated clear cell) was observed (not shown).
The clear cell contains an irregularly shaped, non-uniformely
granulated nucleus and a low number of heteromorphous granules, predominantly with an appearance similar to the α-cell granule (see
[26]).

DISCUSSION

We report here for the first time single K$^+$ channel current
measurements from non-mammalian islet cells, i.e., endocrine cells
from the principal islet of the trout. Since these measurements have
been carried out on clusters of cells extracted from the central
region of the islet, where the predominant endocrine cell types have
been identified as β- and δ-cells, it is likely that the recorded K$^+$

channel activity has originated from either insulin or somatostatin-
secreting cells. In addition, this study represents an original
attempt to characterize K^+ channels keeping the cell intact and associated with other islet cells in their primitive arrangement. In this
respect, the characteristics of the trout principal islet, namely its
giant size, makes this endocrine organ an attractive model system for
the study of single channel activity in quasi-intact conditions.

K^+ Channel Characteristics

The amplitude distribution of the single channel currents appeared to be normally distributed, suggesting that the channel does
not exhibit sub-conductance states. Its unit conductance, estimated
from linear current-voltage relationships, was 74 pS with 120 mM KCl
in the pipette, constant for pipette potentials in the range from -40
to 30 mV. The kinetics of the K^+ channel openings exhibited a marked
voltage dependence. In addition, channel activity seemed to occur in
bursts, the channel shuting for a prolonged length of time at the end
of each burst.

In rat pancreatic islet-cells, at least two distinct large
conductance K^+ channels have been identified. One of them shows at
20-25°C a single conductance of about 50 pS, with $[K^+]_0=140$ mM, or
about 17 pS, with $[K^+]_0=5$ mM in quasi-physiological solution, is
blockable by ATP and is totally insensitive to membrane
potential.[1,6,13] The second type of K^+ channel has, at 22-26°C, a
single conductance of about 250 pS, in symetrical K^+ solutions, or 90
pS in quasi-physiological solutions, and its mean open time is modulated by Ca^{2+} and voltage.[7, 12, 13] The K^+ channel recorded from fish
islet cells resembles the rat islet cell Ca^{2+}-activated K^+ channel
with regard to its voltage sensitivity. It is possible that its lower
conductance may be explained in part in terms of temperature differences.[4]

Endocrine Cell Types in the Principal Islet

We have observed that β- and δ-cells are confined to the central
region of the principal islet, whereas the α- and PP-cells are mainly
located in the peripheral region. Thus this confirms the immunocytochemical data of Wagner and McKeown on the distribution of the cells

in the islet of _Salmo gairdneri_.[30] In other fish species, a similar
cell topology has been reported.[11,15,28]

We have seen that, in the central region, large aggregates of
β-cells are easily identified by the characteristic laminar rod shape
of the core of its granule, as reported for other teleosts species.[15,18,28]
We have also observed that fish islet δ-cells, identified by
their characteristic pale granules without halo, appear closely associated with β-cells. The features of the somatostatin granule reported here have been identified in most teleosts,[17,18,27,28,29]
while in two species they seem to have a totally different appearance:
reduced size and an electron dense core.[15,25]

In the periphery of the islet, we have shown that the predominant
cell type, identified before as a glucagon producing cell[30] contained
hexagonal shaped granule structures, a finding supported by the observation that glucagon, from _Xiphophorus helleri_, crystallizes in a
rhombic dodecahedron form in _in vitro_ studies.[19] Pancreatic polypeptide (PP) producing cells, characterized by homogeneous granules
provided with round cores and narrow halo have also been described in
other species.[16,27] The PP-cells observed here also resemble a
distinct cell type of unidentified physiological function designated
sometimes as F-cell.[17,28,29]

The less frequent cells described here (clear cells) were not
devoid of granules,[26] in contrast with other reports. In this study,
the few existent granules usually resembled the α-cell granules, a
feature that may reflect a different developmental stage of the α-cell
secretory material or of the α-cells themselves. In keeping with this
possibility, Wagner and McKeown[30] have reported that a non-specific
staining of these cells could be observed after exposure to anti-
glucagon and anti-GIP sera.

ACKNOWLEDGEMENTS

Thanks are given to Dr. B. Suarez-Isla and Dr. L. M. Rosario for
reading the manuscript and to Dr. R. Ornberg for suggestions on the
presentation of the ultrastructural data. This work was supported in
part by the Science Research Council of the United Kingdom.

REFERENCES

1. F. M. Ashcroft, D. E. Harrison, and S. J. H. Ashcroft, Glucose
     induces closure of single potassium channels in isolated rat
     pancreatic β-cells, Nature 312:446 (1984).
2. I. Atwater, A. Goncalves, A. Herchuelz, P. Lebrun, W. J.
     Malaisse, E. Rojas, and A. Scott, Cooling dissociates glu-
     cose-induced insulin release from electrical activity and
     cationic fluxes in pancreatic islets, J. Physiol.
     348:615 (1984).
3. I. Atwater, L. M. Rosario, and E. Rojas, Properties of the
     Ca-activated K-channel in pancreatic β-cells. Cell Calcium
     4:451 (1983).
4. J. N. Barrett, K. L. Magleby, and B. S. Pallotta, Properties of
     single calcium-activated potassium channels in cultured rat
     muscle, J. Physiol. 331:211 (1982).
5. D. Colguhoun and A. G. Hawkes, The principles of the stochastic
     interpretation of ion-channel mechanisms, in: "Single-Chan-
     nel Recording," B. Sakmann and E. Neher, ed., Plenum Press,
     N.Y. and London, pp. 135-176 (1983).
6. D. L. Cook and C. N. Hales, Intracellular ATP directly blocks K^+
     channels in pancreatic β-cells, Nature 311:271 (1984).
7. D. L. Cook, M. Ikeuchi, and W. Y. Fujimoto, Lowering of pHi inhib-
     its Ca^{2+}-activated K^+ channels in pancreatic β-cells, Nature
     311:269 (1984).
8. L. J. DeFelice and J. R. Clay, Membrane current and membrane
     potential from single-channel kinetics, in: "Single-Channel
     Recording," B. Sakmann and E. Neher, ed., Plenum Press, N.Y.
     and London, p. 323 (1983).
9. G. T. Eddlestone, A. Goncalves, J. A. Bagham, and E. Rojas,
     Electrical coupling between cells in islets of Langerhans
     from mouse, J. Membrane Biol. 77:1 (1984).
10. G. T. Eddlestone and E. Rojas, Evidence of electrical coupling
     between mouse pancreatic β-cells, J. Physiol. 303: 76P
     (1980).
11. S. Falkmer and B. Hellman, Identification of the cells in the
     endocrine pancreatic tissue of the marine teleost Cottus
     scorpius by some silver impregnation procedures, Acta
     Morphologica Neerlando-Scandinavica 4:145 (1961).
12. I. Findlay, M. J. Dunne, and O. H. Petersen, High-conductance K^+
     channel in pancreatic islet cells can be activated and

inactivated by internal calcium, _J. Membrane Biol._
83:169 (1985).

13. I. Findlay, M. J. Dunne, and O. H. Petersen, ATP-sensitive inward
    rectifier and voltage- and calcium-activated K^+ channels in
    cultured pancreatic islet cells, _J. Membrane Biol._
    88:165 (1985).

14. O. P. Hammill, A. Marty, E. Neher, B. Sakmann, and F. J.
    Sigworth, Improved patch-clamp techniques for high-resolu-
    tion current recording from cells and cell-free membranes
    patches, _Pflugers Arch._ 391:85 (1981).

15. C. Klein and R. H. Lange, Principal cell types in the pancreatic
    islet of a teleost fish, _Xiphophorus helleri H._, _Cell Tiss.
    Res._ 176:529 (1977).

16. C. Klein and S. Van Noorden, Pancreatic polypeptide (PP)-and
    glucagon cells in the pancreatic islet of _Xiphophorus
    helleri H._ (teleostei), _Cell. Tiss. Res._ 205:187 (1980).

17. K. Kobayashi, S. Shibasaki, and Y. Takahashi, Light and electron
    microscopic study on the endocrine cells of the pancreas in
    a marine teleost, _Fugu rubripes rubripes_, _Cell Tiss. Res._
    174:161 (1976).

18. K. Kobayashi and Y. Takahashi, Light and electron microscope
    observations on the islets of Langerhans in _Carassius
    carassius longsdorfii_, _Arch. Histol. Jap._ 31:433 (1970).

19. R. H. Lange and C. Klein, Rhombic dodecahedral secretory granules
    in glucagon producing islet cells, _Cell Tiss. Res._ 148:561
    (1974).

20. F. Malaisse-Lagae, A. Sener, P. Lebrun, A. Herchuelz, V.
    Leclercq-Meyer, and W. J. Malaisse, Reponse secretoire,
    ionique et metabolique des ₁lots de Langerhans aux anomeres
    du D-mannose, _C.R. Hebd. Seanc. Acad. Sci., Paris_ 294:605
    (1982).

21. A. Marty and E. Neher, Ionic channels in cultured rat pancreatic
    islet cells, _J. Physiol._ 326:36P (1982).

22. P. Meda, I. Atwater, A. Goncalves, A. Bangham, L. Orci, and E.
    Rojas, The topology of electrical synchrony among β-cells in
    the mouse islet of Langerhans, _Quart. J. Exp. Physiol._
    69:719 (1984).

23. P. Meda, R. M. Santos, and I Atwater, Direct identification of
    electrophysiological-monitored cells within intact mouse
    islets of Langerhans, _Diabetes_ 35:232 (1986).

24. R. L. Michaels and J. D. Sheridan, Islets of Langerhans: dye
    coupling among immunocytochemical distinct cell types,
    Science 214:801 (1981).

25. M. Nakamura, M. Yokote, Ultrastructural studies on the islets of
    Langerhans of the carp, Z. Anat. Entwicki.-Gesch. 134:61
    (1971).

26. R. M.Santos, Modulation of glucose-induced electrical activity by
    second messengers in mouse pancreatic β-cells, Ph.D.Thesis
    Department of Biophysics, School of Biological Sciences.
    University of East Anglia Norwich, U.K. (1985).

27. Y. Stefan and S. Falkmer, Identification of four endocrine cell
    types in the pancreas of Cottus scorpius (Teleostei) by
    immunofluorescence and electron microscopy, Gen. Comp.
    Endocrinol. 42:171 (1980).

28. N. W. Thomas, Morphology of endocrine cells in the islet tissue
    of the cod Gadus callarias, Acta Endocrinol. 63:679 (1970).

29. N. W. Thomas, Observations on the cell types present in the
    Principal islet of the Dab Limanda limanda, Gen. Comp.
    Endocrinol. 26:496 (1975).

30. G. F. Wagner and B. A. Mokeown, Immunocytochemical localization
    of hormone-producing cells within the pancreatic islets of
    the rainbow trout (Salmo gairdneri), Cell Tiss. Res. 221:181
    (1981).

POTASSIUM CHANNELS IN ADRENOCORTICAL AND PARATHYROID CELLS

J. Lopez-Barneo, L. Tabares and A. Castellano
Departamento de Fisiologia
Facultad de Medicina
Avda. Sanchez-Pizjuan 4
41009 Seville
Spain

In the last decade numerous studies on the electrophysiology of
endocrine cells have firmly established the important role played by
membrane electrical events in "stimulus-release coupling"[16]. However, very
recent development of patch clamp techniques[8] has made it possible to
undertake a detailed characterization of membrane ionic conductances
present in these cells, and to elucidate their precise role in the
regulation of secretion. Among the different ionic channels so far
identified in endocrine cells, are several types of K-channels that are
important not only for the regulation of action potential duration, and
thus the amount of Ca entry, but also because some of these channels are
the target for the action of secretagogues and intracellular messengers.
Different K- channels have been identified in β-cells (see this workshop)
as well as in chromaffin[13,25] pituitary[6] and adrenocortical cells[21].

In this report we present information about the electrophysiology of
adrenocortical and parathyroid cells with special emphasis on the K-
channels so far identified in their membranes. These cells are rather
unusual and do not seem to follow the general scheme of stimulus-secretion
coupling postulated in other endocrine cells. Adrenocortical cells secrete
steroid hormones in presence of adrenocorticotropic hormone (ACTH) by a
mechanism that may not require exocytosis since they appear to have no
significant amount of stored hormone and it is generally agreed that
stimulation by secretagogues primarily increases steroid biosynthesis which

can freely diffuse across the lipid phase of the membrane[10,19]. However, external Ca^{2+} is required for an adequate response of adrenocortical cells to ACTH[10,11] and reports indicate the existence in the adrenal cortex of steroid-containing organelle similar to the secretory granules seen in other secretory cells[7]. Parathyroid cells secrete parathyroid hormone (PTH) in response to a decrease in external Ca^{2+} concentration and maximal secretion is attained in the absence of Ca^{2+}[24]. In addition, membrane hyperpolarization is associated with increased secretion and depolarization with decreased secretion[12]. These observations contrast with findings in other endocrine cells where membrane depolarization and Ca^{2+} entry are critical events in the process of exocytosis[16].

MATERIALS AND METHODS

Experiments were performed on cultured adrenocortical and parathyroid cells as well as in intact parathyroid glands in vitro. Y-1 cells, which are a transformed line of murine adrenocortical cells that secrete steroid hormones in presence of ACTH[4], were cultured in Ham's F-10 medium with 5% fetal bovine serum and antibiotics added. Cells were used for electrical recording 2-24 h after plating in small culture dishes. Parathyroid cells were obtained from bovine glands following a procedure previously described[23]. Glands were cleaned, treated with collagenase (20 mg/gr of tissue) and the cells dissociated and centrifuged in a Percoll gradiente. Parathyroid cells were cultured in Ham's F-10 medium with 5% fetal serum, 1% BSA and antibiotics. Aliquots of this cell suspension were placed in culture dishes and maintained in an incubator until use (aprox. 24-72 h later).

Intracellular recordings in parathyroid cells were done on intact glands obtained from adult male rats maintained in vitro using the techniques previously described[12]. The recording amplifier has a FET operational amplifier wired as a voltage follower, a constant current pump to inject current through the recording microelectrode and electronic compensation of stray capacitance and microelectrode resistance. The composition of solutions used in the experiments is given in the figure legends.

Single channel currents were recorded from cell-attached and inside-out excised membrane patches of cultured adrenocortical and parathyroid cells using the patch-clamp techniques[8,21]. Patch pipettes were fabricated from glass capillaries and had a resistance between 2-6 Mohm. The patch-clamp amplifier was built in our laboratory based on the design by Sigworth[20]. The current-to-voltage converter is a FET operational amplifier (Burr Brown 3523, U.S.A.) with a 8.5 Mohm feedback resistor. The cut-off frequency of the recording system is about 250 Hz, therefore events lasting less than 3-4 ms are partially filtered. The composition of pipette and bath solutions is given in the figure legends.

RESULTS AND DISCUSSION

Potassium Channels in Adrenocortical Cells

In membrane patches from adrenocortical cells we have identified three types of potassium channels mainly based on single channel conductance values, although they also seem to differ in their activation properties. Two of these channels have conductances of about 20 and 38 pS respectively in solutions with 130 mM K^+ facing the external side of the membrane and 5 mM K^+ in contact with the internal side of the membrane. We have concentrated on studying a large conductance (about 180 pS in symmetrical 130 mM K^+), Ca^{2+} and voltage-activated K-channel that is uniformly distributed in Y-1 cell membranes[21].

The basic properties of the large K-channels recorded in cell-attached membrane patches are illustrated in Fig. 1A. Opening and closing of the channel included in the patch appear as steps of outward current that increase in amplitude with membrane depolarization as the electrochemical gradient for K ions increases. With depolarization there is a gradual increase of the time the channel spends in the open state which reflects the voltage dependence of channel activation. This fact is summarized in Fig. 1B (filled circles) Figure legend only refers to it as circles. The percent of time open is near zero at negative voltages but increases with depolarization. The curve fitting the data points is very steep between -20 and +10 mV and approaches channel open probability of 1 at large depolarizations. At low levels of activity, an e-fold increase in the open-state probability is achieved by a 10 mV increase in membrane

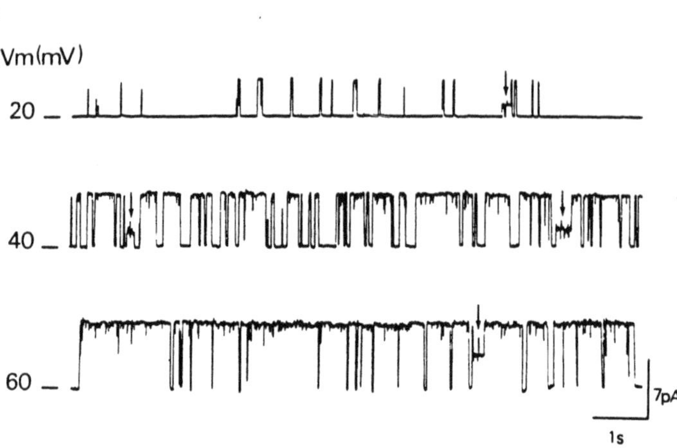

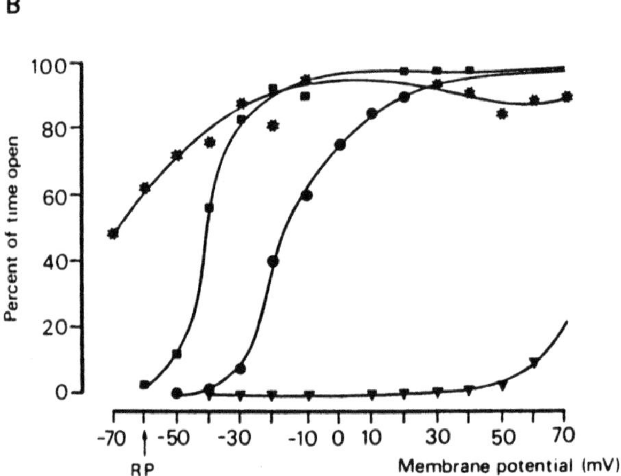

Fig. 1. A. Single K-channel currents recorded from an adrenocortical cell membrane patch <u>in situ</u>. Numbers on the left of each trace indicate, in mV, the amplitude of the depolarization from the cell resting potential. Bars indicate zero current level. Arrows have been drawn over intermediate conductance sub-states of the channel. Pipette and bath solutions had, in mM: NaCl, 140; KCl, 2.7; CaCl$_2$, 2; Hepes, 10. B. Percent of time a channel is open as a function of membrane potential and internal ionic Ca. Circles: cell-attached recordings; other symbols: inside-out excised patches. Pipette and bath solution compositionwas, in mM. KCl, 130; NaCl, 10; MgSO$_4$, 2; Hepes, 10. Variable amounts of EGTA were used to obtain the following concentration of ionic Ca at the inner surface of the membrane: 50 μM (stars); 2-3 μM(squares); 0.01 μM (triangles). pH of all solutions was 7.3. Temperature 20-23°C. Adapted from Tabares et al., 1985.

potential which is a value close to that calculated for the same channel type in other cells[14].

Both, in cell-attached and excised membrane patches, we have observed at least two different intermediate conductance sub-states of the channel

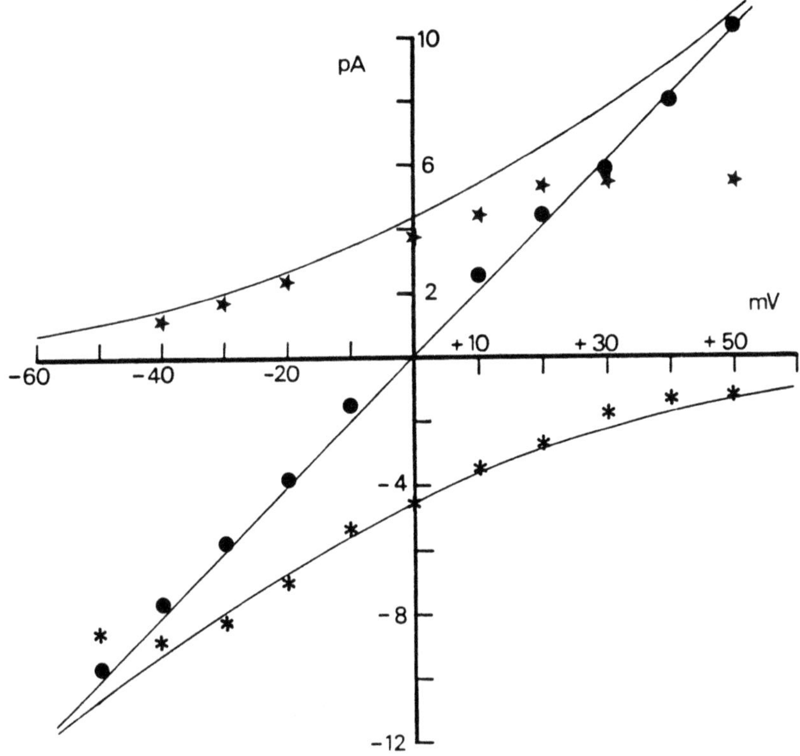

Fig. 2.   Current-voltage plots from single Ca-activated K-channels
          in adrenocortical cell inside-out excised membrane patches.
          Continuous lines are the values predicted by the Goldman, Hodgkin,
          Katz equation[9]. Symbols are data points obtained in: symmetrical
          130 mM K$^+$ solutions (circles); 130 mM external, 2.7 mM internal K$^+$
          (asterisks) and 5 mM external, 130 mM internal K$^+$ (stars).
          Concentration of other ions was the same as in Fig. 1B. Equimolar
          substitution of KCl and NaCl were used to maintain constant the
          osmolarity. Internal Ca^{2+} concentration was 1 µM but in the
          experiment represented by stars that was 2 mM. Temperature
          20-23°C. Tabares and Lopez-Barneo, unpublished.

(Fig. 1A). These sub-states have conductances about 35% (records at 20 and

40 mV) and 55% (record at 60 mV) of the value seen in the normal conducting

state. Nevertheless, we cannot discard the existence of conductance sub-

states of short lifetime which are not detected due to the limited

frequency response of our recording system. Intermediate conductance

sub-states have been seen in the case of large Ca^{+2} dependent K-channels

recorded in other preparations[1].

The large conductance of the K-channel recorded in adrenocortical

cells is also activated by intracellular Ca ions in a critical

concentration range between 0.01-10 µM. The effects of internal Ca^{2+} are

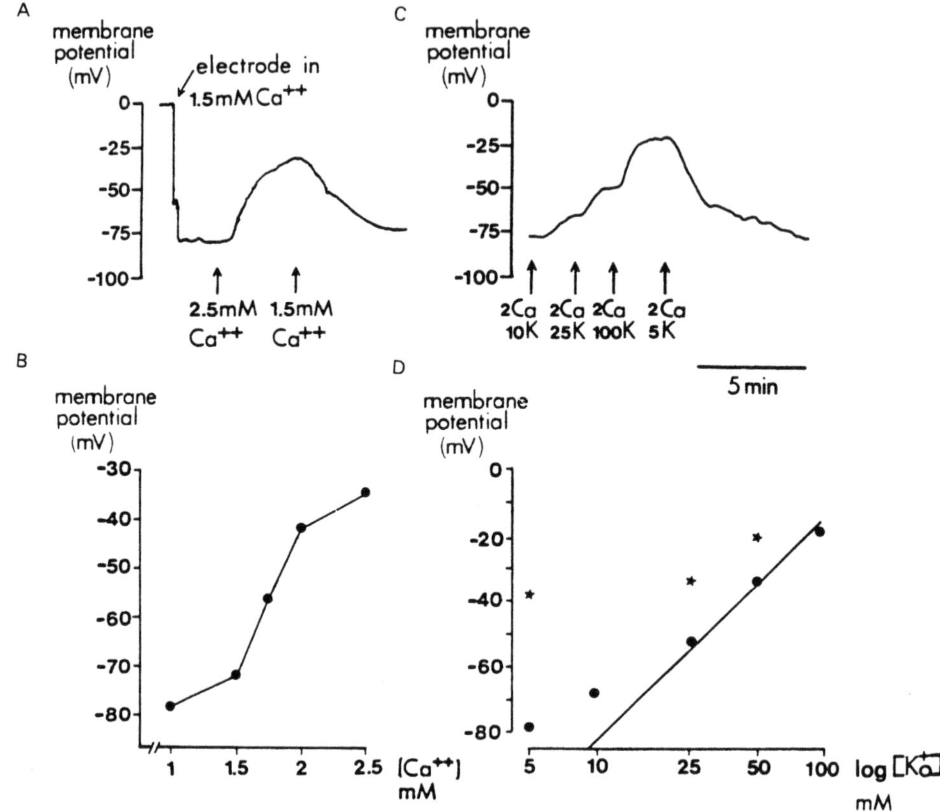

Fig. 3. A and B. Modifications of membrane potential in parathyroid cells
by changes in the concentration of external calcium. Solution
composition was, in mM: NaCl, 120; KCl, 5; Na-fumarate, 2.7;
Na-pyruvate, 4.9; Na-glutamate, 4.9; $MgSo_4$, 1.1; Glucose, 2.8;
Hepes, 10. $CaCl_2$ as indicated in the figure. C and D. Effect of
external K concentration on parathyroid cell membrane potential.
Solution composition was the same as in parts A and B of the
figure but chloride salts were substituted by glutamate.
Equimolar substitution of Na and K salts were used to maintain
constant the osmolarity. pH of all solutions was 7.4. Temp.
37°C. Adapted from Lopez-Barneo and Armstrong, 1983.

summarized in Fig. 1B. Open-channel probability as a function of membrane
potential is plotted at different Ca^{2+} concentrations at the inner surface
of the membrane. The curves fitting the data points can be compared with
the one obtained from cell-attached recordings (filled circles) refered to
in this only as <u>circles</u>. In a low internal Ca^{2+} solution (approx. 0.01 μM,
triangles) channels are usually closed even at large depolarizing voltages.
In the presence of high Ca^{2+} (50 μM, stars) channels are open most of the
time in a broad voltage range; channel open-state probability only
decreases at potentials below -40 mV. With positive holding potentials,

130

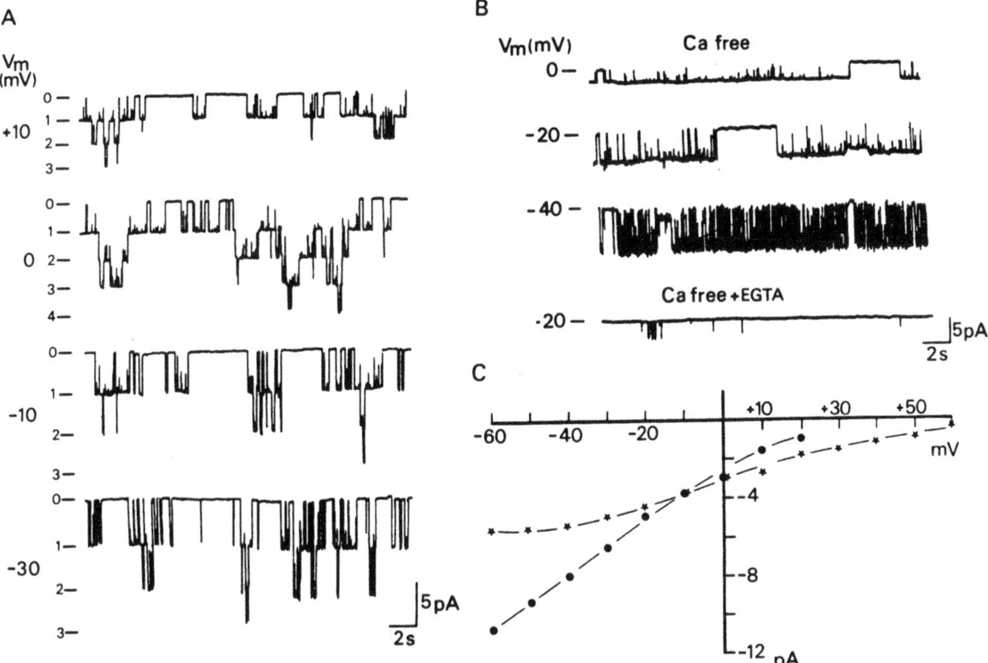

Fig. 4. A and B. Single K-channel currents recorded from inside-out
excised patches of cells dispersed from bovine parathyroid glands.
Pipette solution had 130 mM K^+ and 100 μM Ca. Bath solution had
2.7 mM K^+ and approx. 4 μM Ca^{2+}. Other ions were the same as in
figure 1 and 2. In the record at -20 mV the concentration of EGTA
was 5 mM and the estimated concentration of ionic Ca^{2+} <0.01 μM.
C. Current-voltage plots of the channels shown in parts A (stars)
and B (circles) of the figure. Temp. 20-23°C. Lopez-Barneo and
Castellano, unpublished.

and in presence of high internal Ca^{2+}, long-lasting shut intervals are
present in the records which are probably due to channel inactivation or to
blockage by internal Ca^{+2} [15,22]

Single channel I-V curves obtained with variable concentrations of
external and internal K^+ are summarized in Fig. 2. Symbols are data points
obtained in different experimental conditions and the continuous lines were
calculated with the Goldman-Hodgkin-Katz flux equation [2,9]. In symmetrical,
130 mM K^+, solutions (filled circles) the mean single channel conductance
is 181 ± 54 pS (mean ± SD, n=29). This large dispersion around the mean
has also been noted by other workers [25]. The data points follow the
theoretical straight line between ± 50 mV, however we have previously shown
that current saturates at larger depolarizing or hyperpolarizing
voltages [21]. In excised patches bathed in solutions with 130 mM K^+ facing

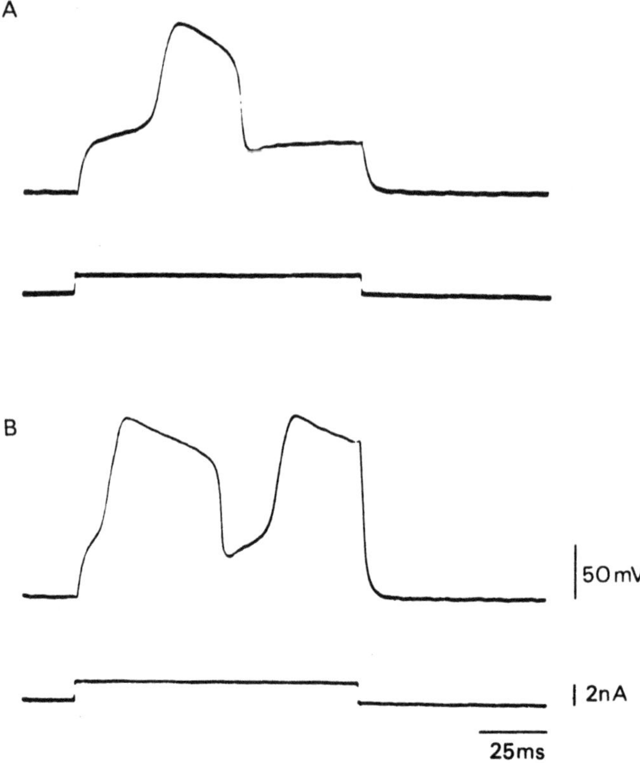

Fig. 5. A and B. Action potentials elicited by direct stimulation of
         intacellularly recorded Y-1 adrenocortical cells. The composition
         of external solution was, in mM: NaCl, 140; KCl, 2.7; $MgSO_4$ 1.1,
         $CaCl_2$ 4; glucose 5; Hepes, 10. pH = 7.4. Temp. 20-23°C. Tabares
         and Lopez-Barneo, unpublished.

the external side of the membrane and 2.7 mM K^+ facing the internal side,
the data points (asterisks) follow the theoretical curve as expected for a
highly selective K-channel in spite of the presence of 10 and 140 mM Na^+ in
the external and internal solutions respectively. This high selectivity of
the channel has been recorded in chromaffin[25] and other cells[2]. Deviations
from the curve at negative voltages may be due to diffusion-limited ion
flow or to the blocking effect of external Na^+ [25]. In solutions containing
5 mM external and 130 mM internal K^+ (stars) single channel current
approaches the theoretical curve only at negative voltages. The deviation
at positive voltages can be explained assuming the presence of a voltage-
dependent blockade by internal Na and Ca ions.[2,22,25] Thus, the Ca
activated K-channel seen in adrenocortical cells, similar to channels found
in other preparations, highly selects K over Na or Cl ions. However, there
are deviations from independence, as illustrated in Fig. 2, due to voltage-
dependent blockade by impermeant (or less permeant) Na and Ca ions.

Potassium Channels in Parathyroid Cells

Patch-clamp experiments have not been performed on parathyroid cells until very recently (see below) and therefore information about K^+ conductances in this preparation is nearly absent. However, intracellular recordings strongly suggest that membrane K^+ permeability participates in the electrical response of parathyroid cells to alterations in external Ca^{+2} concentration[12]. This response, as illustrated by Fig. 3A, consists of a depolarization of about 40 mV when external Ca^{2+} concentration is increased by 1 mM. Mean membrane potential calculated from several cells at different external Ca^{2+} concentrations is shown in Fig. 3B. Extreme values of potential are 78 ± 2 mV in 1 mM Ca^{2+} and 34 ± 6 mV in 2.5 mM Ca^{2+} (mean ± SD, n=66).

Membrane potential and the depolarizing response of parathyroid cells to high external Ca^{2+} are not deeply affected by substitution of external Na^+ and Cl^- by other impermeant ions or by Ca-channel blockers[12]. However, these cells, as shown in Fig. 3C, are very permeable to K^+ in presence of low Ca^{2+}. Cell impalament was done in a solution with 5 mM K^+ and with glutamate substituting Cl ions. In this solution 2 mM Ca^{2+} was needed to match the Ca^{2+} activity measured in the 1.5 mM, control solution. At this Ca^{2+} concentration, external K^+ causes a reversible depolarization, the amplitude of which depends on the K^+ concentration tested (Fig. 3D, filled circles). Above 25 mM K^+ the cells depolarize 62 mV by a ten-fold increase in external K^+ which is the value predicted by the Nernst equation for a potassium electrode at 37°C. With a high external Ca^{2+} concentration, similar increases in external K^+ have a much smaller effect on the membrane potential (stars in Fig. 3D).

These observations have led us to hypothesize the existence in parathyroid cells of K-channels regulated by external $Ca.^{2+}$ [12] High Ca^{2+} may cause depolarization by decreasing membrane K^+ permeability. On returning to low Ca^{2+}, K-channels may open and contribute to the maintenance of the membrane potential at a value near the equilibrium potential for K ions.

Although direct proofs for the existence of these channels are not available, we have recently identified two different K-channels in excised patches of cells dissociated from bovine parathyroid glands that could participate in their electrical response to alterations in external Ca^{2+}. Unitary conductances, at 0 mV and with 130 mM K^+ in the solution facing the external side of the membrane and 2.7 mM K^+ facing the internal side, are 60 and 110 pS respectively. Inward current steps due to the opening of the K-channels with lower conductance are shown in Fig. 4A and the I-V curve in Fig. 4C (stars). The open-state probability of the four channels present in this membrane patch is only moderately dependent on membrane potential. Open-channel probability increases e-fold by a positive change in the membrane potential of 45 mV. By contrast, the channel with larger conductance (Fig. 4B and C, filled circles) is activated by membrane depolarization and closes when the internal Ca^{2+} is lowered by addition of EGTA. The properties of this channel are very similar to those of the large Ca-activated K-channel of Y-1 cells.

Possible Role of K Channels in the Regulation of Secretion

The precise role played by K-channels in the regulation of adreno-cortical and parathyroid cell secretion is not yet known since a detailed knowledge of membrane ionic conductances and their modification by secretagogues is missing. Ca-activated K-channels contribute to the shape of the action potential in several endocrine cells and thus participate in the fine regulation of Ca^{2+} entry[14,17]. These channels are also activated in secretory systems where the action of secretagogues results in a release of Ca^{2+} from internal stores[6]. In adrenocortical cells, increased Ca^{+2} uptake as well as release of Ca^{2+} from intracellular stores have been suggested to occur in presence of ACTH[5,10,11]. This rise of intracellular Ca^{2+} seems to be critical for certain key enzymatic steps and cytoskeleton modifications required for steroidogenesis[3,4]. Ca^{+2} and voltage activated K-channels found in adrenocortical cells may play a crucial role as a link between intracellular Ca^{2+} and membrane voltage-dependent Ca^{2+} permeability. In fact, we have recently recorded large Ca^{2+} action potentials in Y-1 cells with intracellular microelectrodes (Fig. 5). Modulation of Ca^{2+} and voltage-activated K-channels by intracellular messengers other than Ca ions is a possibility that cannot be ruled out.

Parathyroid cells are unusual among endocrine cells because low external Ca^{2+} stimulates, while high external Ca^{2+} inhibits, PTH secretion[18,24]. Stimulation of secretion is accompanied by a hyperpolari-zation, possibly mediated by an increase of membrane K^+ permeability[12], and can occur at cytosolic Ca^{2+} levels that fail to support exocytosis in other secretory systems[18]. Parathyroid cell membrane potential closely parallels secretory rates at different external Ca^{2+} concentrations[12], therefore it seems highly likely that membrane potential changes are related to changes in secretory activity. Secretion in these cells may involve unknown mechanisms independent of detectable changes in cytosolic Ca^{2+} where modifications of membrane voltage play an important role. Single channel currents reported here could participate in the voltage response of parathyroid cells to changes in external Ca^{+2}. However support for this hypothesis is still lacking and must await for future experimental work.

ACKNOWLEDGEMENTS

Authors wish to thank Drs. C.M. Armstrong and D. Mir for their continuous encouragement and support. We are also indebted to Drs. E. Pintado and C. de Miguel for their valuable help in the dissociation and culture of parathyroid and Y-1 cells. Bovine parathyroid glands were kindly supplied by the Matadero Industrial de Mercasevilla. Research has been supported by grants from CAICYT (3464-83) and Areces Foundation.

REFERENCES

1. J.N. Barrett, K.L. Magleby, and B.S. Pallotta, Properties of single calcium-activated potassium channels in cultured rat muscle, J. Physiol. (Lond.) 331:211 (1982).
2. A.L. Blatz and K.L. Magleby, Ion conductance and selectivity of single calcium-activated potassium channels in cultured rat muscle, J. Gen. Physiol. 84:1 (1984).
3. R. Cheitlin and J. Ramachandran, Regulation of actin in rat adrenocortical cells by corticotropin, J. Biol. Chem. 256:3156 (1981).
4. M.A. Clark and J.W. Shay, The role of tubulin in the steroidogenic response of murine adrenal and rat Leydig cells, Endocrinology 109:2261 (1981).
5. M. Culty, I. Vilgrain, and E.M. Chambaz, Steroidogenic properties of phorbol ester and a Ca^{2+} ionophore in bovine adrenocortical cells suspension, Biochim. Biophys. Res. Comm. 121:499 (1984).

6. J.M. Dubinsky and G.S. Oxford, Dual modulation of K^+ channels by
    thyrotropin-releasing hormone in clonal pituitary cells, Proc.
    Natl. Acad. Sci. USA 82:4284 (1985).

7. R.T. Gemmell, S.G. Laychock, and R.P. Rubin, Ultrastructural and
    biochemical evidences for a steroid containing secretory organelle
    in the perfused cat adrenal gland, J. cell. Biol. 72:209 (1977).

8. O.P. Hamill, A. Marty, E. Neher, B. Sakmann, and F. Sigworth, Improved
    patch-clamp techniques for high-resolution current recording from
    cells and cell-free membrane patches, Pflügers Arch. 391:85 (1981).

9. A.L. Hodgkin and B. Katz, The effect of sodium ions on the electrical
    activity of the giant axon of the squid, J. Physiol. (Lond.) 108:37
    (1949).

10. S.D. Jaanus, M.J. Rosenstein, and R.P. Rubin, On the mode of action of
    ACTH on the isolated perfused adrenal gland, J. Physiol. (Lond.)
    209:539 (1970).

11. D.J. Leier and R.A. Jungmann, Adrenocorticotropic hormone and
    dibutyryl adenosine cyclic monophyosphate-mediated Ca uptake by rat
    adrenal glands, Biochim. Biophys. Acta 329:196 (1973).

12. J. Lopez-Barneo and C.M. Armstrong, Depolarizing response of
    parathyroid cells to divalent cations, J. Gen. Physiol. 82:269
    (1983).

13. A. Marty, Ca^{2+}-dependent K^+ channels with large unitary conductance in
    chromaffin cell membranes, Nature 291:497 (1981).

14. A. Marty, Ca^{2+}-dependent K^+ channels with large unitary conductance,
    Trends Neurosci. 6:262 (1983).

15. B.S. Pallota, Calcium-activated potassium channels in rat muscle
    inactivate from a short-duration open state, J. Physiol. (Lond.)
    363:501 (1985).

16. O.H. Petersen, The electrophysiology of gland cells, in "Monographs of
    the Physiological Society", Academic Press, London, (1980).

17. O.H. Petersen and Y. Maruyama, Calcium-activated potassium channels
    and their role in secretion, Nature 307:693 (1984).

18. D.M. Shoback, J. Tatcher, R. Leombruno, and E.M. Brown, Relationship
    between parathyroid hormone secretion and cytosolic calcium
    concentration in dispersed bovine parathyroid cells, Proc. Natl.
    Acad. Sci. USA 81:3113 (1984).

19. C.P. Sibley, B.J. Whitehouse, G.P. Vinson, C. Goddard, and E.
     McCredie, Studies on the mechanism of secretion of corticosteroids
     by the isolated perfused adrenal gland of the rat, $\underline{J}$. $\underline{Endocrinol}$.
     91:313 (1981).

20. F.J. Sigworth, Electronic design of the patch-clamp, in
     "Single-Channel Recording", B. Sakmann and E. Neher, eds., Plenum
     Press, New York, pp. 3-35 (1983).

21. L. Tabares, J. Lopez-Barneo, and C. De Miguel, Calcium- and voltage-
     activated potassium channels in adrenocortical cell membranes,
     $\underline{Biochim}$. $\underline{Biophys}$. $\underline{Acta}$. 814:96 (1985).

22. C. Vergara and R. Latorre, Kinetics of Ca^{2+}-activated K^+ channels from
     rabbit muscle incorporated into planar bilayers. Evidences for
     Ca^{2+} and Ba^{2+} blockage, $\underline{J}$. $\underline{Gen}$. $\underline{Physiol}$. 82:543 (1983).

23. J. Wallace and A. Scarpa, Regulation of parathyroid hormone secretion
     in vitro by divalent cations and cellular metabolism, $\underline{J}$. $\underline{Biol}$.
     $\underline{Chem}$. 257:10613 (1982).

24. J. Wallace, E. Pintado, and A. Scarpa, Parathyroid hormone secretion
     in the absence of external free Ca and transmembrane Ca influx,
     $\underline{FEBS}$ $\underline{Lett}$. 151:83 (1983).

25. G. Yellen, Ionic permeation and blockage in Ca^{2+}-activated K^+ channels
     of bovine chromaffin cells, $\underline{J}$. $\underline{Gen}$. $\underline{Physiol}$. 84:157 (1984).

PHARMACOLOGICAL PROPERTIES OF THE CHROMAFFIN CELL

CALCIUM CHANNEL

A.G. Garcia, C.R. Artalejo, R. Borges
J.A. Reig and F. Sala

Departamento de Farmacologia
Facultad de Medicina
Universidad de Alicante, Espana

INTRODUCTION

In 1961, Douglas and Rubin[21] concluded that "the role of
acetylcholine as a transmitter at the adrenal medulla is to cause some

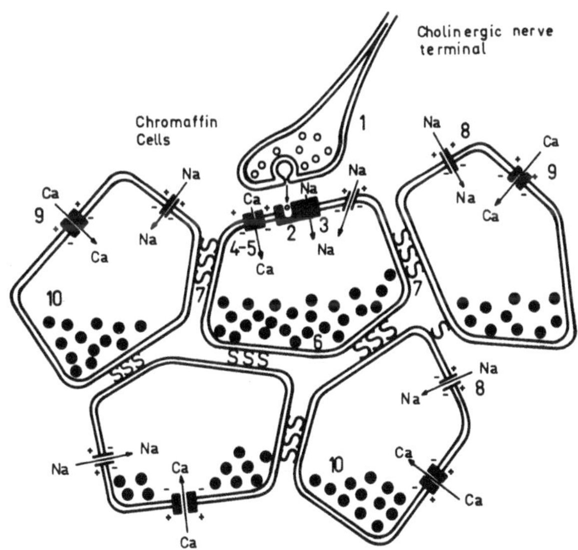

Fig. 1. Sequence of events during the process of stimulus-secretion
coupling. 1. Acetylcholine (ACh) release from splanchnic nerve

terminals. 2. Activation of cholinoceptors. 3. Na entry through ACh
receptor ionophore. 4. Membrane depolarization. 5. Activation of
voltage-sensitive Ca^{2+}-channels. 6. Catecholamine release.
7. Amplification of the secretory signal by intercellular electrical
coupling. 8. Generation of Na-dependent action potentials (Na channels, gap junctions). 9. Voltage-dependent Ca^{2+}-channels. 10.
Catecholamine release.

brief change in medullary cells which allows extracellular Ca^{2+} to
penetrate them and trigger the catecholamine ejection process." The
formation accumulated since then can be summarized in the diagram of
the sequence of events taking place during the secretory cycle
depicted in Fig. 1.

Acetylcholine, released at the splachnic nerve-chromaffin cell
junction, reacts with muscarinic and nicotinic receptors present in
the chromaffin cell membrane. Activation of the nicotinic-receptor
channel induces depolarization of the membrane which in turn activates
voltage-gated Ca^{2+} channels and leads to Ca^{2+} entry necessary for
catecholamine secretion.

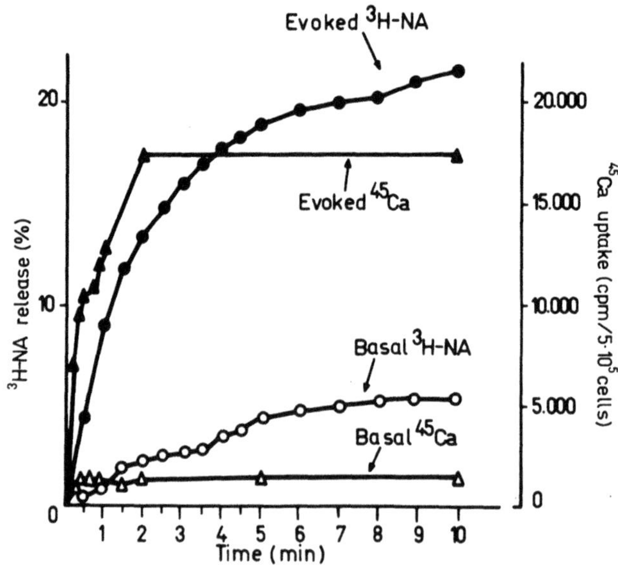

Fig. 2. Time course of ^{3}H-noradrenaline (^{3}H-NA) release and ^{45}Ca uptake
during exposure of cultured bovine adrenal chromaffin cells to high K$^+$.
Cells of individual wells preloaded with ^{3}H-NA, were incubated at 37°C in
Krebs solution (O, Δ) or in high K (59K) different time periods (abscissa).
The net release was calculated by subtracting the K-evoked release from the
basal release and expressed as % of the total ^{3}H-retained by the cells
(fractional release). Data from two experiments made in triplicate.

Acetylcholine stimulation of the chromaffin cell evokes
exocytotic release of catecholamines to the extracellular space and
this process depends critically on cytosolic Ca^{2+}.[5]

RESULTS AND DISCUSSION

The extracellular medium appears to be the main source of the
calcium required for secretion and Ca^{2+}-uptake preceeds the release of
catecholamines (Fig. 2).

Bovine adrenal chromaffin cells, cultured for three days, were
exposed to a high potassium (50 mM) Krebs solution and both, the
uptake of $^{45}Ca^{2+}$ and the release of ^{3}H-noradrenaline (^{3}H-NA) were
simultaneously measured. $^{45}Ca^{2+}$ uptake clearly preceeded the release
of ^{3}H-NA (see Fig. 2).

Since previous studies have established that Ca^{2+} enters the
chromaffin cell through voltage-gated channels,[27] it seems logical to
assume that the regulation of the secretory response must be directed
towards the regulation of Ca^{2+} entry. This could be modulated by the

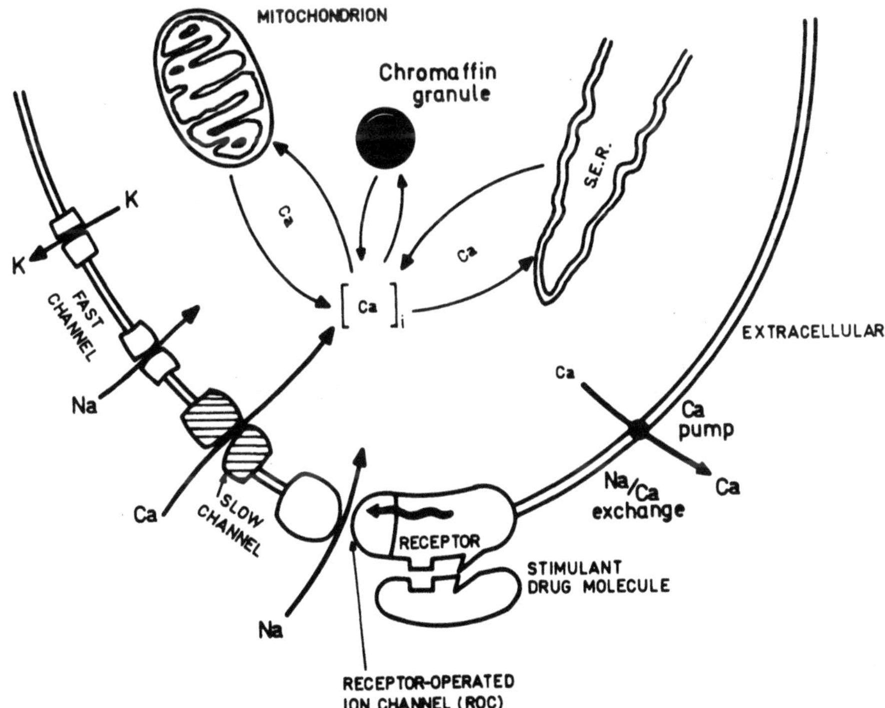

Fig. 3. Ionic channels present in the chromaffin cell membrane and intracellular Ca-sequestering/Ca-extruding systems. From ref. 55.

same mechanisms that control Ca^{2+}-channel activation and deactivation (slow Ca^{2+}-channel in Fig. 3). These studies have also revealed the presence in the chromaffin cell membrane of at least two additional voltage-gated channels, namely, two fast, voltage-gated Na^+- and K^+-channels, and a $[Ca^{2+}]$i-activated K^+-channel.[52,16,26,27,39,41,50,67]

Characterization of these channels requires the use of specific blockers and/or agonists of high affinity. In the radioactive form, these drugs can be used to estimate the density and the topology of the channels on the membrane of the chromaffin cell. For example, ^{3}H-nitrendipine selectively blocks voltage-gated Ca^{2+}-channels and binds to high affinity receptors present in the chromaffin cell membrane.[32,33,34] Measurements of the specific binding of ^{3}H-nitrendipine to membrane fragments from bovine adrenal medulla showed a saturable component with an apparent K_D of 1.18 nM and a B_{max} of 325 fmol/mg protein.[34] These results indicate the presence in the chromaffin cell membrane of high affinity binding sites probably related to voltage-gated Ca^{2+}-channels. Addition of EGTA to the assay medium drastically reduced ^{3}H-nitrendipine binding. This result suggests that Ca^{2+}, structurally related to the Ca^{2+}-cahnnel complex,

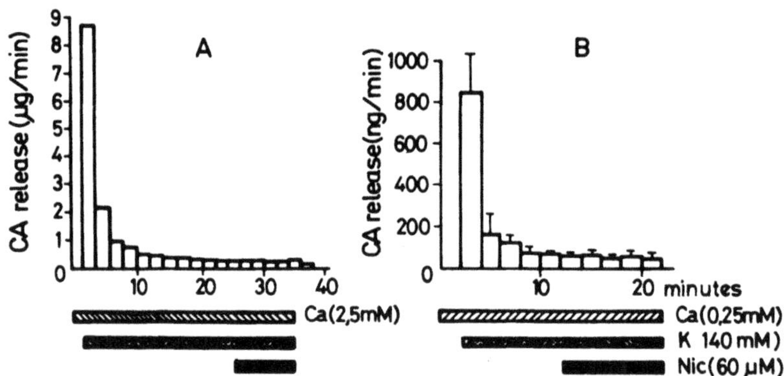

Fig. 4. Catecholamine release evoked by high K from perfused cat adrenal glands in the presence of 2.5 mM Ca (B) or 0.25 mM Ca (A). Nicotine was introduced in the perfusion system once the secretory response to high K was fully inactivated; no enhancement of the secretory rate was observed upon Nic addition. After 1 h washout, a new test with high K solution demonstrated that the gland was able to evoke a large secretory response. Vertical bars in B represent mean ± S.E.

might be required for the binding of the dihydropyridine derivative. Assuming that nitrendipine binds only to Ca^{2+}-channel receptors, the channel density ranges between 1 to 10 channels/um^2.

It has been suggested that acetylcholine, nicotine and high
external K^+ promote Ca^{2+}-entry and catecholamine release mainly by
activation of voltage-gated Ca^{2+}-channels.[15,32,33] However, early
experiments showed that acetylcholine stimulated catecholamine secretion from perfused cat[22] and rat adrenal glands[40] in the presence of
high K^+, a situation in which chromaffin cells are already depolarized. These results suggested that, while K^+-induced depolarization
activated Ca^{2+}-entry through voltage-gated channels, cholinergic
stimulation and activation of the nicotinic-receptor channels, might
independently promote an extra Ca^{2+}-entry through the nicotinic
receptor channel. As the nicotinic-receptor channel is permeant to
Ca^{2+},[10] it is likely that in the depolarized cell activation of the
nicotinic-receptor channel leads to an extra Ca^{2+}-entry.[38,40,41]

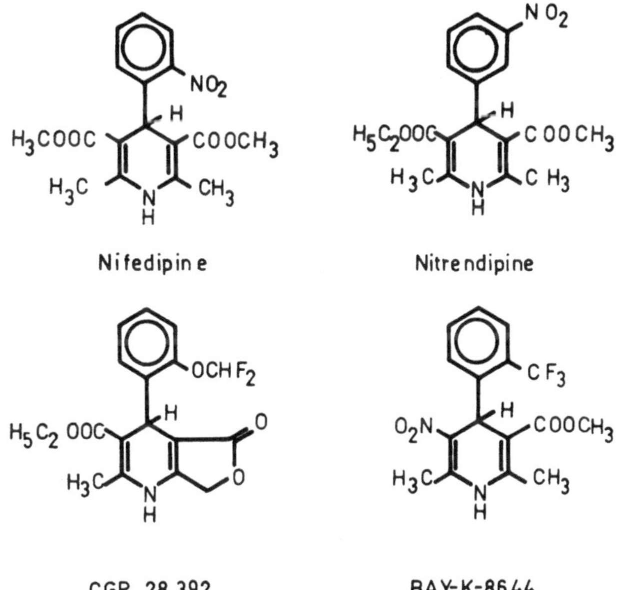

Fig. 5. Chemical structures of some dihydropyridine Ca channel antagonists
(top) or agonists (bottom).

Acetylcholine is known to cause the release of catecholamines
from cat adrenal glands by activation of both nicotinic and muscarinic
receptors.[20,45] In presence of high K^+, the secretory response to
muscarine is preserved.[45] On the other hand and in contrast to
results in bovine adrenal cells, the secretory response to nicotine is
abolished (Fig. 4A). This is probably not due to intracellular
depletion of catecholamines as in the presence of low Ca^{2+} (0.25 mM),

high K^+ evoked the release of 10% of the amount released in normal Ca^{2+} (2.6 mM) yet nicotine failed to further stimulate release (Fig. 4B).

If nicotinic-receptor stimulation and K^+-induced membrane depolarization cause release of catecholamines by activation of the voltage-gated Ca^{2+}-channel, nitrendipine and Mn^{2+} should inhibit secretion in response to these secretagogues. Mn^{2+} inhibited catecholamine release evoked by acetylcholine, nicotine or high K^+ from perfused cat adrenal.[32] Nitrendipine blocked ^{3}H-NA release from cultured bovine adrenal chromaffin cells evoked by nicotine or high K^+.[15]. Conversely, the Ca^{2+}-channel agonist Bay-K-8644, a dihydropyridine derivative, potentiated the secretion evoked by either nicotine or high K^+ in perfused cat adrenal glands.[35] Although these results clearly show that both nicotine- and high K^+-evoked catecholamine release depend on the activation of the Ca^{2+}-channel, they cannot be used to test the idea that Ca^{2+} enters the cell through the nicotinic-receptor channel.

Catecholamine release evoked by several secretagogues is inhibited by agents that act selectively on voltage-gated Ca^{2+}-channels. This is true for Mg^{2+}, Mn^{2+}, Co^{2+}, Ni^{2+}, Cd^{2+} and for the Ca^{2+}-channel antagonists verapamil, D-600, nifedipine and nitrendipine (Table 1). Although the classification of these Ca^{2+}-channel blockers may vary from one author to another, they are generally grouped into three classes.[60] Class I: dihydropyridines (nifedipine, PY 108-068, nimopidine, nitrendipine, niludipine, nisoldipine); Class II: verapamil, D-600, diltiazem, diclofurime; Class III: diphenylalkylamines (pimozide, fendiline, lidoflazine, perhexine, cinnarizine, prenylamine, flunarizine) (Table 1).

One objection to the use of the organic Ca^{2+}-channel antagonists is wide variation in potency exhibited when used in different tissues. For example, while verapamil was reported to be a potent blocker of cardiac and smooth muscle,[28,49,54,65] high doses were required to inhibit high K^+ or acetylcholine-evoked catecholamine release. The lack of specificity is illustrated by the finding that verapamil blocks Na^+-channels in synaptosomes,[53] has muscarinic-like effects,[25,] has $_2$ adrenergic receptor affinity[30] and inhibits the Ca^{2+}-permeability component of the endplate acetylcholine-receptor channel.[10]

Table 1. Effects of inorganic and organic Ca-channel blockers on
chromaffin cell function.

Experimental

System	Compound	Concentration	Inhibitory Effects	Reference
CBCC	Cd^{2+}	0.001-1 mM	Inhibits CaU evoked by 59 mM K (IC50 0.03 mM) or carbachol (IC50 0.06 mM). CAR inhibited equally	38
IBCC	Co^{2+}	0.02-1 mM	Inhibition of CAR (IC50 0.33 mM) and cGMP levels (IC50 0.0 mM evoked by ACh (50 +M)	49
PRAG	Co^{2+}	5 mM	Block of spike frequency and CAR induced by high K	42
PDAG	Co^{2+}	3 mM	Block of CAR induced by GABA	46
PCAG	La^{3+}	1-3 mM	Block of CAR evoked by TES. Inhibition by 80% of CAR evoked by ACh (130 µM)	66
PBAG	La^{3+}	0.5 mM	ACh (1 mM)-induced CAR blocked by 74% and K (150 mM)-evoked by CAR by 44%	8
PCAG	Mg^{2+}	20 mM	Inhibition by 60% of CAR evoked by TES. ACh-evoked release was unaffected	66
IBCC	Mg^{2+}	10-20 mM	ACh (100 µM)- or K (65 mM)-evoked CAR blocked by 40-50%. Block of cGMP increase induced by ACh (50 µM)	49 56
CBCC	Mg^{2+}	0.20 mM	Inhibition of CaU and CAR evoked by 56 mM K (IC 50 5 mM) and carbachol (IC50 3 mM)	38

CBCC	Mg^{2+}	0.20 mM	Block CAR and CaU induced by veratridine and K, but not carbachol	65
PCAG	Mn^{2+}	1-3 mM	Block CAR evoked by ouabain and CA reintroduction. Inhibition by 50-85% of CAR evoked by TES or ACh (130 µM)	59 66
PBAG	Mn^{2+}	2,2 mM	Block CAR evoked by ACh	4
IBCC	Ni^{2+}	10 mM	Block CAR evoked by ACh (100 µM) and K (65 mM)	56
CBC	Ni^{2+}	1 mM	43% inhibition of CAR and 96% inhibition of cGMP levels induced by ACh (50 µM)	49
PCAG, PBAG	D-600	0.1-0.3 mM	Inhibition of CaU and CAR evoked by Ca reintroduction	1,2,59
PBAG	D-600	0.3 mM	85% inhibition on CAR evoked by ACh (10^{-4}M). No effect on CAR evoked by Na deprivation	54
IBCC	D-600	0.01 mM	66% inhibition of CAR and 24% of cGMP levels evoked by ACh (50 µM)	49
PDAG	Dibucaine	8.1 mM	75% inhibition on CAR evoked by GABA	46
IBCC	Diltiazem	3 µM	Inhibition of carbachol-induced CAR	64
CBCC	Nitrendipine	0.1nM-10µM	Inhibition on CAR evoked by by Nicotine (IC50=300 nM) and K(IC50=10 nM)	15
PCAG	Nifedipine	0.1 µM	Block CAR evoked by Ca reintroduction and antagonism with Bay-K-8644	51

CBCC	Trifluoperazine	10 µM	Inhibition of Ca currents	16
IBCC	Verapamil	1-10 µM	35-70% inhibition of CAR evoked by ACh, carbachol or K. 21% inhibition of cGMP levels	49,55,56
PBAG	Verapamil	0.1nM-1mM	ACh-induced CAR (IC50=1 uM	3
PCAG	Verapamil	0.1 mM	CAR evoked by TES or ACh unaffected	66
PDAG	Verapamil	50 µM	Block of GABA-induced CAR	46

Abbreviations: ic50: concentration for 50% inhibition; CBCC: cultured bovine chromaffin cells; IBCC: isolated bovine chromaffin cells; PRAG: perifused rat adrenal glands; PCAG: perifused cat adrenal glands; PDAG: perifused dog adrenal glands; PBAG: perifused bovine adrenal glands; Cau: Calcium uptake; CAR: catecholamine release; TES: transmural electrical stimulation; ACh: acetylcholine

Dihydropyridine derivatives such as nitrendipine are potent blockers of nicotine-evoked ^{3}H-NA release from PC12 cells and of nicotine or high K^+-induced release from bovine adrenal chromaffin cell at concentrations in the range from 20 to 40 nM. In contrast, high concentrations of verapamil (500 nM) are needed to block by 75% high-K^+-evoked catecholamine release from cat adrenal glands[65] and, 10 uM verapamil or D-600 to inhibit acetylcholine-evoked release from isolated bovine adrenal chromaffin cells.[49]

Electrical recordings showed that acetylcholine or high K^+ depolarizes the membrane of gerbils or rat chromaffin cells in culture[17,18,9,7,43] and adrenal cells from intact perfused rat glands.[40] The membrane potential decreased with the logarithm of the concentration of K^+ together with the rate of release of catecholamines supporting the view that Ca^{2+}-entry increased with depolarization as well.

Ca^{2+} currents in adrenal chromaffin cells have been studied with the patch clamp technique in the whole cell recording mode. As expected, whole cell Ca^{2+} currents were found to be blocked by D-600 and Co^{2+}.[27] As in other cell types (this volume), the peak value of

the current occurred when the membrane was completely depolarized and
the apparent reversal potential for the Ca^{2+} current was estimated to
be close to 60 mV. Since the theoretical equilibrium potential for
the Ca^{2+} currents is greater, the low reversal potential must be due
to permeation of the Ca^{2+}-channel by an intracellular cation, perhaps
K^+. Since in presence of 2.6 mM Ca^{2+} in the external medium, single
channel Ca^{2+} current is only a fraction of a pA (10^{-12} A),
Ca^{2+}-channel properties are often studied in presence of Ba^{2+} inplace
of Ca^{2+} in the external medium. Under these conditions, bursts of
Ca^{2+}-channel openings are recorded. Furthermore, a slow decrease in
the rate of Ca^{2+}-channel openings has been observed.[27] Both the rise
in free- Ca^{2+} as measured with quin-2 and secretion evoked by a sudden
increase in external potassium concentration are also transient

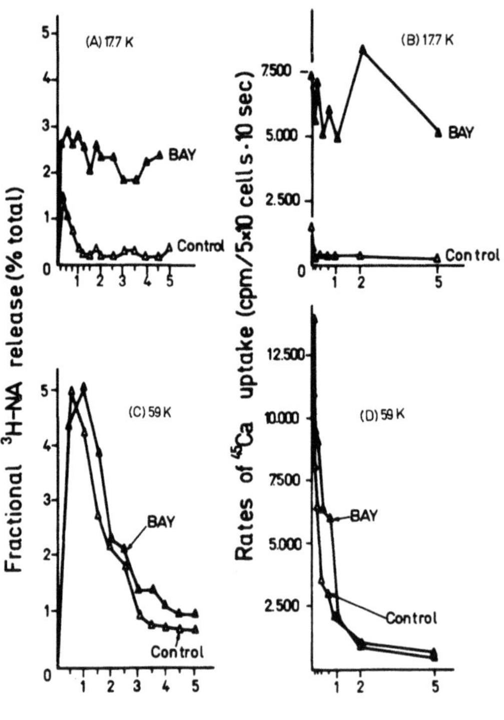

Fig. 6. Effects of Bay-K-8644 (Bay; 1 µM on the rates of release of
^{3}H-noradrenaline (^{3}H-NA) from and ^{45}Ca uptake into K^+ cultured bovine
adrenal chromaffin cells evoked by 17.7 (top) or 59 mM K (bottom).
The experiment to monitor ^{3}H-NA release was performed in individual
single wells, replacing the bathing solution at 10 sec intervals. The rate
of ^{45}Ca uptake at different times was measured after a 10-sec exposure to
^{45}Ca (8 Ci/0.5 ml). Data expressed as fractional ^{3}H-NA release (left side)
or as cpm of ^{45}Ca uptake per half million cells in 10 sec (right side).

events.[47,13] Ca^{2+}-channel inactivation may be caused by the rise in internal Ca^{2+}.[11,62] Ca^{2+}-channel density has been estimated to lie between 5 and 15 channels/um^2,[27] in the same range as the density estimated from ^{3}H-nitrendipine binding.[34]

Small modifications of the structure of the nifedipine molecule produced the Ca^{2+}-channel agonist Bay-K-8644 (Fig. 5). Another interesting dihydropyridine derivative with the properties of a Ca^{2+}-channel agonist is CGP28392 (Ciba-Gygy).

Basal catecholamine release from perfused cat adrenal gland or cultured bovine chromaffin cells remained unmodified after the application of either Bay-K-8644 or CGP28392. Other ionophores such as A 23187 and ionomycin increase basal release.[31,14,15]

Ca^{2+}-channel agonists potentiate the secretory response evoked by a sudden increase in external K$^+$. For example, in cultured bovine chromaffin cells Bay-K-8644 markedly potentiated the release of ^{3}H-NA

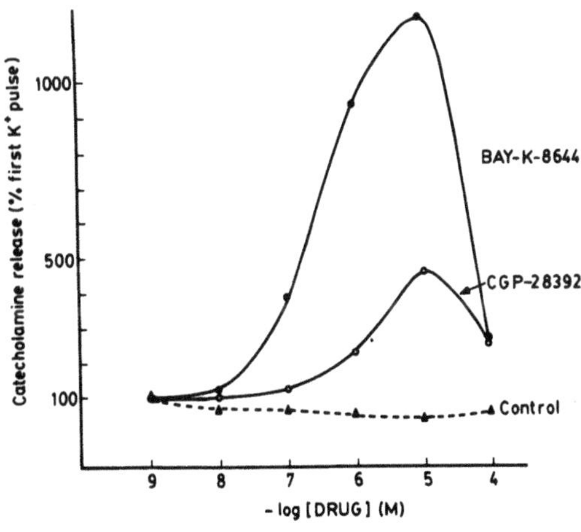

Fig. 7. Effects of increasing concentrations (abscissa) of Bay-K-8644 and GCP28392 on the release of catecholamines from cat adrenal gland (ordinate) evoked by 2-min pulses of 17.7 mM K at 30 min intervals. Net catecholamine release during the first pulse amounted to 0.9 µg (n=9).

and $^{45}Ca^{2+}$-uptake evoked by low (17.7 mM) but not high (59 mM)
external K^+ (Fig. 6). The Ca^{2+}-channel agonists Bay-K-8644 and
CGP28392 enhanced catecholamine release induced by 17.7 mM K^+ in a
concentration-dependent manner, Bay-K-8644 being two orders of
magnitude more potent than CGP28392 (Fig. 7).

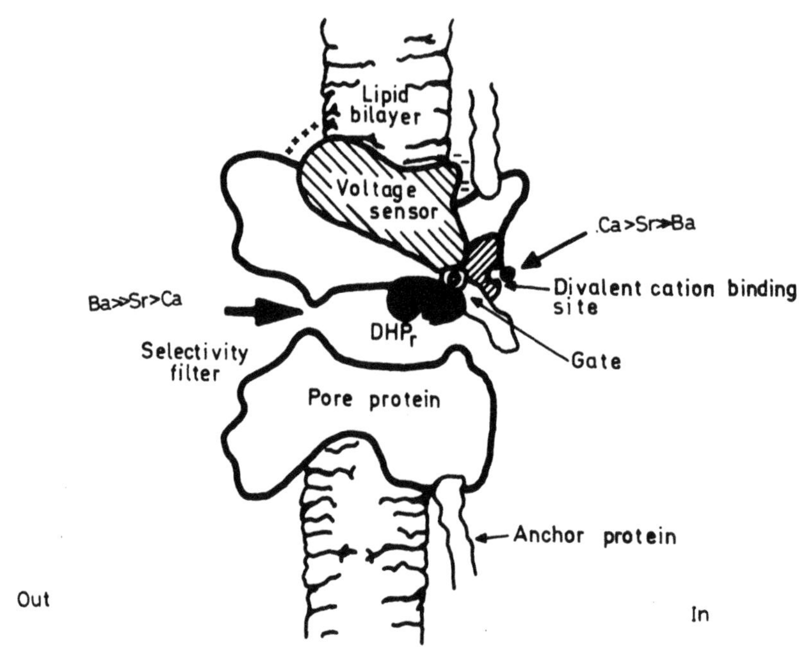

Fig. 8. Model for the regulation of the chromaffin cell Ca-channel. See
text for explanation.

Removal of external Ca^{2+} followed by Ca^{2+} reintroduction is known
to favour catecholamine release. Bay-K-8644 markedly potentiates this
secretory response. Removal of K^+, known to activate Na^+/Ca^{2+}
exchange, enhanced catecholamine release and this effect was not
affected by Bay-K-8644. These results show that the drug is highly
selective for the Ca^{2+}-channel.

Ba^{2+} and Sr^{2+} can substitute for Ca^{2+} in stimulated catecholamine
releasle from cat adrenal gland.[23] In perfused cat adrenal gland,

Bay-K-8644 potentiates high K^+-evoked catecholamine release with Ca^{2+} present in the medium but not with Sr^{2+} or Ba^{2+} in place of Ca^{2+}.

If the Ca^{2+} ionophore A-23187 is applied in the absence of Ca^{2+}, reintroduction of the divalent cation gives rise to an enhanced catecholamine release.[34,35] Since Bay-K-8644 does not modify the rate of release evoked by A-23187 nor the release evoked by Ca^{2+} from cultured chromaffin cells permeabilized with digitonin, it seems that the receptor to the dihydropyridine is not intracellular.

At concentrations greater than 10 uM, Bay-K-8644 inhibits secretion from chromaffin cells. This result suggests a common site of action for Bay-K-8644 and nitrendipine. This idea is supported by the observations that Bay-K-8644 displaces nitrendipine from membranes isolated from bovine adrenal cells and Bay-K-8644 potentiating effects are antagonized by nifedipine in a competitive manner.[34,35]

The potentiating effects of Bay-K-8644 on secretion are best seen at low extracellular Ca^{2+}. Even though the potentiating effects of Bay-K-8644 are more pronounced at 17.7 mM K^+ (Fig. 6), provided that $[Ca^{2+}]_o$ is maintained low (<0.25 mM), the Ca^{2+}-channel agonist will potentiate catecholamine release even in presence of higher concentrations of potassium. As proposed by Joost and Atwater (ch. , this volume) the inhibition of insulin release during continued exposure to high K^+ may be due to Ca^{2+}-channel inactivation caused by accumulation of Ca^{2+} in the cell cytosol. Such an interpretation applied to our results (Fig. 6) in chromaffin cells would indicate that the active site of Bay-K8644 is not the same site as for $[Ca^{2+}]_i$-induced Ca^{2+}-channel blockade.

The antagonistic effect of nifedipine on secretion is reversed by increasing $[Ca^{2+}]_o$ suggesting a link between the receptor site for dihydropyridines and the Ca^{2+}-channel.

The Ca^{2+}-channel model depicted in Fig. 8 includes a dihydropyridine receptor located in the pore itself, a voltage sensor with its associated gate and an internal Ca^{2+} binding site. This internal Ca^{2+} site is proposedto be responsible for $[Ca^{2+}]_i$-induced Ca^{2+}-channel blockade.[11,24,62] It is further proposed that the divalent cations bind to this site following the sequence

$$Ca^{2+} > Sr^{2+} > Ba^{2+}.$$

In summary, catecholamine release from the medullary chromaffin
cells is intimately associated with activation of Ca^{2+}-channels in the
membrane. Some pharmacological properties of the channel have been
discussed.

ACKNOWLEDGEMENTS

Thanks are given to Fundacion Areces, CAICYT (# 0626/81) and FISS
of Spain for support and to Prof. F. Hoffmeister for the sample of
Bay-K-8644 and nifedipine. Experiments for Fig. 6 were performed by
L. Gandia and those for Figs. 2 and 6 were carried out in collaboration with D. Aunis.

REFERENCES

1. J. Aguirre, J. Falute, J.E.B. Pinto, and J.M. Trifaro, Effects of
   methoxyverapamil on the stimulation by Ca^{2+}, Sr^{2+} and inhibition by
   Mg^{2+} on catecholamine release from the adrenal medulla, <u>Br</u>. <u>J</u>.
   <u>Pharmac</u>. 66:591 (1979).

2. J. Aguirre, J.E.B. Pinto, and J.M. Trifaro, Calcium movements during
   the release of the catecholamines from the adrenal medulla: effects
   of methoxyverapamin and external cations, <u>J</u>. <u>Physiol</u>. 269:371
   (1977).

3. L. Arqueros and A.J. Daniels, Analysis of the inhibitory effect of
   verapamil on the adrenal medullary secretion, <u>Life</u> <u>Sci</u>. 23:2415
   (1978).

4. L. Arqueros and A.J. Daniels, Manganese as agonist and antagonist of
   calcium ions: dual effect upon catecholamine release from adrenal
   medulla, <u>Life</u> <u>Sci</u>. 28:1535 (1981).

5. P.F. Baker and D.E. Knight, Calcium-dependent exocytosis in bovine
   adrenal medullary cells with leaky plasma membranes, <u>Nature</u> 276:620
   (1978).

6. P.F. Baker and D.E. Knight, Calcium control of exocytosis in bovine
   adrenal medullary cells, <u>TINS</u> 7:120 (1984).

7. B. Biales, M. Dichter, and A. Tischler, Electrical excitability of
   adrenal chromaffin cells, <u>J</u>. <u>Physiol</u>. 262:743 (1976).

8. J.L. Borowitz, Effect of lanthanum on catecholamine release from
   adrenal medulla, <u>Life</u> <u>Sci</u>, Part <u>I</u> 11:959 (1972).

9. B.L. Brandt, S. Hagiwara, Y. Kodokoro, and S. Miyazaki, Action
   potentials in the rat chromaffin cell and effects of acetylcholine,
   <u>J</u>. <u>Physiol</u>. 263:417 (1976).

10. P.D. Bregestovski, R. Miledi, and I. Parker, Calcium conductance of acetylcholine-induced endplate channels, _Nature_ 279:638 (1979).

11. P. Brehm and R. Eckert, Calcium entry leads to inactivation of calcium channel in Paramecium, _Science_ 202:1203 (1978).

12. A.M. Brown, D.L. Kunze, and A. Yatani, The agonist effect of dihydropyridines on Ca channels, _Nature_ 311:570 (1984).

13. R.D. Burgoyne and T.R. Cheek, Is the transient nature of the secretory response of chromaffin cells due to inactivation of calcium channels?, _FEBS Lett._ 182:115 (1985).

14. M.H. Carvalho, J.C. Prat, A.G. Garcia, and S.M. Kirpekar, Ionomycin stimulates secretion of catecholamines from cat adrenal gland and spleen, _Am. J. Physiol._ 242:E137 (1982).

15. V. Cena, G.P. Nicolas, P. Sanchez-Garcia, S.M. Kirpekar, and A.G. Garcia, Pharmacological dissection of receptor-associated and voltage-sensitive ionic channels involved in catecholamine release, _Neuroscience_ 10:1455 (1983).

16. D.E. Clapham and E. Neher, Trifluoperazine reduces inward ionic currents and secretion by separate mechanisms in bovine chromaffin cells, _J. Physiol._ 353:541 (1984).

17. W.W. Douglas and T. Kanno, The effect of amethocaine on acetylcholine induced depolarization and catecholamine secretion in the adrenal chromaffin cell, _Br. J. Pharmac._ 30:612 (1967).

18. W.W. Douglas, T. Kanno, and S.R. Sampson, Effects of acetylcholamine and other medullary secretagogues and antagonists on the membrane potential of adrenal chromaffin cells: an analysis employing techniques of tissue culture, _J. Physiol._ 118:107 (1967).

19. W.W. Douglas and A.M. Poisner, On the mode of action of acetylcholine in evoking adrenal medullary secretion: Increased uptake of calcium during the secretory response, _J. Physiol._ 162:385 (1962).

20. W.W. Douglas and A.M. Poisner, Preferential release of adrenaline from the adrenal medula by muscarine and pilocarpine, _Nature_ 208:1102 (1965).

21. W.W. Douglas and R.P. Rubin, The role of calcium in the secretory response of the adrenal medulla to acetylcholine, _J. Physiol._ 159:40 (1961).

22. W.W. Douglas and R.P. Rubin, The mechanism of catecholamine release from the adrenal medulla and the role of calcium in stimulus-secretion coupling, _J. Physiol._ 167:288 (1963).

23. W.W. Douglas and R.P. Rubin, The effects of alkaline earths and other divalent cations on adrenal medullary secretion, _J. Physiol._ 175:231 (1964).

24. R. Eckert and J.E. Chad, Inactivation of Ca channels, *Prog. Biophys. Molec. Biol.* 44:215 (1984).

25. E. El-Fakahany and E. Richelson, Effect of some calcium antagonists on muscarinic receptor-mediated cyclic GMP formation, *J. Neurochem.* 40:705 (1983).

26. E.M. Fenwick, A. Marty, and E. Neher, A patch clamp study of bovine chromaffin cells and of their sensitivity to acetylcholine, *J. Physiol.* 331:577 (1982).

27. E.M. Fenwick, A. Marty, and E. Neher, Sodium and calcium channels in bovine chromaffin cells, *J. Physiol.* 331:599 (1982).

28. A. Fleckenstein, Die Bedentung der energiereichen Phosphate fur Kontraktilitat und Tonus des Myokards, *Verh. DtschGes. Inn. Med.* 70:81 (1964).

29. A. Fleckenstein, Specific pharmacology fo calcium in myocardium cardiac pacemakers and vascular smooth muscle, *Ann. Rev. Pharmac. Toxicol.* 17:149 (1977).

30. A.M. Galzin and S.Z. Langer, Presynaptic $alpha_2$ adrenoceptor antagonism by verapamil but not by diltiazem in rabbit hypothelamic slices, *Br. J. Pharmac.* 78:571 (1983).

31. A.G. Garcia, S.M. Kirpekar, and J.C. Prat, A calcium ionophore stimulating the secretion of catecholamines from the cat adrenal, *J. Physiol.* 244:253.

32. A.G. Garcia, V. Cena, and J. Frias, Pharmacological dissection of ionic channels involved in catecholamine release from the chromaffin cell, *Actual. Chim. Ther. 11° serie*, 165 (1984).

33. A.G. Garcia, F. Sala, M.G. Ladona, V. Cena, and C. Montiel, Analysis of the catecholamine secretory process by using a novel dihydropyridine calcium "agonist" and potassium or calcium gradients, *in*: "Regulation of Transmitter Function", W.S. Vizi and K. Magyar, eds., Amsterdam, Elsevier (1984).

34. A.G. Garcia, F. Sala, J.A. Reig, S. Viniegra, J. Frias, R. Fonteriz, and L. Gandia, Dihydropyridine Bay-K-8644 activates chromaffin cell calcium channels, *Nature* 309:69 (1984).

35. A.G. Garcia, F. Sala, V. Cena, C. Montiel, and M.G. Ladona, Modulation by calcium of the kinetics of the chromaffin cell secretory response, *in*: "Stimulus-secretion Coupling n Chromaffin Cells", K. Rosenheck, ed., Boca Raton, Florida, CRC Press (in press).

36. S. Hagiwara and L. Byerly, Calcium Channel, *Ann. Rev. Neurosci.* 4:69 (1981).

37. P. Hess, J.B. Lansman, and R.W. Tsien, Different modes of Ca channels gating behaviour favoured by dihydropyridine Ca agonists and antagonists, *Nature* 311:538 (1984).

38. R.W. Holz, R.A. Senter, and R.A. Frye, Relationship between Ca^{2+} uptake and catecholamine secretion in primary dissociated cultures of adrenal medulla, J. Neurochem. 39:635 (1982).

39. T. Hoshi, J. Rothlein, and S.J. Smith, Facilitation of Ca^{2+}-channel currents in bovine adrenal chromaffin cells, Proc. Natl. Acad. Sci. USA 81:5871 (1984).

40. K. Ishikawa and T. Kanno, Influences of extracellular calcium and potassium concentrations on adrenaline release and membrane potential in the perfused adrenal medulla of the rat, Jap. J. Physiol. 28:275 (1978).

41. Y. Kidokoro, Electrophysiology f adrenal chromaffin cells, in: "The Electrophysiology of the Secretory Cell", A.M. Poisner and J.M. Trifaro, eds., Elsevier Science Publishers B.V., Amsterdam (1985).

42. Y. Kidokoro and A.K. Ritchie, Chromaffin cell action potentials and their possible role in adrenaline secretion from rat adrenal medulla, J. Physiol. 307:199 (1980).

43. Y. Kodokoro, S. Miyazaki, and S. Ozawa, Acetylcholine-induced membrane depolarization and potential fluctuations in the rat adrenal chromaffin cell, J. Physiol. 324:203 (1982).

44. D.L. Kilpatrick, R.J. Slepetis, J.J. Corcoran, and N. Kirshner, Calcium uptake and catecholamine secretion by cultured bovine adrenal medulla cells, J. Neurochem. 38:427 (1982).

45. S.M. Kirpekar, J.C. Prat, and M.T. Schiavone, Effect of muscarine on the release of catecholamines from the perfused adrenal gland of the cat, Br. J. Pharmac. 77:455 (1982).

46. S. Kitayama, K. Morita, T. Dohi, and A. Tsujimoto, The nature of the stimulatory action of gamma-aminobutyric acid in the isolated perfused dog adrenals, Naunyn-Schmiedeberg's Arch. Pharmacol. 326:106 (1984).

47. D.E. Knight and N.T. Kesteven, Evoked transient intracellular free Ca^{2+} changes and secretion in isolated bovine adrenal medullary cells, Proc. Roy. Soc. London B. 218:177 (1983).

48. S. Kokubun and H. Reuter, Dihydropyridine derivates prolong the open state of Ca channels in cultured cardiac cells, Proc. Natl. Acad. Sci. USA 81:4824 (1984).

49. S. Lemaire, G. Derome, R. Tseng, P. Mercier and I. Lemaire, Distinct regulations by calcium of cyclic GMP levels and catecholamne secretion in isolated bovine adrenal chromaffin cells, Metabolism 30:462 (1981).

50. A. Marty, Ca^{2+}-dependent K^{+} channels with large unitary conductance in chromaffin cell membranes, Nature 291:497 (1981).

51. C. Montiel, A.R. Artalejo, and A.G. Garcia, Effects of the novel
    dihydropyridine Bay-K-8644 on adrenomedullary catecholamine release
    evoked by calcium reintroduction, <u>Biochem. Biophys. Res. Comm.</u>
    120:851 (1984).

52. D.A. Nachshen and M.R. Blaustein, Influx of calcium, strontium and
    barium in presynaptic nerve endings, <u>J. Gen. Physiol.</u> 79:1065 (1982).

53. P.J. Norris, D.K. Dhaltiwal, D.P. Druce and H.F. Bradford, The
    suppression of stimulus-evoked release of amino acid
    neurotransmitters from synaptosomes by verapamil, <u>J. Neurochem.</u>
    40:514 (1983).

54. J.E.B. Pinto and J.M. Trifaro, The different effects of D-600
    (methoxy-verapamil) on the release of adrenal catecholamines
    induced by acetylcholine, high potassium or sodium deprivation, <u>Br.
    J. Pharmac.</u> 57:127 (1976).

55. H. Sasakawa, K, Kumamura, S. Yamamoto, and R. Kato, Effects of W-7 on
    catecholamine release and $^{45}Ca^{2+}$ uptake in cultured adrenal
    chromaffin cells, <u>Life Sci.</u> 33:2017 (1983).

56. A.S. Schneider, H.T. Cline, K. Rosenheck, and M. Sonenberg, Stimulus-
    secretion coupling in isolated adrenal chromaffin cells: calcium
    channel activation and possible role of cytoskeletal elements, <u>J.
    Neurochem.</u> 37:567 (1981).

57. M. Schramm, G. Thomas, R. Towart, and G. Franckowiak, Novel
    dihydropyridines with positive iontropic action through activation
    of Ca^{2+} channels, <u>Nature</u> 303:535 (1983).

58. S.M. Simon and R.R. Llinas, Compartmentalization of the submembrane
    calcium activity during calcium influx and its significance in
    transmitter release, <u>Biophys. J.</u> 48:485 (1985).

59. J. Sorimachi and S. Nishimura, Operation of internal Na-dependent Ca
    influx mechanism associated with catecholamine secretion in the
    adrenal chromaffin cells, <u>Jap. J. Physiol.</u> 34:19 (1984).

60. M. Spedding, Calcium antagonist subgroups, <u>TIPS</u> 6:109 (1985).

61. M. Takahashi and A. Ogura, Dihydropyridines as potent calcium channel
    blockers in neuronal cells, <u>FEBS Lett.</u> 152:191 (1983).

62. D. Tillotson, Inactivation of Ca conductance dependent on entry of Ca
    ions in molluscan neurons, <u>Proc. Natl. Acad. Sci. USA</u> 76:1497
    (1979).

63. R.W. Tsien, Calcium channels in excitable cell membranes, <u>Ann. Rev.
    Physiol.</u> 45:341 (1983).

64. A. Wada, N. Yanagihara, F. Izumi, S. Sakurai, and H. Kobayashi,
    Trifluoperazine inhibits $^{45}Ca^{2+}$ uptake and catecholamine secretion
    and synthesis in adrenal medullary cells, <u>J. Neurochem.</u> 40:481
    (1983).

65. A. Wada, H. Takara, F. Izumi, H. Kobayashi, and N. Yanagihara, Influx of ^{22}Na through acetylcholine receptor-associated Na channels: Relationship between ^{2}Na influx, ^{45}Ca influx and secretion of catecholamines in cultured bovine adrenal medulla cells, Neuroscience 15:283 (1985).

66. A.R. Wakade, Studies on secretion of catecholamines evoked by acetylcholine or transmural stimulation of the rat adrenal gland, J. Physiol. 313:463 (1981).

67. G. Yellen, Ionic permeation and blockade in Ca^{2+}-activated K channels of bovine chromaffin cells, J. Gen. Physiol. 84:157 (1984).

INSULIN RELEASE AND K^+-INDUCED DEPOLARIZATION IN MOUSE

PANCREATIC β-CELLS

S. Joost and I. Atwater

Laboratory of Cell Biology and Genetics
Building 8, Room 403
NIADDK
National Institutes of Health
Bethesda, MD 20892

INTRODUCTION

Ca influx through the β-cell membrane represents an important step in the triggering of glucose-induced insulin secretion. The way in which β-cells recognize the glucose concentration and subsequently regulate Ca^{2+} influx has been the focal point of much of the research in islet physiology over the last two decades. We have studied Ca-channel activity in the β-cell using high external potassium $[K^+]_o$ to depolarize the membrane.[1] The analysis of the resulting membrane noise revealed that the conductance of a single channel was about 6 pS with a mean open time of about 20 msec at 0 mV.[1] Thus, β-cell Ca-channels behave similarly to Ca-channels described in other cells.[6]

Ca currents in β-cells have been recently studied using the patch clamp method in the whole cell configuration.[14,17] Little time dependent inactivation of the Ca current was observed and, in one of the studies,[17] a lack of sensitivity to glucose _per se_ was described.

However, in patch clamp studies, dispersed single β-cells, which are known to respond to glucose only very poorly, were used.[11] Also, the experiments were usually done at room temperature, a condition which was reported to completely inhibit insulin release.[2]

Therefore, we chose a different approach in order to investigate Ca-channel activity in the intact islet. An increase in the extra-

cellular K^+ concentration depolarizes the β-cell membrane and induces
a transient insulin secretion.[7,8] Assuming that the K^+-induced release is dependent upon the recurrent opening of Ca-channels during
the depolarization, the time course of the hormone release is supposed
to reflect Ca-channel activity. In this study we have compared the
time course of insulin output during constant membrane depolarization
by high K^+ in the absence and presence of glucose.

METHODS

Chemicals

Labelled ^{125}I insulin was obtained from New England Nuclear
(Boston, MA); Anti-insulin serum was a gift from Dr. Leclerq-Meyer
(Brussels, Belgiuim). Pork insulin was obtained from Elanco
(Indianapolis, IN). Bovine serum albumin was from Calbiochem (San
Diego, CA). All other reagents were of analytical grade.

Experimental Procedures

Female albino mice, 4-5 months old, fed ad libitum were used.
For each experiment, a single large partially microdissected islet
from the tail region of the pancreas was pinned in a small perifusion
chamber as described in detail elsewhere[15] and perifused at 37°C with
modified Krebs-Ringer-bicarbonate (KRB) buffer. The KRB contained (in
mM): 110 NaCl, 5 KCl, 25 $NaHCO_3$, 2.56 $CaCl_2$, and 1.1 $MgCl_2$, equilibrated with 95% O_2/5% CO_2 and 0.5% bovine serum albumin. Glucose was
added when indicated. When the potassium concentration was raised the
sodium concentration was decreased to maintain constant osmolarity.
The perifusion rate was constant during the course of each experiment
and varied between 1.4 - 1.8 ml/min for different experiments. Samples for insulin determination were collected at 30 sec or 1 min
intervals.

Insulin Assay

Insulin release from single mouse pancreatic islets was measured
by radio-immunoassay (RIA) using a modified method previously
described.[10] Pork insulin was used as a standard; ^{125}I labeled
porcine insulin was used as the radioactive tracer and guinea-pig
anti-porcine insulin serum was used as antibody. Dilutions of radioactive tracer and antibody were carried out in 0.1 M Tris-buffer (pH

7.4) containing 0.5% bovine serum albumin. Calibration curves were
run with standards diluted either in Tris-buffer or in KRB-buffer used
for the experiments. None of the modifications of the perifusion media
interfered with the assay procedure. All samples were measured in
duplicate; 200 ul of standard or undiluted sample were incubated with
400 ul Tris-buffer, containing radioactive tracer and antibody (final
dilution of 1:300000), for 18 to 24 hours at room temperature. The
bound fraction was separated from the free hormone by ammonium sulfate
(final concentration of 22%) precipitation. The supernatant was
removed and the radioactivity in the precipitate was counted in a
Micromedic gamma counter. The lower limit of detection of the assay
was 16 pg insulin/ml. Intra-assay variation was 11.1%, inter-assay
variation 10.7%.

RESULTS

We have developed a technique for studying insulin release from
single, microdissected mouse islets during continuous perifusion.[15]

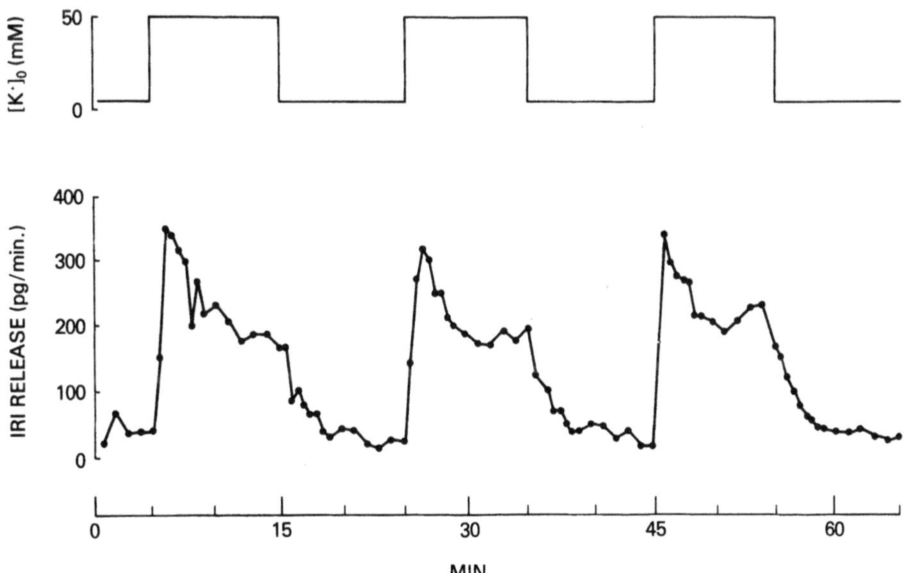

Fig. 1. Insulin release induced by high K$^+$ in the presence of low
levels of glucose. Three consecutive 10 min exposures to 50 mM K$^+$ in
the presence of 5.6 mM glucose. Each point represents the median
between duplicates.

Because of variability from islet to islet, control and test conditions must be compared consecutively in the same islet. We have
previously shown that several repetitive exposures for 3 min to 50 mM
K[+] in the absence of glucose induced comparable release responses.[4] It
was also shown that repetitive exposures to 22.2 mM glucose induced
comparable release responses if 5.6 mM glucose was present between the
exposures.[3] Fig. 1 clearly shows that 3 consecutive 10-minute exposures to 50 mM K[+] in the presence of 5.6 mM glucose caused a secretion
pattern which was quite reproducible.

Figure 2 illustrates an experiment in which 2 exposures to 50 mM
K[+] were tested in the same islet, first in the presence of 5.6 mM
glucose, second in the absence of glucose. As can be seen, peak
insulin release was not changed in both cases, but the time course of
decay during the K[+]-exposure was quite different. In three experiments, the half time of the decay averaged 6.5 min in the absence and
33.5 min in the presence of glucose.

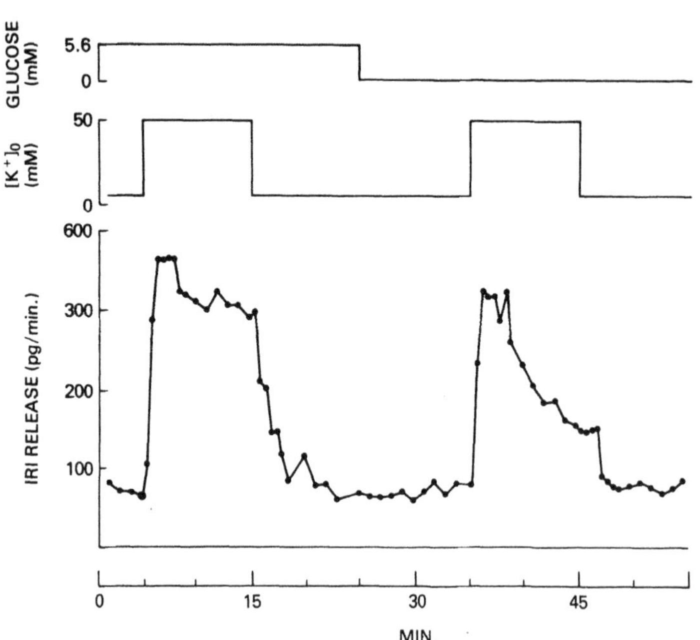

Fig. 2. Comparison of K[+]-induced insulin release in the presence and
absence of glucose. Glucose was changed from 5.6 mM to 0 between
consecutive exposures to 50 mM K[+], as indicated in the upper panel.
Each point represents the median between duplicates.

In order to determine the relationship between glucose concentration and time course of the decay in insulin release during exposure
to 50 mM K$^+$, single islets were stimulated with high K$^+$ in the presence of various glucose concentrations. One such experiment is illustrated in Fig. 3. In the presence of glucose (11 mM), no decay in
insulin output evoked by high K$^+$ stimulation could be observed during
the 10 min exposure. In the presence of low, subthreshold, glucose
concentrations the time course of insulin release showed a fairly
rapid decay. The results were similar if the glucose concentration
was progressively increased instead of decreased. It is also shown in
Fig. 3 that the background insulin release between the exposures to 50
mM K$^+$ was enhanced in the presence of 11.1 mM and 8.3 mM glucose,while
at 5.6 mM and 2.8 mM of the sugar, a lower basal insulin level was
measured.

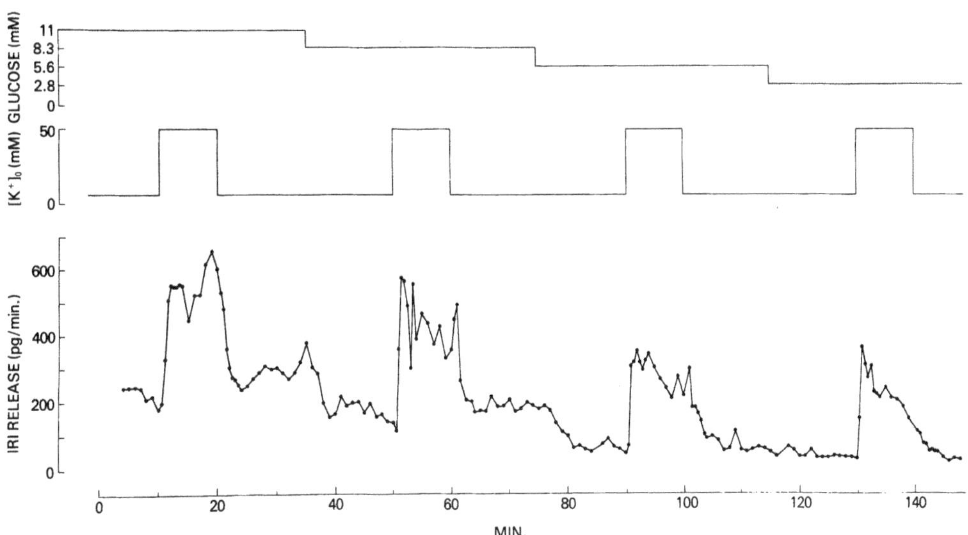

Fig. 3. Effects of glucose on K$^+$-induced insulin secretion. Consecutive exposures to 50 mM K$^+$ in the presence of 11.1, 8.3, 5.6, and 2.8
mM glucose, as indicated in the upper panel. Points represent the
median between duplicates.

DISCUSSION

The results presented here showed that depolarization of the
β-cell membrane by high K$^+$ leads to a transient increase in insulin

output, confirming previous reports.[4] The fast decay of insulin release in response to a K[+]-stimulus is unlikely to be due to depletion of insulin granules since no attenuation of the response to 50 mM K[+] could be observed with repeated exposures to high K[+].[4]

It is well documented that the increased Ca^{2+} influx through voltage-gated Ca-channels causes accumulation of cytosolic Ca^{2+} which in turn causes inactivation of the channels.[9] A similar mechanism may be operating in β-cells, and this might be responsible for the transient stimulation of insulin release in response to K[+]. However, whole cell Ca^{2+} currents in patch clamp experiments showed little or no inactivation.[14,17] The different behavior could be caused by the EGTA used in the internal solutions during the whole cell recording.

We have shown that glucose alters the time course of K[+]-induced insulin release by reducing the decay of the insulin secretion (Fig. 2). This effect of glucose appears to be concentration dependent since 5.6 mM glucose only delayed the decay of K[+]-stimulated insulin output while 11 mM glucose prevented the inactivation of release. This inactivation process was not voltage dependent, since at all glucose concentrations tested, 50 mM K[+] depolarized the membrane to the same value, $^{-12}$mV.[1]

Although at this point we cannot rule out other mechanisms, we rather favor the idea that glucose or a factor produced by its metabolism interferes with Ca-channel inactivation in the β-cell. The insensitivity of the Ca^{2+} currents to glucose in whole cell patch clamp experiments[17] might be explained by the lack of glucose metabolism due to the fact that in this technique the intracellular medium is replaced by the perfusion solution of the pipette.

If our interpretation of the results is correct, i.e., that glucose prevents inactivaiton of Ca^{2+} currents, then there are two possibilities: the sugar might induce a change in the molecular properties of the channels or it might enhance Ca^{2+} buffering within the β-cell and thus attenuate or suppress Ca^{2+}-induced Ca-channel inactivation. Although cytosolic free Ca^{2+} measurements using fluorescent indicators showed an increase in cytosolic Ca^{2+} evoked by glucose in an insulin-secreting cell line[18] as well as in obese mouse β-cells,[12] this does not necessarily provide evidence against the latter hypothesis. This is particularly true in the case of the experiments with β-cells from obese mouse islets where it could be

demonstrated that an early effect of glucose was to lower cytosolic
Ca^{2+} concentration. The decrease in the fluorescent signal is masked
by the large increase in Ca^{2+} due to Ca^{2+} entry through the voltage-
gated Ca-channels. Furthermore, these fluorescent techniques indicate
average Ca^{2+} concentrations, but both experimental data[13] and, more
recently, quantitative modeling[16] have demonstrated that the increase
in cytosolic Ca^{2+} may be highly localized spatially as well as tempo-
rally.

REFERENCES

1. I. Atwater, C. M. Dawson, G. T. Eddlestone, and E. Rojas, Voltage
      noise measurements across the pancreatic β-cell membrane:
      calcium channel characteristics, J. Physiol. (London)
      314:195 (1981).
2. I. Atwater, A. Goncalves, A. Herchuelz, P. Lebrun, W. J.
      Malaisse, E. Rojas, and A. Scott, Cooling dissociates glu-
      cose-induced insulin release from electrical activity and
      cation fluxes in rodent pancreatic islets, J. Physiol.
      (London) 384:615 (1984).
3. I. Atwater and A. M. Scott, Measurements of insulin release and
      β-cell membrane potential in mouse islets of Langerhans: the
      effects of repetitive glucose pulses, in "Proc. Internatl.
      Diabetes Federation," Elsevier Press, Madrid (in press,
      1986).
4. C. Dawson, I. Atwater, and E. Rojas, Potassium-induced insulin
      release and voltage noise measurements in single mouse
      pancreatic islets, J. Membr. Biol. 64:33 (1982).
5. G. T. Eddlestone, A. Goncalves, J. A. Bangham, and E. Rojas,
      Electrical coupling between β-cells in islets of Langerhans
      from the mouse, J. Membr. Biol. 77:1 (1984).
6. S. Hagiwara and L. Byerly, Calcium channel, Ann. Rev. Neurosci.
      4:69 (1981).
7. J. C. Henquin and A. Lambert, Cationic environment and dynamics
      of insulin secretion: II. Effect of a high concentration of
      potassium, Diabetes 23:933 (1974).
8. A. Herchuelz, N. Thonnart, A. Sener, and W. J. Malaisse, Regula-
      tion of calcium fluxes in pancreatic islets: the role of
      membrane depolarization, Endocrinology 107:491 (1980).
9. B. Hille, Ionic channels of excitable membranes, Sinauer Associ-
      ates Inc., Sunderland, MA, Ch. 87, pg. 4 (1984).

10. H. G. Joost, Effects of a possible β-cell membrane label, metahexamide-isothiocyanate, on insulin release, _Horm. Metab. Res._ 11:104 (1979).

11. A. Lernmark, The preparation of, and studies on, free cell suspensions from mouse pancreatic islets, _Diabetologia_ 10:431 (1974).

12. P. Rorsman, H. Abrahamsson, E. Gylfe, and B. Hellman, Dual effects of glucose on the cytosolic Ca activity of mouse pancreatic β-cells, _FEBS Lett._ 170:196 (1984).

13. B. Rose and W. R. Loewenstein, Calcium ion distribution in cytoplasm visualized by aequorin: diffusion in cytosol restricted by energized sequestering, _Science_ 190:1204 (1975).

14. L. Satin and D. L. Cook, Voltage-gated Ca-current in pancreatic β-cells, _Pfluegers Arch._ 404:385 (1985).

15. A. Scott, I. Atwater, and E. Rojas, A method for the simultaneous measurement of insulin release and β-cell membrane potential in single mouse islets, _Diabetologia_ 21:470 (1981).

16. S. M. Simon and R. R. Llinas, Compartmentalization of the submembrane calcium activity during calcium influx and its significance in transmitter release, _Biophys. J._ 48:485 (1985).

17. G. Trube and P. Rorsman, Calcium and potassium currents recorded from pancreatic β-cells under voltage-clamp control, see this volume (1986).

18. C. B. Wollheim and T. Pozzan, Correlation between cytosolic free Ca and insulin release in an insulin-secreting cell line, _J. Biol. Chem._ 259:2262 (1984).

CALCIUM AND POTASSIUM CURRENTS RECORDED FROM PANCREATIC β-CELLS UNDER
VOLTAGE CLAMP CONTROL

G. Trube and P. Rorsman[*]

Max-Planck-Institut fur biophysikalische Chemie
POB 2841, D-3400 Gottingen
Federal Republic of Germany
[*]Permanent address: Dept. Medical Cell Biology
POB 571, S-751 23 Uppsala
Sweden

It has been known for over 15 years that glucose and most other
stimuli for insulin secretion induce electrical activity in the pancreatic
β-cell[4]. The secretory and electrical responses are furthermore known to
be closely correlated[13]. From membrane potential recordings obtained with
conventional intracellular electrodes and from flux measurements at least
three types of currents have been postulated[10].

(i) A K^+ conductance which is active in unstimulated β-cells and
inhibited by glucose[17].

(ii) A voltage-dependent Ca^{2+} current accounting for the depolarizing
phase of the action potential[5].

(iii) A voltage-dependent K^+ conductance underlying the repolarization
of the action potential[15].

It has long been desired to study these currents under voltage-clamp
conditions. However, because of their small size, β-cells have been
unaccessible for such measurements until the development of the patch-
clamp technique. Using the whole cell-configuration of this method[8] we
have recorded voltage-activated Ca^{2+} and K^+ currents in the same membrane

potential range as the action potentials. In addition, we describe another
K^+ conductance which is not voltage-dependent, but regulated by
intracellular ATP. This conductance is also active around the resting
potential and is identical to the ATP and glucose-sensitive K-channels
previously described[1,3,16].

MATERIALS AND METHODS

Procedures for the preparation of cells and measurements of ionic
currents are reported elsewhere[16]. Briefly, islets of Langerhans were
isolated from the pancreases of NMRI-mice by collagenase digestion and
dissociated into single cells by shaking in a Ca^{2+} free medium[12]. The
cells were maintained in tissue culture for 1-3 days prior to the
experiments. The whole-cell configuration of the patch-clamp technique was
used for the measurements of voltage-clamp currents[6,8]. During the
experiments the cells were immersed in a solution (sol. A) containing 138
mM NaCl, 5.6 mM KCl, 1.2 mM $MgCl_2$, 2.6 mM $CaCl_2$, 10 mM HEPES-NaOH (pH 7.4)
and 10 mM D-glucose. The electrode filling solution in whole-cell
experiments dialyzes the interior of the cell. Electrodes (pipettes) were
therefore filled with a medium (sol. B) mimicking the composition of the
cytoplasm and containing 125 mM KCl, 30 mM KOH, 4 mM $MgCl_2$, 2 mM $CaCl_2$, 10
mM EGTA, 3 mM Na_2ATP and 5 mM HEPES-KOH (pH = 7.15; pCa = 7). In some
experiments Na_2ATP was replaced by 6 mM NaCl and $MgCl_2$ reduced to 1 mM
(sol. C). Outward K^+ currents had to be blocked in experiments on the
inward Ca^{2+} current. This was achieved by either equimolar substitution of
CsCl for the KCl in solution B (sol. D) or using the impermeant cation
N-methyl-D-glucamine[7]. In the latter case the intracellular medium (sol.
E) contained 145 mM N-methyl-D-glucamine, 120 mM HCl, 4 mM $MgCl_2$, 2 mM
$CaCl_2$, 10 mM EGTA, 3 mM Na_2 ATP, 5 mM HEPES-KOH (pH = 7.15). The
osmolarity of all media ranged between 290 and 300 mOsm. Experiments were
performed at room temperature (22°C).

RESULTS AND DISCUSSION

Membrane currents recorded from a single β-cell under voltage-clamp
conditions are shown in Fig. 1A. Depolarizing voltage steps to potentials
between -40 mV and +10 mV were applied to the cell from a holding potential

of -70 mV. A pulse to -30 mV evoked a rapidly activating inward current
(time to peak 7 ms). At more positive potentials a slowly rising outward
current of larger amplitude was superimposed on the inward current. This
current is similar to the delayed rectifying K^+ current in other types of
cells (cf. 11, chs. 2 and 5). The inward current is carried by calcium as
will be shown below. The peak amplitude of the inward and outward currents
are plotted versus the potential in Fig. 1B. The outward current increased
linearly with the potential at voltages above -20 mV with an average
amplitude of 300 ± 60 pA (mean ± S.E.M. of 13 experiments) at 0 mV. It was
completely and reversibly blocked by adding 20 mM tetraethylammonium (TEA)
to the bath solution (Fig. 1D). The inward current had a V-shaped current-
voltage relation and was maximal during pulses to -10 mV (Fig. 1B, bottom).
The size of the inward currents in different experiments was more variable
than that of the outward currents. We found that a high input resistance
(>10 Gohm), which indicates little leak between the cell interior and the
bath, was necessary to record inward currents such as those in Fig. 1A.

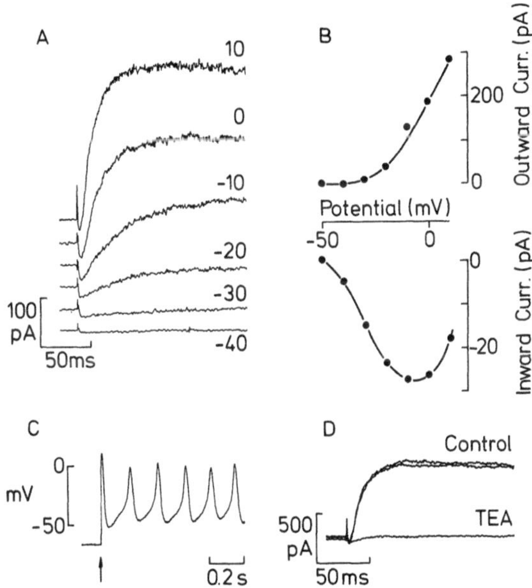

Fig. 1. A, membrane currents during voltage-clamp pulses going from -70 mV
to potentials indicated (in mV) at the end of the traces. Pipette
solution B. Pulse rate: 0.35 Hz. Deflections by capacitive
currents were retouched with the exception of those in the top
trace. B, current-voltage relations of peak inward and outward
currents in A. Note different ordinate scales. Curves were
fitted to the values (circles) by eye. C, action potentials
recorded from the same cell as in A and B after releasing the
voltage-clamp (arrow). D, different cell. Control: currents
before and after the application of 20 mM TEA. Voltage-clamp
pulses going from -70 mV to 10 mV. Pipette solution B.

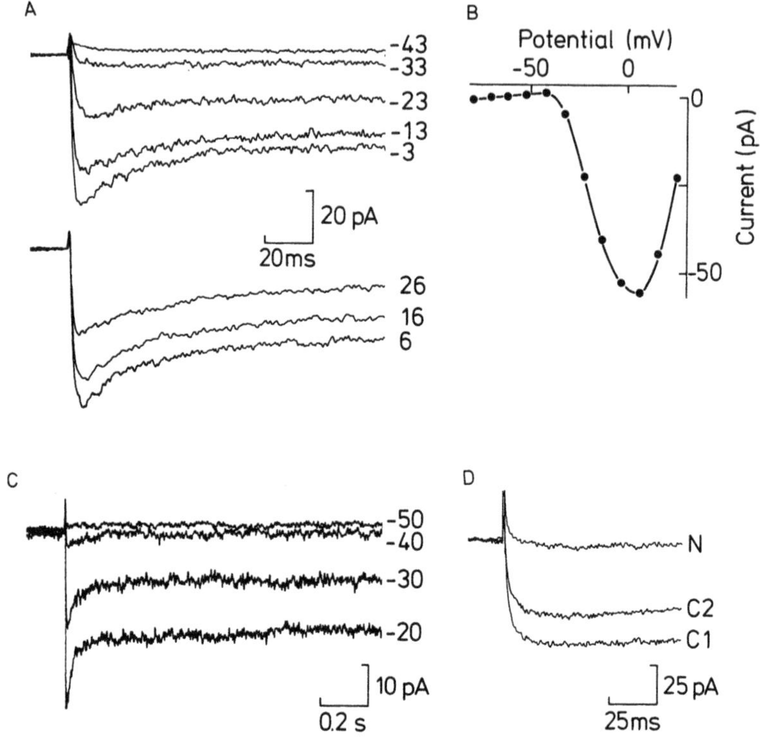

Fig. 2. Ca^{2+} currents. A, currents during depolarizing steps from -70 mV
to potentials indicated at the right end of the traces. All
records are from the same cell but are displayed in two groups for
clarity. Pipette solution D. Pulse rate 0.45 Hz. B,
current-voltage relation of peak currents in A. C, currents
during long (2s) depolarizing steps. Holding potential -70 mV,
pulse potentials at the right end of the traces. Pulse rate: 0.16
Hz. Pipette solution E. D, current responses during voltage
steps going from -70 mV to -10 mV. C1 and C2: records before and
after (5 min washout) exposure to nifedipine. N: current in
presence of 5 μM nifedipine. Pipette solution E. Bath solution A
containing 10.24 mM $CaCl_2$ in all experiments.

When a tight seal between the pipette and the cell had been obtained,
spontaneous action potentials (spikes) could be observed after releasing
the voltage-clamp (Fig. 1C). The spikes usually started from a potential
between -50 mV and -40 mV, which is in the same range as that where time-
and voltage-dependent currents became detectable. They peaked between -20
mV and 0 mV as has been reported for the action potentials in intact
islets[14]. In most experiments the frequency of action potentials was less
regular than in the example of Fig. 1C. For unknown reasons, slow waves
and bursts of action potentials, which are typically found in electrical
recordings from whole islets, were not observed in single cells but only in
experiments on cell clusters (diameter about 100 μm, not shown). Single

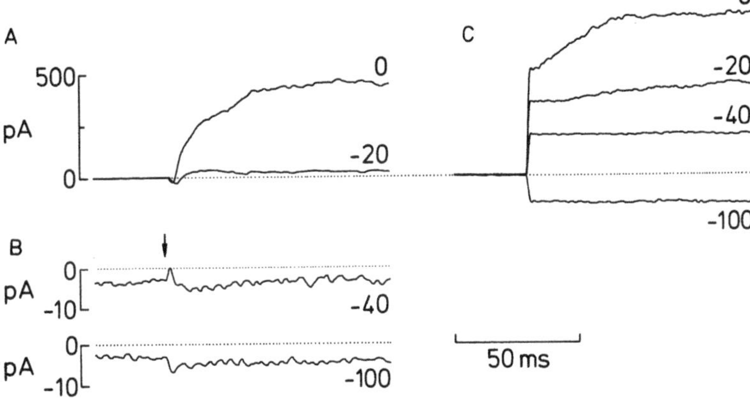

Fig. 3. Washout of intracellular ATP and development of a K^+ conductance.
A and B, current response to voltage pulses going to the indicated
potentials. Records taken 40 s after going into the whole-cell
recording mode. Responses in B shown at higher gain; onset of
pulses marked by arrow. C, same experiment 9 min later. Dotted
lines in A-C indicate zero current level. Holding potential -70
mV throughout.

cells had to be used for voltage-clamp experiments because electrical

coupling between cells was not sufficient to obtain a good clamp of an

entire cluster.

It is evident from Fig. 1A that the large outward K^+ current tended to

obscure the inward current. The K^+ current therefore had to be blocked to

permit a study of the inward current. K^+ in the pipette (intracellular

solution) was consequently replaced by either Cs^+ or the impermeant cation

N-methyl-D-glucamine (NGM: see Materials and Methods, solutions D and E).

It was also advantageous to increase extracellular $[Ca^{2+}]$ from 2.6 to 10.2

mM, which increased the magnitude of the inward current by 60%. Fig. 2A

shows a series of superimposed current traces recorded with a pipette

containing Cs^+. Inward currents were first seen when the cell was

depolarized from -70 mV to -33 mV and increased in size until +6 mV. The

current-voltage relation of the peak inward currents is shown in Fig. 2B.

In 5 similar experiments the Ca^{2+} current reached a maximum amplitude of 48

± 8 pA at potentials around +10 mV. If NMG was used instead of Cs^+ the peak of the current-voltage relation was shifted to more negative potentials by about 20 mV. Different liquid junction potentials between the pipette and the bath can account for about half of this deviation[7]. The inward currents were not affected by the Na^+ channel blocker tetrodotoxin, but were suppressed by Co^{2+} (5 mM) or nifedipine (5 μM, Fig. 2D) as is expected for a Ca^{2+} current(cf 14 and 11, chs. 4 and 12).

The Ca^{2+} currents seemed to inactivate partially in the experiment of Fig. 2A. In most experiments with pipettes containing Cs^+ the decay of the current was stronger than in Fig. 2A. We believe that part of this time course results from an incomplete blockage of K^+ channels and an outward current carried by Cs^+. When NMG was used instead of Cs^+ (sol. E), the current stayed constant for at least 100 ms in 50% of the cells (e.g. Fig. 2C). The decay in the remaining experiments could be explained by assuming the existence of two classes of Ca-channels which are mixed at variable proportions from cell to cell[2]. Alternatively, the solution exchange between the intracellular space and the pipette may have been less effective in these experiments resulting in a lower Ca^{2+} buffering capacity, and an accumulation of Ca^{2+} close to the membrane may have caused Ca-mediated inactivation of the Ca^{2+} current (cf. 11, ch. 4). When inactivation was present, the current reached a steady level within about 100 ms and stayed constant for more than 1 s (Fig. 2C). The fact that a small inward current was maintained at -40 mV suggests that the Ca^{2+} channels may also contribute to the plateau and the slow waves induced by elevated glucose.

Resting islet cells usually have a stable resting potential close to -70 mV and depolarize to a plateau potential around -40 mV when exposed to stimulatory concentrations of glucose[10]. The depolarization is believed to be primarily due to a decreased K^+ conductance[17], however, the currents described above are not influenced by the extracellular glucose concentration (0-30 mM). This is expected since it is generally believed that the glucose dependent responses in β-cells are mediated by intracellular products of glucose metabolism[9]. In the whole-cell configuration of the patch-clamp technique the composition of the intracellular medium is determined by the pipette solution. Therefore, it is necessary to vary its composition when investigating membrane currents which are normally influenced by glucose. Fig. 3 shows an experiment where ATP was omitted

from the pipette solution (sol. C). During the first minute after breaking
the patch in the pipette tip the currents were indistinguishable from those
seen in the presence of ATP. Whereas a delayed outward current was
activated during voltage jumps to 0 mV, currents during pulses from -70 mV
to -40 mV or -100 mV were hardly detectable (Fig. 3A, B). At later times
the current amplitudes increased progressively until a new steady state was
reached after about 10 min (Fig. 3C). A slowly activating outward current
was still seen at 0 mV on top of a large current activating
instantaneously. Such time-independent currents were also evoked at the
other potentials, where almost no response had been seen before. The
polarity of these currents reversed at -70 mV and the holding current at
this potential was still close to zero (dotted lines) indicating that the
time-independent conductance is primarily selective for K^+ and not due to
unspecific leak (K^+-equilibrium potential = -85 mV, calculated from pipette
$[K^+]$ = 155 mM and bath $[K^+]$ = 5.6 mM). Changes such as those in Fig. 3
were observed in four similar experiments, but never occurred in the
presence of ATP. We therefore conclude that another K^+ conductance was
induced when intracellular ATP was washed out by the pipette solution.
Cook and Hales[3] recently found ATP-sensitive K^+ channels in isolated
patches of β-cell membrane, and it is pertinent that the same channels are
influenced by glucose in cell-attached patches[1,16].

CONCLUSION

The present study demonstrates the existence of voltage-activated K^+
and Ca^{2+} currents as well as an ATP-sensitive K^+ conductance which could
regulate the resting potential in pancreatic β-cells. When the ATP-
sensitive channels were blocked by millimolar concentrations of
intracellular ATP, the β-cell was sufficiently depolarized to produce
spontaneous action potentials. The voltage-dependent currents are very
similar to the delayed rectifying K^+ current and the Ca^{2+} inward current
present in many other cell types[11]. Since these currents activated in the
potential range of the spikes and no other voltage dependent currents were
observed, it seems likely that they are responsible for the generation of
the action potentials.

ACKNOWLEDGEMENT

We wish to thank Dr. S. DeRiemer for reading and commenting on the
manuscript. This work was supported by the Deutsche Forschungsgemeinschaft
and the Swedish Medical Research Council. P.R. was a recipient of a travel
grant from the Wallenberg Fund at Uppsala University.

REFERENCES

1. F.M. Ashcroft, D.E. Harrison, and S.J.H. Ashcroft, Glucose induces
    closure of single potassium channels in isolated rat pancreatic
    β-cells, _Nature_ 312:446 (1984).

2. E. Carbone and H.D. Lux, A low voltage-activated calcium conductance
    in embryonic chick sensory neurons, _Biophys. J._ 46:413 (1984).

3. D.L. Cook and C.N. Hales, Intracellular ATP directly blocks K^+
    channels in pancreatic β-cells, _Nature_ 311:271 (1984).

4. P.M. Dean and E.K. Matthews, Electrical activity in pancreatic islet
    cells, _Nature_ 219:389 (1968).

5. P.M. Dean and E.K. Matthews, Electrical activity in pancreatic islet
    cells: effects of ions, _J. Physiol._ 210:265 (1970).

6. E.M. Fenwick, A. Marty, and E. Neher, A patch-clamp study of bovine
    chromaffin cells and of their sensitivity to acetylcholine, _J.
    Physiol._ 635:577 (1982).

7. J.M. Fernandez, A.P. Fox, and S. Krasne, Membrane patches and whole-
    cell membranes: a comparison of electrical properties in rat clonal
    pituitary (GH_3 cells), _J. Physiol._ 356:565 (1984).

8. O.P. Hamill, A. Marty, E. Neher, B. Sakmann, and F. Sigworth, Improved
    patch-clamp techniques for high-resolution current recordings from
    cells and cell-free membrane patches, _Pflügers Arch._ 391:85 (1981).

9. C.J. Hedeskov, Mechanisms of glucose-induced insulin secretion,
    Physiol. Rev. 60:442 (1980).

10. J.C. Henquin and H.P. Meissner, Significance of ionic fluxes and
    changes in membrane potential for stimulus-secretion coupling in
    pancreatic β-cells, _Experientia_ 40:1043 (1984).

11. B. Hille, "Ionic Channels of Excitable Membranes", MA: Sinauer
    Associates Inc., Sunderland (1984).

12. A. Lemmark, The preparation of, and studies on, free cell suspensions
    from mouse pancreatic islets, _Diabetologia_ 10:431 (1974).

13. H.P. Meissner and I. Atwater, The kinetics of electrical activity of
    beta cells in response to a square wave stimulation of glucose or
    glibenclamide, _Horm. Metab. Res._ 8:11 (1976).

14. H.P. Meissner and W. Schmeer, The significance of calcium ions for the
    glucose-induced electrical activity of pancreatic β-cells, in "The
    Mechanisms of Gated Calcium Transport Across Biological Membranes,
    S.T. Ohnishi and M. Endo, eds., Academic Press, New York, pp
    157-165 (1981).

15. B. Ribalet and P.M. Beigelman, Calcium action potentials and potassium
    permeability activation in pancreatic β-cells, _Am. J. Physiol._
    239:C124 (1980).

16. P. Rorsman and G. Trube, Glucose dependent K^+ channels in pancreatic
    β-cells are regulated by intracellular ATP, _Pflügers Arch._ 405:305
    (1985).

17. J. Sehlin and I.-B. Taljedal, Glucose induced decrease in Rb^+
    permeability in pancreatic β-cells, _Nature_ 253:635 (1975).

VOLTAGE-ACTIVATED Ca^{2+} AND K^+ CURRENTS IN AN INSULIN-SECRETING CELL LINE
(RINm5F)

I. Findlay and M.J. Dunne

M.R.C. Secretory Control Research Group
The Physiological Laboratory
University of Liverpool
P.O. Box 147
Liverpool L69 3BX
United Kingdom

A cloned and continuous line of cells of islet derivation is a useful
tool for studies which require large quantities of homogenous cellular
material. The RINm5F cell line was developed from an insulinoma induced
initially in a rat and then transplanted through mice before being
established as a continuous cell line[7]. The secretion of insulin by RINm5F
cells can be stimulated by secretagogues such as glyceraldehyde and
L-alanine and by exposure to high K^+ solutions[11,13]. Though interestingly
insulin secretion is not enhanced by the β-cell secretagogue, glucose[11].
Although numerous biochemical studies of these cells have been made their
electrophysiology is unknown and there is therefore no comparison with the
body of literature concerning the electrophysiology of β-cells. Elsewhere
in this volume[10] we have shown, using patch-clamp electrophysiological
techniques, that RINm5F cells maintain a stable negative membrane potential
and that upon exposure to a secretagogue RINm5F cells depolarise and fire
spike-potentials. In this study we examine in more detail the spike
potentials and show them to result from voltage-activation of Ca-channels.

MATERIALS AND METHODS

Stock cultures of RINm5F cells were provided by Professor C.B.
Wollheim (Institut de Biochemie Clinique, University of Geneva). Cells
were maintained in RPMI 1640 medium (Flow Labs) containing 11 mM glucose
and supplemented with 10% (v/v) heat-inactivated foetal calf serum, 100
U/ml penicillin, 100 µg/ml streptomycin and 2 mM glutamine. Cells were
seeded in Falcon type 3001 plastic petri dishes (35 mm diameter) and
incubated at 37°C in a humidified atmosphere of 95% O_2, 5% CO_2.

Patch clamp experiments were performed upon single cells in the
'whole-cell' configuration[8]. The series resistance and linear capacitance
of the patch pipettes (whose resistances measured with standard 'intracellular' solution were between 2 and 5 Mohm) and the cells were compensated
with a List EPC-7 patch-clamp amplifier. Cell currents were recorded on
tape (Racal 4DS) and later replayed for analysis. Individual current
records have not been corrected for leakage conductance or residual
capacitive transients. Calculated current-voltage curves have been
corrected for leakage conductance.

The standard extracellular solution contained (mM): 140 NaCl, 4.7 KCl,
1.13 $MgCl_2$, 2 $CaCl_2$, 2.5 glucose, 10 HEPES, pH 7.2. Cobalt ($CoCl_2$),
cadmium ($CdCl_2$), additional Ca, EGTA or TEA (TEA, Br; Fluka) were added
directly to the bath solution when required. Stock solutions of nifedipine
and verapamil were prepared in DMSO and small aliquots were added to
standard extracellular solution just prior to use. The standard
'intracellular' solution contained in the patch pipette consisted of (mM):
140 KCl, 10 NaCl, 1.13 $MgCl_2$, 10 glucose, 1 ATP, 10 HEPES and 1 EGTA. No
Ca^{2+} was added and the pH was 7.2.

Before experiments were begun the RPMI 1640 tissue culture medium was
removed and the cells which remained attached to the petri dish were washed
several times in standard extracellular solution. All experiments were
performed at room temperature (20-23°C).

RESULTS

The membrane potential of -52 mV ± 1 mV (n = 109) of RINm5F cells in
the 'whole-cell' configuration was recorded by clamping the cell membrane
current. Fig. 1 shows a typical membrane voltage recording made from a

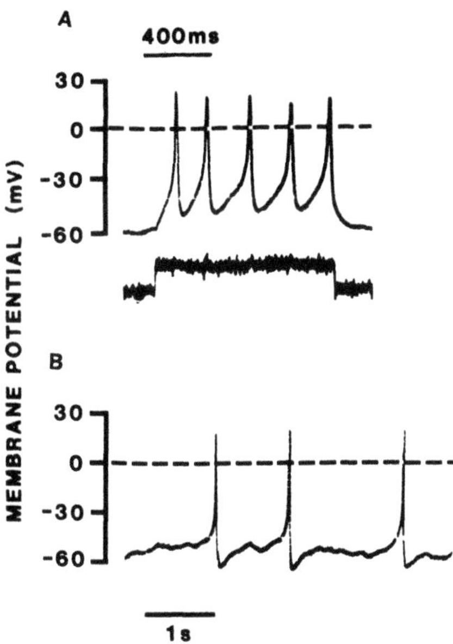

Fig. 1. Whole-cell membrane voltage recorded by clamping the membrane
current at 0 pA. A. A positive current pulse (lower trace) evokes
repetitive overshooting spike potentials in the membrane voltage
record (upper trace). B. Spontaneous overshooting spike
potentials in the membrane voltage record. From reference 5.

RINm5F cell under current clamp. In Fig. 1A a positive (depolarising)
current step applied to the cell in the absence of any secretagogue evoked
a number of spike-potentials. Each spike had three distinct phases.
Initially the potential rose slowly, the rate of depolarisation then
increased dramatically to peak at approximately +20 mV and the membrane
then rapidly depolarised. Fig. 1B shows a voltage record obtained when the
cell was spontaneously firing spike-potentials and clearly shows that the
repolarisation of the spike was followed by a distinct after-hyperpolarisa-
tion.

Under voltage-clamp control the potential of each RINm5F cell was held
at -50 mV and depolarising or hyperpolarising voltage steps of 30 ms
duration were applied to the cell at a fixed rate of 0.2 Hz. More frequent
stimulation caused a progressive decline in amplitude of the recorded
currents. The membrane currents which were evoked by depolarising voltage
steps from a clamped potential of -50 mV are shown in Fig. 2A. Stepping
the voltage to -40 mV resulted in no change in membrane current while
depolarisation to -30 mV evoked a small inward membrane current which
declined gradually during the 30 ms duration voltage step. When the
membrane voltage was stepped to either -20 or -10 mV much larger inward

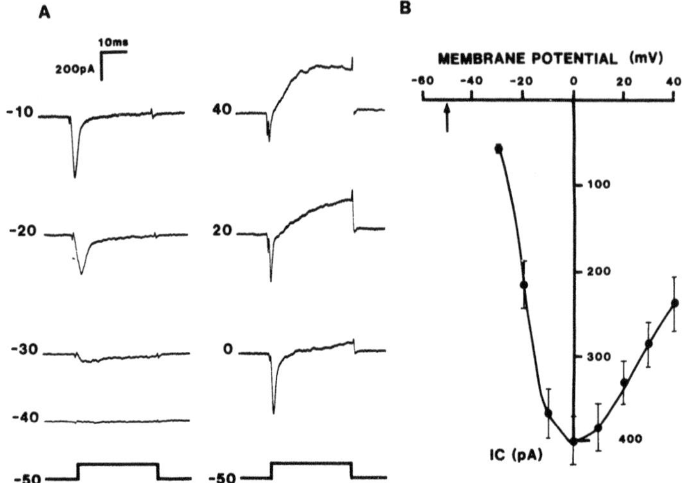

Fig. 2. Membrane currents evoked by voltage steps from a holding potential
of -50 mV. A. Individual membrane current traces recorded during
voltage steps from -50 mV to the potential indicated next to each
trace. Downwards deflections represent current moving into the
cell, upward deflections represent current moving out of the cell.
B. Current-voltage (I-V) relationship of the maximum inward
current recorded from experiments such as that illustrated in A.
Mean and standard error values were obtained from 20 experiments.
The arrow indicates the holding potential of -50 mV. IC = inward
current. From reference 5.

currents were evoked which then declined rapidly, within 10 ms of the onset
of the voltage step (Fig. 2A). Stepping the membrane voltage to more
positive membrane potentials resulted in a decline in the amplitude of the
initial inward membrane current. At the same time after the initial inward
current these voltage steps evoked a clear outward membrane current which
increased with further depolarisation of the cell (Fig. 2A).

The current-voltage relationship for the initial membrane current is
shown in Fig. 2B. It can be seen that the amplitude of the voltage-
activated inward currents increased from the activation threshold of
approximately -30 mV to maximum values at -10 and 0 mV. Reversal of the
inward currents has not been observed during these experiments but it would
clearly be at a value more positive than +55 mV which is the equilibrium
potential for Na^+ under these experimental conditions. Under the quasi-
physiological cation gradients used in these experiments Ca^{2+} is the only
ion with an equilibrium potential more positive than +55 mV. Further
evidence that the voltage-activated inward currents are carried by Ca^{2+}
ions is provided in Fig. 3. Chelation of Ca^{2+} bathing the cell by the
addition of an excess of EGTA abolished the inward current (Fig. 3A),

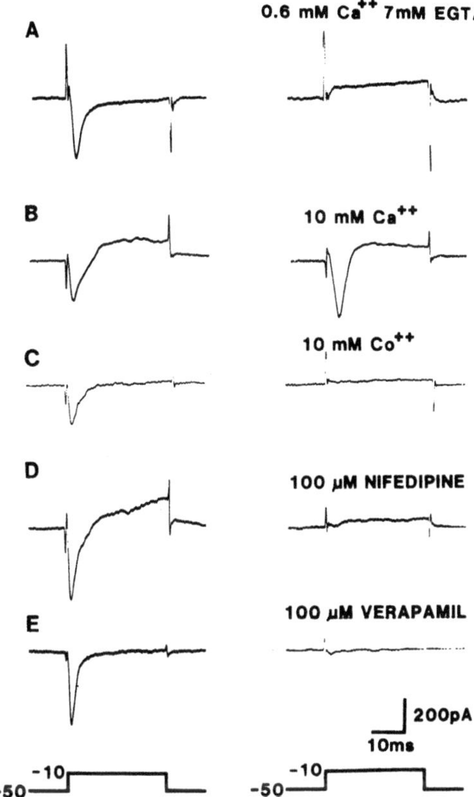

Fig. 3. Membrane currents evoked by voltage steps from a holding potential
of -50 mV to -10 mV. The left-hand column illustrates individual
records obtained under control conditions with 2 mM $[Ca^{2+}]_o$
(except in B. where $[Ca^{2+}]_o$ = 1 mM). The right-hand column
illustrates records obtained from the same cells but now recorded
under the test conditions which are identified by each individual
trace. From reference 5.

whereas increasing the concentration of Ca^{2+} bathing the cell increased the
amplitude of the inward current (Fig. 3B). Addition of inorganic Ca-
channel blockers Co^{2+} (Fig. 3C) and Cd^{2+} (Not shown) as well as organic
Ca-channel blockers nifedipine (Fig. 3D) and verapamil (Fig. 3E) inhibited
the inward current.

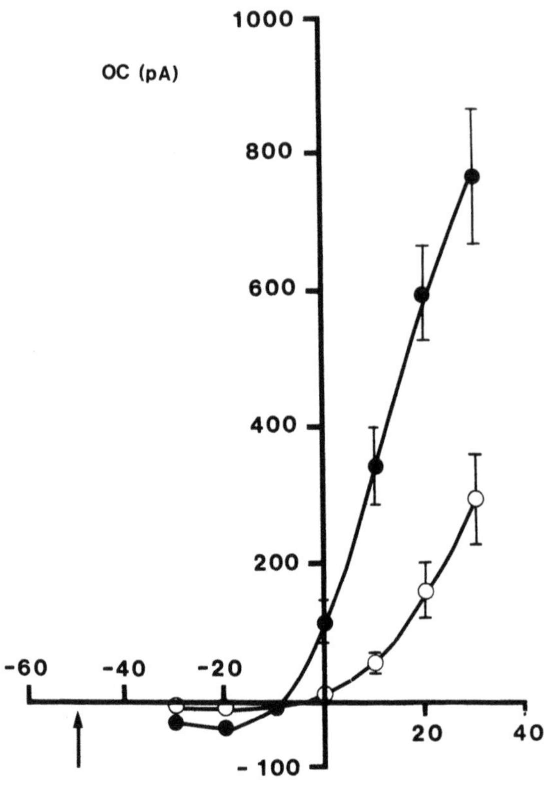

Fig. 4. Current-voltage (I-V) relationship of membrane currents recorded
at the end of each 30 ms duration depolarising voltage step from a
holding potential of -50 mV (arrow). Closed circles (●) represent
control experiments with 2 mM $[Ca^{2+}]_o$ (n = 20). Open circles (○)
represent the combined results of experiments which utilised Co^{2+}
(5 and 10 mM), nifedipine (50 and 100 µM) and verapamil (100 µM)
to block inward currents carried by Ca^{2+} (n = 11). OC = outward
current. From reference 5.

In control experiments an outward membrane current could be observed
after the initial transient inward current when the membrane voltage was
stepped to values more positive than -10 mV. This current was voltage-
activated, increasing in amplitude with the magnitude of applied depolarisation (Fig. 2A). That this outward current was also at least partially
activated by an increase in intracellular Ca^{2+} was indicated by the
consistent observation that in the presence of each of the Ca-channel
blockers when the inward Ca^{2+} current was inhibited the outward membrane
currents were severely reduced (Fig. 4). Figs. 5 and 6 illustrate the
effects of the inclusion of the K-channel blocker tetraehylammonium (TEA)
in the medium bathing the RINm5F cells. In the presence of 2 mM TEA the

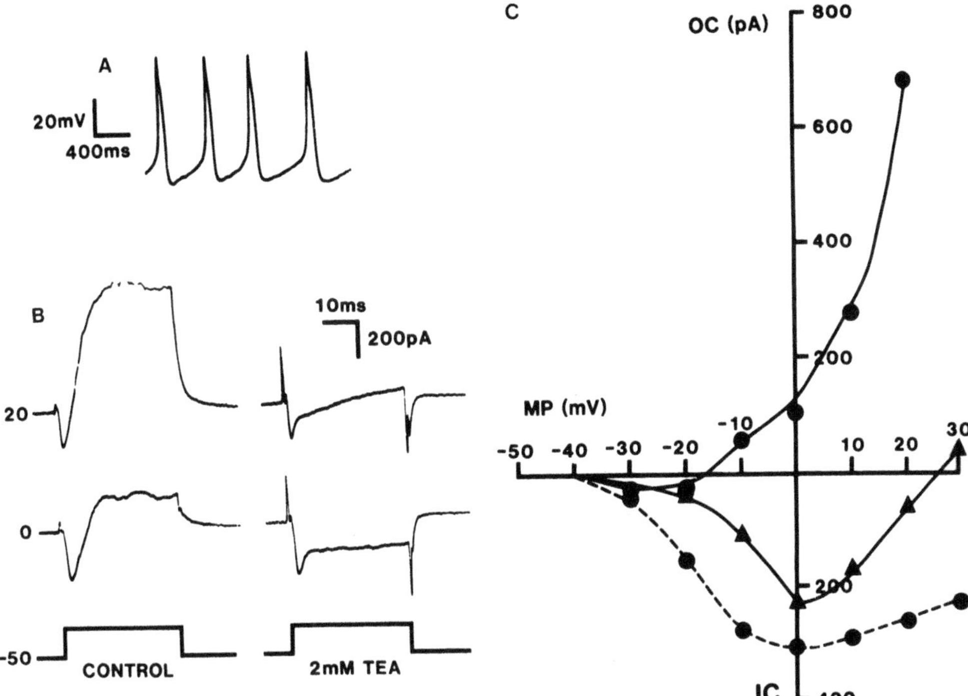

Fig. 5. The effect of 2 mM TEA upon RINm5F cell membrane voltage and
currents. A. Depolarising current-evoked spike potentials in the
presence of 2 mM TEA in the bathing solution. B. Membrane
currents evoked by voltage steps from a holding potential of -50
mV to the potential indicated next to each trace. The left hand
column illustrates records obtained under control conditions. The
right hand column illustrates records obtained from the same cell
recorded after the inclusion of 2 mM TEA in the bathing medium.
C. Current-voltage relationship of membrane currents recorded from
the same cell as B. Dotted line represents the peak inward
current. Solid lines represent the current recorded at the end of
each 30 ms depolarising voltage step (circles = control: triangles
= 2 mM TEA).

duration of current-evoked spike potentials was slightly increased (compare
Fig. 5A with Fig. 1A) and the outward membrane currents recorded under
voltage clamp were depressed, being evoked only by voltage steps to
membrane potentials of +20 mV and greater (Fig. 5B, C). 20 mM TEA
completely abolished outward membrane currents within the tested voltage
range (Fig. 6C) and markedly increased the duration of current-evoked spike
potentials (Fig. 6A).

The inclusion of TEA in the medium bathing RINm5F cells exposed a
component of inward membrane current that had been masked by the voltage-

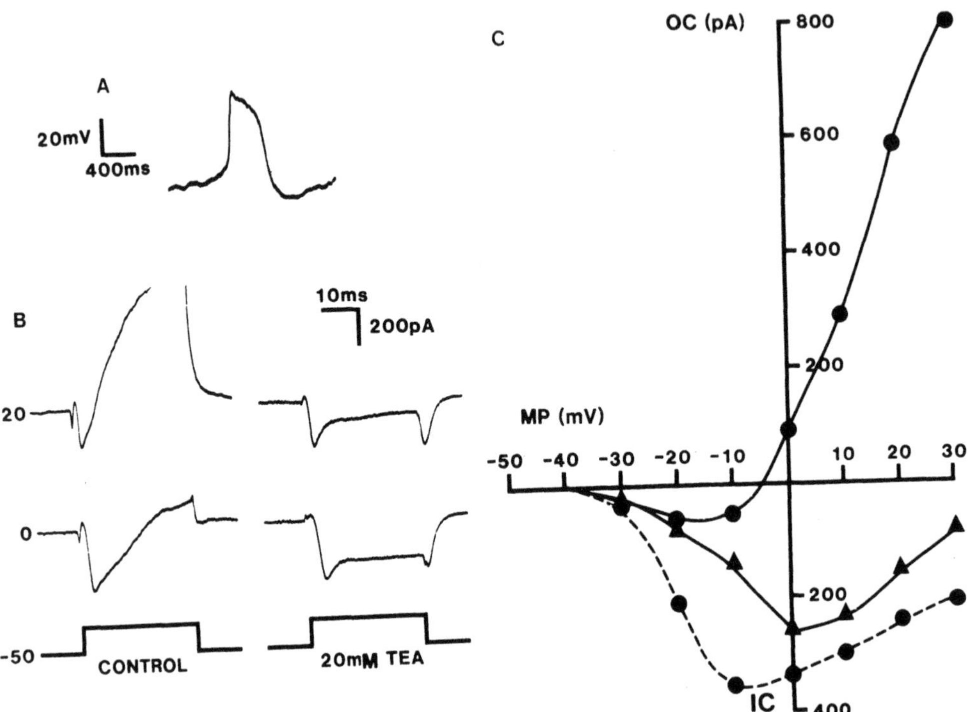

Fig. 6. The effect of 20 mM TEA upon RINm5F cell membrane voltage and currents. Explanation as for Fig. 5.

and Ca^{2+} activated outward membrane current observed under control conditions. The Ca^{2+} current could now be seen to consist not only of an initial transient peak of inward membrane current which was unaffected by TEA but also of an inward current which was sustained for the duration of the voltage step (Figs. 5B, 6B, and 6C) and formed a tail current when the membrane voltage was returned to the holding potential of -50 mV (Fig. 6B).

DISCUSSION

Insulin secretion from β-cells stimulated by glucose is associated with cell depolarisation and the generation of Ca-dependent spike potentials[9]. Similar behaviour is shown by cells of the insulin-secreting cell line RINm5F when exposed to the secretagogue glyceraldehyde[10]. Spike potentials can be directly stimulated in RINm5F cells by membrane depolarisation (Fig. 1A) as has also been shown for β-cells[1,9], and the characteristics of these voltage-activated spike potentials can be directly compared

with the membrane currents recorded from RINm5F cells under voltage-clamp
control.

The rising phase of the spike potential evoked by membrane depolarisation consisted initially of a slow rate of depolarisation which accelerated
when the potential reached -30 to -25 mV (Fig. 1A). The voltage-clamp
records obtained under control conditions showed a similar pattern where
depolarisation to about -30 mV evoked only a small, sustained inward
current which as the cell potential was stepped to more positive voltages
was seen as a large and rapidly activating current (Fig. 2A). It is clear
that the inward membrane currents evoked by membrane depolarisation were
carried by Ca^{2+} ions since they were not observed in the absence of extracellular Ca^{2+} (Fig. 3A), they were enhanced when the extracellular concentration of Ca^{2+} was increased (Fig. 3B) and they were inhibited by both
inorganic (Fig. 3C) and organic Ca-channel blockers (Figs. 3D, 3E).

The spike potentials recorded from RINm5F cells peak at approximately
+20 mV, rapidly repolarise and undershoot the previous resting membrane
potential (Fig. 1). The voltage-clamp record under control conditions
showed that the evoked inward membrane current declined rapidly as the
membrane developed a voltage-activated outward membrane current (Fig. 2A).
The outward membrane current most probably reflects activation of voltage-
and Ca-activated K-channels (K^+_{MAXI}) by both membrane depolarisation and an
increase in intracellular Ca^{2+} since the outward current was reduced when
Ca^{2+} entry was prevented (Fig. 4) and inhibited by extracellular TEA (Figs.
5, 6)[6].

Removal of the K^+ current component from whole cell recording revealed
the inward Ca^{2+} current to consist of two distinct phases. Initially a
rapidly activating current which reached a peak value and then declined
within 10 ms of the onset of the depolarising voltage step. Secondly a
sustained current which showed a little decline during the 30 ms test pulse
(Fig. 6). It is possible that these separate phases represent activation
of 2 classes of Ca-channel in RINm5F cells. One responsible for the
generation of the spike potentials and the other contributing to a slow
plateau current[4]. Records of both rapidly and slowly inactivating Ca-
currents have been observed also in sensory neurons[3] and atrial cells[2].
Satin and Cook[12] have recently demonstrated inward membrane currents
carried by both Ba and Ca ions in neonatal rat β-cells where the currents

carried by Ca ions shows rapid inactivation which was not apparent when Ba^{2+} was used as the charge carrier. Whether Ba^{2+} would prevent inactivation of the initial inward Ca-current in RINm5F cells must await further experimentation.

ACKNOWLEDGEMENTS

We thank Professor O.H. Petersen (University of Liverpool) and Professor C.B. Wollheim (University of Geneva) for many helpful discussions. This work was supported by a grant from the Medical Research Council (U.K.).

REFERENCES

1. I. Atwater, A.A. Goncalves, and E. Rojas, Electrophysiological measurements of an Oscillating Potassium Permeability during the Glucose-Stimulated Burst Activity in Mouse Pancreatic β-Cell, Biomed. Res. 3:645 (1982).

2. B.P. Bean, Two kinds of calcium channels in canine atrial cells. Differences in kinetics, selectivity and pharmacology, J. Gen. Physiol. 86:1 (1985).

3. J.L. Bossu, A. Feltz, and J.M. Thomann, Depolarisation elicits two distinct calcium currents in vertebrate sensory neurons, Pflügers Arch. 403:360 (1985).

4. D.L. Cook, W.E. Crill, and D. Porte, Jr., Plateau potentials in pancreatic islet cells are voltage-dependent action potentials, Nature 286:404 (1980).

5. I. Findlay and M.J. Dunne, Voltage-activated Ca^{2+} currents in insulin-secreting cells, FEBS Lett. 189:281 (1985).

6. I. Findlay, M.J. Dunne, S. Ullrich, C.B. Wollheim, and O.H. Petersen, Quinine inhibits Ca^{2+}-independent K^+ channels whereas tetraethylammonium inhibits Ca^{2+}-activated K^+ channels in insulin-secreting cells, FEBS Lett. 185:4 (1985).

7. A.F. Gazdar, W.L. Chick, H.K. Oie, H. Sims, D.L. King, G.L. Weir, and V. Lauris, Continuous, clonal, insulin- and somatostatin-secreting cell lines established from a transplantable rat islet cell tumor, Proc. Natl. Acad. Sci. USA 77:3519 (1980).

8. O.P. Hamill, A. Marty, E. Neher, B. Sakmann, and F.J. Sigworth,
    Improved patch-clamp techniques for high-resolution current
    recording from cells and cell-free membrane patches, Plfügers Arch.
    391:85 (1981).

9. E.K. Matthews and Y. Sakamoto, Electrical characteristics of
    pancreatic islet cells, J. Physiol. 246:421 (1975).

10. O.H. Petersen, I. Findlay, and M.J. Dunne, Three different potassium
    channels in insulin-secreting cells, This Volume (1985).

11. G.A. Praz, P.A. Halban, C.B. Wollheim, B. Blondel, A.J. Strauss, and
    A.E. Renold, Regulation of immunoreactive-insulin release from a
    rat cell line (RINm5F), Biochem. J. 210:345 (1983).

12. L.S. Satin and D.L. Cook, Voltage-gated Ca^{2+} current in pancreatic
    β-cells, Pflügers Arch. 404:385 (1985).

13. C.B. Wollheim, S. Ullrich, and T. Pozzan, Glyceraldehyde, but not
    cyclic AMP-stimulated insulin release is preceded by a rise in
    cytosolic free Ca^{2+}, FEBS Lett. 177:17 (1984).

VOLTAGE-GATED Ca CURRENT IN PANCREATIC ISLET β-CELLS

L.S. Satin and D.L. Cook

Department of Physiology and Biophysics and Medicine
University of Washington and Seattle V.A. Medical Center
Seattle, Washington   98108

INTRODUCTION

Glucose induces rhythmic bursts of electrical spiking in mouse
pancreatic islet β-cells. Several groups have shown that the spikes are
dependent on extracellular Ca and are blocked by Ca channel blockers[5,6].
This electrical activity is believed to mediate Ca uptake necessary for
insulin secretion[1,8]. In order to study the underlying Ca spike
conductances, we used the whole-cell recording technique of Hamill et al.[4].

MATERIALS AND METHODS

The whole-cell method was implemented using isolated β-cells obtained
from monolayer cultures of neonatal rat islets. These cultures have
electrical spiking activity and secrete insulin in a glucose and Ca
dependent manner. Current and voltage-clamp recordings were made using a
Dagan 8900 patch clamp amplifier. Seals obtained ranged from 5-30
gigaohms. Series resistance was not compensated since our peak currents
were less than 60 pA and pipette resistances were typically 5 mohms or
less.

Solutions were chosen to block the outward current through K⁻channels
since this current normally obscured the inward current. We used 10 mM TEA
outside and replaced all internal K with Cs, which is impermeable through K-

channels. We used internal ATP, at a concentration of 2 mM, to block the
ATP-sensitive K current[3]. Internal free calcium was kept low with 5 mM
EGTA to prevent activation of Ca-activated K current[2]. To enhance inward
current through Ca channels, we used Ba instead of Ca as a charge carrier,
since in many systems, Ba permeates the Ca channel more readily than Ca
itself.

All experiments were performed at 20-22°C. These results have
recently appeared in print[7].

RESULTS

Using pipettes containing 140 mM KCl, 3μM free Ca and with no K -
channel blockers present, we recorded spontaneous spikes, as shown at the
top of Figure 1. This spike resembled normal mouse β-cell spikes in its
amplitude and time course. A spike elicited with injected current
(recorded from a different cell) using solutions containing K channel
blockers and extracellular Ca is shown in the middle panel. The spike is

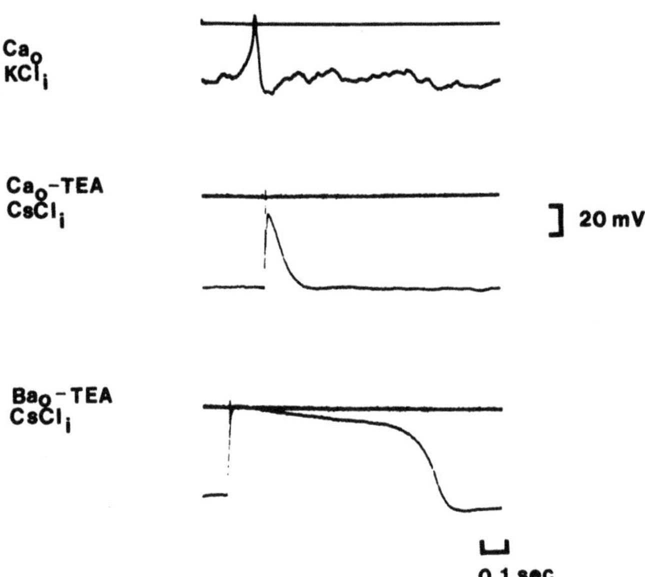

Fig. 1. (TOP) Spike recordings from cultured neonatal β-cells.
Spontaneous spike recorded using KCl-filled pipette containing
3 μM free Ca and 3 mM Ca in the bath. Straight line indicates
ground potential. (MIDDLE) Evoked spike recorded with Cs-filled
pipette and 5 mM EGTA (no added Ca) and 3 mM Ca plus 10 mM TEA in
the bath. Recorded from a different cell than above. Note
spike-broadening and decrease in baseline noise. (LOWER) Evoked
spike from same cell as middle panel after Ca was replaced by Ba.

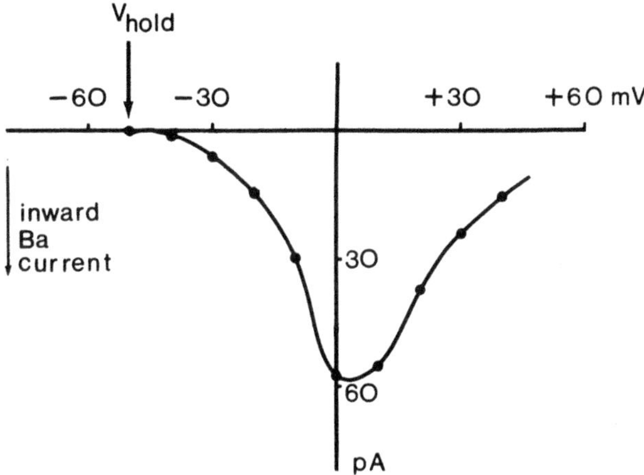

Fig. 2. Representative current-voltage relation for β-cell Ca-channels,
obtained using 3 mM Ba as a Ca analogue. Arrow denotes holding
potential. Leakage current has already been subtracted.

broader due to block of repolarizing K current. Note that the baseline
voltage noise in the top trace was decreased by K-channel blockers,
suggesting that it was due to random K-channel activity. When Ca was
replaced by Ba, a spike of augmented amplitude and duration was obtained,
as shown in the lower panel.

In order to study the underlying membrane ionic currents we voltage-
clamped isolated β-cells. Holding potential was set to -50 or -60 mV. A
40 msec hyperpolarizing command to -80 mV evoked an inward leakage current.
However, stepping positive beyond -40 mV resulted in an inward current,
which became maximal at 0 mV. These currents activated rapidly (about 15
msec) and showed little decay or 'inactivation' during the command pulse.
When the pulse was terminated, the current deactivated rapidly.

By measuring the peak inward currents and plotting them as a function
of membrane potential, we obtained the current-voltage (I-V) relation,
shown in Figure 2, after leakage current subtraction. The slope of the
leakage current line yielded an input resistance of 4.1 gigaohms. The
inward current progressively activated with depolarization beyond about -50
mV as more Ca-channels opened and peaked about 0 mV. Cells were variable
in this regard, showing peaks between -20 and 0 mV. Peak current was less
than about 60 pA in 3 mM Ba and decreased at positive voltages, in part due
to decreased driving force acting on Ba ions. Most non leak-subtracted
currents reversed at potentials less than +50 mV.

To test whether Na or Ba was the primary current carrier, we replaced
external Na with choline. Neither this nor the inclusion of 3 µM TTX
abolished the inward current. However, when Ba was replaced by Mg in the
external solution, the inward current was reversibly abolished leaving only
leakage current. Recovery from Ba removal was sometimes hampered by a
shift in the I-V and an increase in outward current that appeared with
time.

To test whether Ba was indeed moving through Ca-channels, we added 2
mM Co to the Ba-containing saline. Cobalt at this concentration is a
well-known Ca-channel blocker and it blocks bursting electrical activity
and insulin release in β-cells. We found that Co abolished the inward
current in a reversible manner. The Ca-channel blocker Ni also abolished
the current at a concentration of 2 mM, and the blockade was more readily
reversible.

In other systems Ca-channel currents are typically larger in Ba than
in Ca, as mentioned before. To test this we replaced Ba with Ca and found
that the current was smaller. Calcium currents showed more marked
'inactivation', although in most cells a more slowly decaying component
still remained after 40 msec.

The dihydropyridine Ca-channel activator BAY-K 8644 has been recently
shown to increase glucose-dependent insulin release and ^{45}Ca uptake in
mouse islets (see Malaisse-Lagae et al., this volume). We found that 5 µM
BAY-K reversibly increased the Ba current and thus, may augment insulin
release by directly interacting with the Ca-channel in the β-cell membrane.

CONCLUSION

This report is a direct demonstration and an initial characterization
of voltage-dependent Ca-channels in the β-cell membrane. Using the whole-
cell voltage clamp, we are now examining other inward currents that may
contribute to the pattern of electrical activity seen in β-cells.

ACKNOWLEDGEMENTS

We thank Dr. Wilfred Fujimoto and his staff for supplying the cells
and Mr. L.D. Stamps for his excellent technical assistance. L. Satin was
supported by NIH NS07097 and D.L. Cook by NIH AM29816 and the Veterans
Administration.

REFERENCES

1. D.L. Cook, Electrical pacemaker mechanisms of pancreatic islet cells,
     Fed. Proc. 43:2368 (1984).
2. D.L. Cook, M. Ikeuchi, and W. Fujimoto, Lowering of pH_i inhibits
     Ca-activated K channels in pancreatic β-cells, Nature 311:269
     (1984).
3. D.L. Cook and C.N. Hales, Intracellular ATP directly blocks K channels
     in pancreatic β-cells, Nature 311:271 (1984).
4. O.P. Hamill, A. Marty, E. Neher, B. Sakmann, and F.J. Sigworth,
     Improved patch-clamp techniques for high-resolution current
     recording from cells and cell-free membrane patches, Pflügers Arch.
     391:85 (1981).
5. H.P. Meissner and W. Schmeer, The significance of calcium ions for the
     glucose-induced electrical activity of pancreatic β-cells, in: "The
     Mechanism of Gated Calcium Transport Across Biological Membranes",
     S. Ohnishi and M. Endo, eds., Academic Press, New York, NY (1981).
6. B. Ribalet and P.M. Beigelman, Calcium action potentials and potassium
     permeability activation in pancreatic β-cells, Am. J. Physiol.
     239:C124 (1980).
7. L.S. Satin and D.L. Cook, Voltage-gated Ca current in pancreatic islet
     β-cells, Pflügers Arch. 404:385 (1985).
8. C.B. Wollheim and W.G. Sharp, Regulation of insulin release by
     calcium, Physiol. Revs. 61:914 (1981).

EFFECTS OF VERAPAMIL AND NIFEDIPINE ON GLUCOSE-INDUCED ELECTRICAL ACTIVITY
IN PANCREATIC β-CELLS

M. Vasseur, A. Debuyser and M. Joffre

Physiologie Animale CNRS U.A. 290
"Biomembranes" U.E.R. Sciences
86022 Poitiers Cedes (France)

It has been known for many years that calcium plays a key role in the
stimulus-secretion coupling of the β-cell. Glucose induced-insulin release
is thought to be triggered by an increase in the cytosolic calcium
concentration which appears partially related to the gating of Ca-channels,
located at the plasma membrane. Experiments with Ca-channel blockers have
confirmed this hypothesis: D600, Verapamil and Nifedipine cause a
dose-dependent inhibition of the glucose-induced insulin release[2,3,4].

It has been well established that D-glucose induces electrical
activity in the β-cell in a typical pattern of slow waves in the membrane
potential and bursts of action potentials, and that a correlation exists
between this activity and insulin release. Matthews and Sakamoto[5] and
Meissner and Schmeer[7] have shown that the calcium channel blocker, D600,
inhibits the glucose induced electrical activity.

The subject of the present work was to compare the effects of two
calcium channel blockers (Verapamil and Nifedipine) on the β-cell membrane
potentials and spike frequency induced by three concentrations of glucose:
2.8, 11, and 22 mM.

MATERIALS AND METHODS

All the experiments were performed with pancreatic islets of female
NMRI mice. They were killed by decapitation two hours after an
intraperitoneal injection of pilocarpine or after overnight fasting.
Pieces of pancreas were fixed in a small chamber and continuously perifused
with a modified Krebs' solution at 37°C. Islets were partially dissected
directly in the chamber under a binocular microscope.

The perifusion medium contained, in mM: 122 NaCl, 4.7 KCl, 1.1 $MgCl_2$,
2.56 $CaCl_2$, 20 $NaHCO_3$, and was continuously equilibrated with a mixture of
O_2/CO_2 (95/5%) to maintain a pH of 7.4. Glucose was added at concentrations
of 2.8, 11 or 22 mM as desired.

Microelectrodes, made from glass capillary tubes and filled with
potassium citrate solution (2 M, pH 7.2) were used for recording the
membrane potential. They had resistances ranging between 100 and 200 Mohm.
Membrane potentials were recorded and are denoted as: V_r, maximal
repolarization potential; V_t, threshold potential between slow and fast
depolarization; V_p, plateau potential; V_d, depolarization potential at the
top of the spikes. The prepotential is defined as the slowly depolarizing
potential occurring just before the onset of an action potential during
repetative spike activity.

RESULTS

Effects of Verapamil on Glucose-induced Electrical Activity

In 2.8 mM glucose, Verapamil (above 10^{-5} M) slowly depolarizes the
cell (about 10 mV) without any induced electrical activity.

In 11 mM glucose, low concentrations of Verapamil (below 5.10^{-6} M) do
not change the pattern of the slow waves. There is only a slight decrease
of the plateau and the silent phase durations. At higher concentrations,
Verapamil changes the typical slow waves into a continuous spike activity
which is obtained by a progressive depolarization of V_r and V_t to V_p.
Simultaneously the duration of the silent phases decreases (Fig. 1).

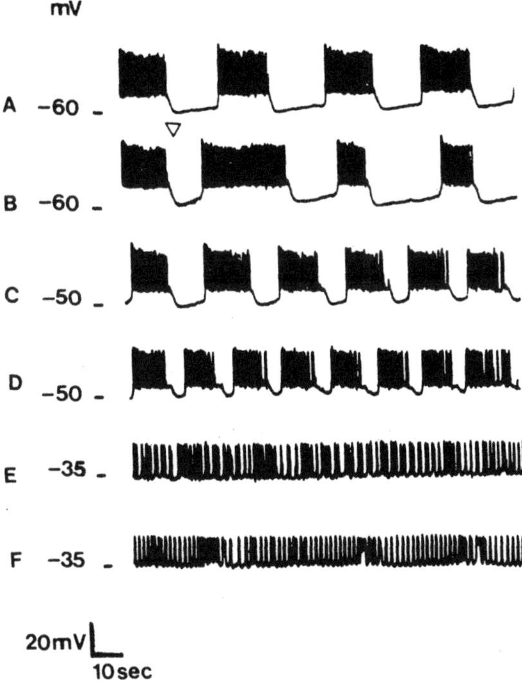

Fig. 1. Effects of Verapamil (10^{-5} M) on the electrical activity of a
single mouse β-cell, exposed to 11 mM glucose. A to F were
recorded from the same cell. Verapamil was added at the time
indicated by the pointer over trace B. A, control; B, 0; C, 3 min
30 sec; D, 5 min; E, 10 min; F, 12 min 30 sec after addition of
Verapamil.

Verapamil (10^{-7} to 10^{-4} M) induces a dose-dependent decrease of spike
frequency without any change in spike amplitude. The decrease in spike
frequency occurs concomitantly with an increase in spike duration and in
spike interval duration and with a decrease in the slope of the
prepotential.

In 22 mM glucose, Verapamil (5.10^{-8} to 10^{-5} M) does not inhibit the
continuous spike activity (potential levels are not modified) but
progressively decreases the spike frequency until a new steady state is
reached. At concentrations above 5.10^{-6} M, Verapamil often inhibits spike
activity, first by a decrease in frequency then by a decrease in amplitude.

Effect of Nifedipine on Glucose-induced Electrical Activity

In 2.8 mM glucose, Nifedipine (10^{-7} M) has no effect on the membrane
potential.

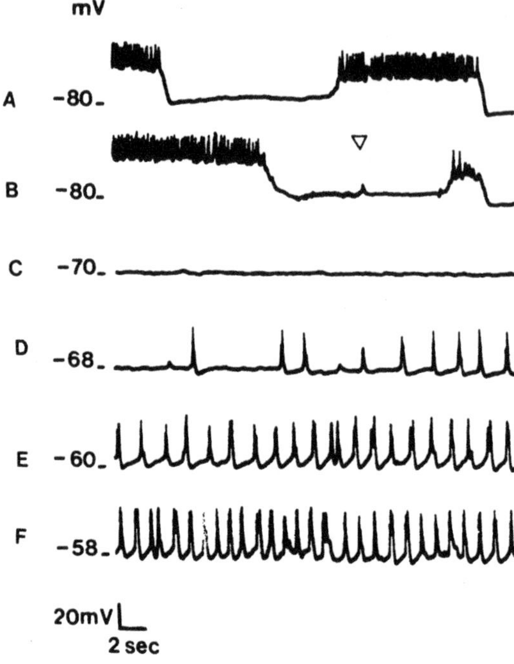

Fig. 2. Effect of Nifedipine (10^{-7} M) on the electrical activity of a
single mouse β-cell exposed to 11 mM glucose. A to F were
recorded from the same cell. Nifedipine was added at the time
indicated by the pointer over trace B. A, control; B, 0; C, 2
min; D, 2 min 30 sec; E, 4 min; F, 14 min after addition of
Nifedipine.

In 11 mM glucose, low concentrations of Nifedipine (below 10^{-8} M) do
not change the pattern of the slow waves. For higher concentrations (above
10^{-7} M) a few seconds after the exposure to Nifedipine a silent phase at
the repolarization potential appears (Fig. 2). Then the cell rapidly
depolarizes to the plateau potential, without the appearance of any slow
waves. Compared with Verapamil the continuous spike activity is reached
very quickly (5 min instead of 15 min).

Nifedipine (10^{-9} to 10^{-6} M) decreases the spike frequency in a
dose-dependent way without any effect on the spike amplitude in 11 mM
glucose.

In 22 mM glucose, Nifedipine (10^{-8} M to 10^{-7} M) induces a transient
slow wave pattern (in that case the membrane potential levels are the same
as in 11 mM glucose) but the cells rapidly return to a continuous spike
activity by a decrease of V_r and V_t which rapidly join V_p. With Nifedipine

198

(10^{-6} M) a transient silent phase occurs by repolarization to V_r, then the
cells return to a continuous spike activity. Spike activity rapidly
decreases then increases until a lower steady state is reached. Then the
amplitude of the spikes decreases by an increase of V_d and a decrease of V_p
until the spikes entirely disappear.

In 22 mM glucose, Nifedipine (10^{-9} to 10^{-6} M) decreases the spike
frequency in a dose-dependent way. For concentrations above 10^{-6} M the
spikes disappear first by decreasing in frequency then by decreasing in
amplitude.

DISCUSSION

The dose-dependent effects of Verapamil and Nifedipine on glucose-
induced electrical activity have been studied[8]. In 11 and 22 mM glucose,
the concentrations for half-maximal inhibition, ED_{50}, by Nifedipine are not
significantly different (10^{-8} M and 3.2 x 10^{-8} M), whereas the ED_{50} by
Verapamil is different in 11 or 22 mM glucose (6 x 10^{-6} M and 3 x 10^{-7} M,
respectively). These results show that the β-cells are more sensitive to
Nifedipine than to Verapamil and that glucose stimulation increases the
sensitivity to Verapamil but not to Nifedipine. The modifications of the
electrical activity in the presence of Ca-channel blockers are
qualitatively different from the changes induced by a decrease in glucose
concentration. This suggests that Verapamil and Nifedipine do not act on
the β-cell in the same way as glucose.

Spike frequency in the β-cell has been related to the insulin
release[6]. Thus, the decrease in spike frequency in the presence of
Verapamil or Nifedipine exposure could explain the inhibition of insulin
release with these two drugs[2,4]. In 2.8 mM glucose, basal insulin release
does not involve calcium influx through the Ca-channel[9]. Verapamil (but
not Nifedipine) depolarizes the β-cell. This observation is consistant
with the previously reported direct effect of Verapamil on potassium
permeability.

Verapamil and Nifedipine suppress the rapid depolarization of slow
waves. This indicates that the rapid depolarization may be related to an
entry of calcium through the Ca-channels. These Ca-channel blockers also

depolarize the β-cell to the plateau potential. This action may be more
related to a decrease in calcium entry then to a decrease in cytosolic
calcium concentration. Since calcium efflux is not affected by these
calcium antagonists[2,4], a subsequent decrease in cytosolic calcium
concentration may produce a decrease in calcium dependent potassium
permeability.

Verapamil and Nifedipine induce a very similar final state but the
kinetics of these modifications are different suggesting that Verapamil and
Nifedipine may not act in the same way on β-cells.

REFERENCES

1. P. Lebrun, W.J. Malaisse, and A. Herchuelz, Effect of calcium
   antagonists on potassium conductance in islet cells, Biochem.
   Pharmacol. 30:3291 (1981).
2. W.J. Malaisse and C. boschero, Calcium antagonists and islet function.
   Effect of Nifedipine, Hormone Res 8:203 (1977).
3. W.J. Malaisse, G. Devis, D.G. Pipeleers, and G. Somers, Calcium
   antagonists and islet function. Effect of D600, Diabetologia 12:77
   (1976).
4. W.J. Malaisse, A. Herchuelz, J. Levy and A. Sener, Calcium antagonists
   and islet function. The possible site of action of Verapamil,
   Biochem. Pharmacol. 26:735 (1977).
5. E.K. Matthews and Y. Sakamoto, Electrical characteristics of pancreatic
   islet cells, J. Physiol. 246:421 (1975).
6. H.P. Meissner and H.Schmelz, Membrane potential of beta-cells in
   pancreatic islets, Pflügers Arch 351:195 (1974).
7. H.P. Meissner and W. Schmeer, The significance of calcium ions for the
   glucose-induced electrical activity of pancreatic β-cells, in: "The
   Mechansims of Gatted Calcium Transport Across Biological Membranes",
   Academic Press, Inc., (1981).
8. M. Vasseur, A. Debuyser and M. Joffre, Sensibility of the pancreatic
   beta cells to calcium channel blockers. An electrophysiological
   study of Verapamil and Nifedipine (submitted to Pflügers Arch.).
9. C.B. Wollheim and G.W.G. Sharp, Regulation of insulin release by
   calcium, Physiol. Rev. 61:914 (1981).

STIMULATION OF INSULIN RELEASE BY ORGANIC CALCIUM-AGONISTS

F. Malaisse-Lagae, A. Sener and W.J. Malaisse

Laboratory of Experimental Medicine
Brussels Free University
Brussels, Belgium

The concentration of ionized calcium in the cytosol of the pancreatic
β-cell is thought to play a critical role in the regulation of insulin
secretion. We have recently been able to document in isolated pancreatic
endocrine cells removed from normal rats that an increase in extracellular
D-glucose concentration provokes a rapid and sustained increase in
cytosolic Ca^{2+} activity[1]. Several factors could contribute to this
increase in cytosolic Ca^{2+} concentration, including facilitated Ca^{2+} influx
into the β-cell and, possibly, changes in either Ca^{2+} outflow or Ca^{2+}
intracellular distribution. The present note refers to the stimulation of
insulin release by organic calcium-agonists[3,5]. Our results support the
view that the gating of Ca^{2+} channels represents an efficient modality for
stimulation of insulin secretion.

Effect of Calcium-agonists on ^{45}Ca Handling by Pancreatic Islets

Both BAY K 8644 (methyl 1,4-dihydro-2,6-dimethyl-3-nitro-4-(2-tri-
fluoromethylphenyl)-pyridine-5-carboxylate) and CGP 28392 (4-[2(difluoro-
methoxy)phenyl]-1,4,5,7-tetrahydro-2-methyl-5-oxofuro[3,4-b]pyridine-3-
carboxylic acid ethylester) stimulate the net uptake of ^{45}Ca by islets
incubated in the presence of D-glucose (7.0 mM). In the case of CGP 28392,
the effect is dose-related in the 5 to 100 μM range. However, at a
concentration of 500 μM, CGP 28392 inhibits glucose-stimulated ^{45}Ca net
uptake[3]. Both agents also cause a rapid stimulation of ^{45}Ca release from

prelabelled islets perifused in the presence of both D-glucose (7.0 mM) and Ca^{2+} (1.0 mM). The latter effect is suppressed when the perifusate is deprived of Ca^{2+} and contains EGTA (0.5 mM), supporting the view that it reflects stimulation of a process of exchange between influent $^{40}Ca^{2+}$ and effluent $^{45}Ca^{2+}$.

Effect of Calcium-agonists Upon Insulin Release

In the perifused islets, the stimulation of ^{45}Ca outflow by organic calcium-agonists coincides with a rapid and sustained increase in insulin output[3,5,6]. In static incubations, the magnitude of the secretory response to either BAY K 8644 or CGP 28392 is modulated by both the concentration of the drug and that of D-glucose. No effect of these drugs is seen in the absence of D-glucose or at a low concentration (2.8 mM) of

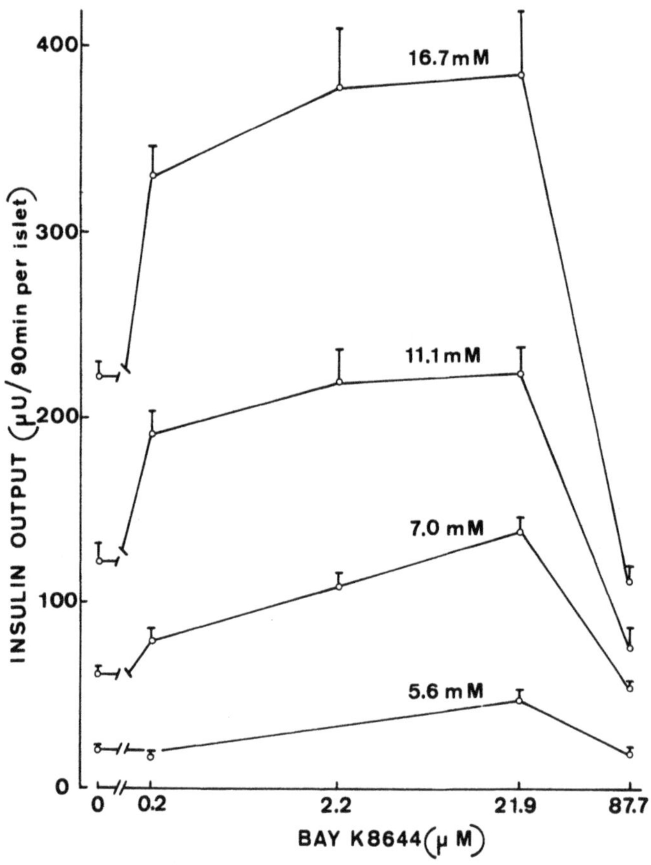

Fig. 1. Effect of increasing concentrations of BAY K 8644 (logarithmic scale) upon insulin release from islets incubated at increasing concentrations of D-glucose (5.6 to 16.7 mM). Mean values (± SEM) refer to 9 or more individual measurements.

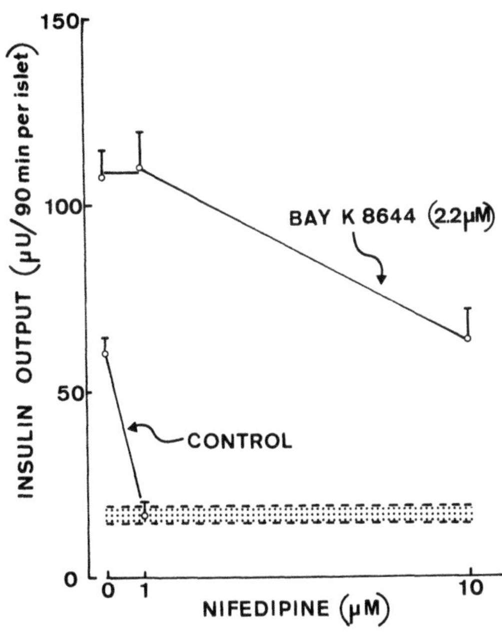

Fig. 2. Effect of increasing concentrations of nifedipine upon insulin release from islets incubated in the presence of D-glucose (7.0 mM) with or without BAY K 8644 (2.2 M). The shaded area corresponds to the basal release of insulin measured in the absence of D-glucose. Mean values (± SEM) refer to 11 or more individual measurements.

the hexose. An obvious stimulation of insulin release is observed at intermediate glucose concentrations (5.6 to 11.1. mM). At a higher glucose concentration (16.7 mM), an increase in insulin output was still observed with BAY K 8644 but not with CGP 28392. The sensitivity to increasing concentrations of BAY K 8644 is greater at high than at low concentrations of D-glucose (Fig. 1). At very high concentrations, both BAY K 8644 and CGP 28392 may cause inhibition of glucose-stimulated insulin release. In all cases, there is a close correlation between the changes in ^{45}Ca net uptake and insulin release, respectively.

Combined Effects of Calcium-agonists and Calcium-antagonists

The calcium-agonists protect the β-cell against the inhibitory action of organic calcium-antagonists. Thus, in the presence of either BAY K 8644 or CGP 28392, the inhibition of glucose-stimulated insulin release by either nifedipine or verapamil is less marked than that observed in the absence of these calcium-agonists (Fig. 2). This secretory behaviour is paralleled by comparable changes in ^{45}Ca net uptake by the islets[5].

Combined Effects of CGP 28392 and Nonnutrient Secretagogues

CGP 28392 fails to affect insulin release evoked in the absence of D-glucose and Ca^{2+} by the combination of Ba^{2+} and theophylline[3]. CGP 28392 augments insulin release evoked in the presence of D-glucose (5.6 mM) by gliclazide. The effects of the hypoglycemic sulfonylurea and the calcium-agonists appear additive. Such is not the case, however, at a higher concentration of D-glucose (11.1 mM). The insulinotropic actions of CGP 28392 and gliclazide, respectively, also differ from one another by the fact that gliclazide, but not CGP, protects the β-cell against the inhibitory action of diazoxide[4].

Concluding Remarks

The data briefly reviewed in this account emphasize the view that the gating of Ca^{2+} channels, as presumably caused by organic calcium-agonists, represents an efficient modality for stimulation, or at least facilitation, of insulin release. The data obtained with CGP 28392 at increasing concentrations of D-glucose are compatible with the view that this agent prolongs the mean open time of Ca^{2+} channels, as documented in cultured cardiac cells[6]. A calcium-antagonistic potential of both BAY K 8644 and CGP 28392 can be revealed at high concentrations of these drugs. The calcium-agonists protect the β-cell against the inhibitory action of organic calcium-antagonists but not diazoxide. The combined used of calcium-agonists and other insulinotropic agents (e.g. hypoglycemic sulfonylurea) may help to identify the primary site or mode of action of these secretagogues in the pancreatic β-cell.

REFERENCES

1. M. Deleers, M. Mahy, and W.J. Malaisse, Glucose increases cytosolic Ca^{2+} activity in pancreatic islet cells, Biochem. Internat. 10:97 (1985).

2. S. Kokubun and H. Reuter, Dihydropyridine derivatives prolong the open state of Ca channels in cultured cardiac cells, Proc. Natl. Acad. Sci. USA 81:4824 (1984).

3. W.J. Malaisse and P.C.F. Mathias, Stimulation of insulin release by an organic calcium agonist, Diabetologia 28:153 (1985).

4. W.J. Malaisse, A. Sener, and F. Malaisse-Lagae, Combined effects of a
    calcium-agonist and hypoglycemic or hyperglycemic sulfonamides upon
    insulin release, Res. Commun. Chem. Pathol. Pharmacol. 49:71
    (1985).

5. F. Malaisse-Lagae, W.J. Malaisse, and P.C.F. Mathias, Gating and
    blocking of calcium channels by dihydropyridines in the pancreatic
    β-cell, Biochem. Biophys. Res. Commun. 123:1062 (1984).

6. U. Panten, S. Zielmann, M.-T. Schrader, and S. Lenzen, The
    dihydropyridine derivative, BAY K 8644, enhances insulin secretion
    by isolated pancreatic islets, Naunyn-Schmiedeberg's Arch.
    Pharmacol. 328:351 (1985).

CONTRIBUTION OF ISOTOPE FLUX STUDIES TO UNDERSTANDING THE MECHANISM OF THE
β-CELL MEMBRANE

P.C. Croghan, C.M. Dawson, A.M. Scott and J.A. Bangham

Department of Biophysics
School of Biological Sciences
University of East Anglia
Norwich, NR4 7TJ, England

The membrane potential of the animal cell membrane is determined
principally by the diffusion of ions. Such a diffusion potential is
determined by the concentrations of the diffusing (electrogenic) ions on
either side of the membrane and by the permeability of the membrane to
these ions. In the resting membrane, the permeability of the K^+ ion is
usually the dominant factor and depolarization of the membrane is caused by
a reduction in K^+ permeability and/or an increase in the permeability of
other ions such as Na^+ and/or Ca^{2+}. These changes in membrane potential
play a key role in the function of various cell types, including the
pancreatic β-cell, where they seem to initiate processes leading ultimately
to insulin release.

Ion permeability of a membrane can be investigated by various
techniques: measurements of electrical potential difference and electrical
conductance of the whole cell or an isolated patch of membrane and by
studying ion fluxes across the membrane using isotopes. This contribution
will be principally concerned with the application of isotope techniques.
As the movement of ions involves an electrical term, it is essential that
ion flux studies are accompanied by determinations of the membrane
potential under similar experimental conditions[8,9,10]. Unfortunately, in
many investigations of the β-cell, this has not been done and this greatly
restricts the information that can be obtained from those ion flux studies.
Flux studies on β-cells are made on groups of cells, usually groups of

collagenase-isolated islets but occasionally on a single islet[27].
Electrical measurements are made on single cells within a microdissected
islet in similar experiments. As, additionally, the collection of a sample
for isotope analysis requires a significant time, rapid changes in
permeability associated with the rapid electrical events cannot be studied.
However, an advantage of the isotope technique is that changes in the
intercellular coupling within an islet, that may occur quite rapidly and
that will affect measurements of input resistance, will have no direct
effect on isotope fluxes[8].

Recently there has been a considerable effort in investigating the
properties of isolated patches of the β-cell membrane[1,5,6,17]. It should
be remembered that the patch-clamp technique is looking at a very small
area of a single cell[1], whereas the ion flux studies are macroscopic,
looking at the behaviour of large groups of cells. It is obviously
important to attempt to relate the findings of these very different
techniques in order to elucidate the function of the β-cell membrane under
physiological conditions.

THEORETICAL BASIS

Isotope Fluxes

The proper way to consider isotope fluxes is to define a rate
constant, most simply

$$\frac{dA_j^*}{dt} = -k_j A_j^* , \qquad (1)$$

where A_j^* is the amount of isotope and k_j is the rate constant. In the
simplest case, the unidirectional flux (J_{12}) is then

$$J_{12} = k_j A_j , \qquad (2)$$

where A_j is the amount of species j in the compartment. The process is
simple where there is only one labelled compartment but becomes more
complex where there are several labelled compartments[31].

208

Fluxes can be measured by either influx or efflux experiments. It is
experimentally simpler to carry out and to interpret efflux experiments
from previously loaded cells. In this type of experiment, the activity
leaving the cells per unit time can be determined by collecting the
perifusate for defined times. Equation (1) can be integrated

$$A_{jt}^{*} = A_{jo}^{*} \exp[-k_j t] \ , \tag{3}$$

where A_{jt}^{*} and A_{jo}^{*} are the amount of isotope in the cells at time t and at t
= 0 respectively. This equation can be used to determine efflux rate
constants[8].

$$k_{jt + \Delta t/2} = -1/\Delta t \ \log_e \ [(A_{jt+\Delta t}^{*})/A_{jt}^{*}] \tag{4}$$

If there are significant amounts of isotope in slowly exchanging
compartments, it may be better to use the rate-of-loss data directly. Then

$$(\Delta A_j^{*}/\Delta t)_t = (\Delta A_j^{*}/\Delta t)_o \ \exp \ [-k_j t] \ , \tag{5}$$

where $\Delta A_j^{*}/\Delta t$ is the amount of isotope lost per unit time. Although this
approach is noisy, it is better for multi-compartment systems as it does
not require information about the amount of isotope present in each
compartment.

Membrane Permeability

The term rate constant does not imply any particular mechanism of
transport across the membrane and thus does not itself define any
particular membrane property. In the case of a passive diffusion flux, the
membrane property can be defined in terms of a permeability coefficient.
It must be stressed that a permeability coefficient is a quantity derived
in terms of a specific model. The most useful model is the Goldman
constant-field model[18]. The assumptions made in this model should always
be borne in mind. A particular assumption is that the ions are moving

solely under the influence of their concentrations and of the electric field and do not interact with each other (independence principle). In ion-selective channels, an element of interaction is inevitable.

In the case of Rb^+ and K^+, it is likely that the efflux will be a passive diffusion process. In this case, the efflux rate constant and the membrane potential can be used directly to define a permeability coefficient (P_j). The unidirectional efflux defined by the efflux rate constant, equation (2), and by the Goldman model can be equated

$$P_j = k_j \; V/A \quad [(1 - \exp \; [-aE] \;)/aE] \tag{6}$$

where V/A is the volume/area of a β-cell (strictly the volume should be the volume of water in the cell), $a = z_j F/RT$ and E is the membrane potential. The term within the outer square brackets indicates the influence of the membrane potential on the ion flux.

In the case of ions such as Na^+ and especially Ca^{2+}, that are present in the cytosol at concentrations below the equilibrium value, the passive component of the efflux is small. Thus in the case of Na^+, taking the membrane potential as -64 mV[10] and the Na^+ concentration in the cytosol as 42 mM[16], the passive component of the efflux, calculated from the Ussing flux-ratio equation[32], is only 3% of the total efflux. In the case of Ca^{2+}, the passive efflux is essentially zero. The efflux of these ions must thus be an active process and the efflux rate constant reflects the rate of active transport. However, if the efflux is determined under steady-state conditions, it is also possible to estimate the permeability coefficient by equating the efflux defined by equation (2) with the passive influx defined by the Goldman model

$$P_j = k_j \; (V/A)(C_i/C_o)[1/aE(\exp \; [aE]) - 1] \tag{7}$$

where C_i and C_o are the concentrations of ion j in the cytosol and external medium respectively. This derivation is independent of any assumption as to the mechanism of active transport (neutral, electrogenic or coupled transport driven by the electrochemical potential difference of another ion).

<u>Membrane Conductance</u>

Measurements of the displacement of membrane potential resulting from injection of small currents into the β-cell have been used to determine input resistance[2,16]. The interpretation of these data is complicated by the presence of coupling between the cells of the islet[15]. Particularly as the extent of the coupling may vary, it is impossible to accurately define membrane specific conductance. However, rapid changes in the measured conductance can presumably be regarded as indicating changes in membrane conductance rather than changes in intercellular coupling. Determinations of ion efflux rate constants and permeabilities are, however, independent of intercellular coupling. Using these data, the membrane conductance for specific ions can be computed using the Goldman model[8,10]. Two types of ion conductance can be defined: slope conductance and chord conductance.

Slope conductance (g_j) is defined as

$$g_j = -dIj/dE. \tag{10}$$

Then using the Goldman model and assuming that P_j is independent of membrane potential

$$g_j = P_j \alpha Z_j F/(1-\exp[aE])^2 [(C_o - C_i)aE \exp[aE] + (C_o - C_i \exp[aE])(1-\exp[aE])]. \tag{11}$$

This equation makes no assumption that ion j is in equilibrium across the membrane.

Chord conductance (G_j) is defined as

$$G_j = -I_j/(E-E_j), \tag{12}$$

where E_j is the Nernst potential of ion j. Then using the Goldman model

$$G_j = P_j Z_j FaE(C_o - C_i \exp[aE])/((E-E_j)(1-\exp[aE])) \tag{13}$$

Under conditions where the net current flowing across the membrane is zero

$$0 = \sum_{j=1}^{N} I_j \quad , \tag{8}$$

where I_j is the current carried by ion j. Then, using the Goldman model, an equation relating the membrane potential to the ion concentrations and permeabilities can be obtained[18]

$$E = -1/a \, \log_e [(K_i + \alpha Na_i + \cdot)/(K_o + \alpha Na_o + \cdot)], \tag{9}$$

where α is the permeability ratio, P_{Na}/P_K. Terms for all significantly permeable ions should be included, except that terms for ions in equilibrium drop out. This equation, of course, includes all the assumptions of the Goldman model. This equation has been applied to β-cells, ignoring ions other than K^+ and Na^+, and α calculated from the relation between membrane potential and K_o^+ and Na_o^+ [2,16,28].

An immediate problem with equation (9) concerns Cl^-. This ion may well be in equilibrium at the normal resting potential of the β-cell but this would not be the case when the potential is depolarized by increasing the ratio K_o^+/Na_o^+. Then if the membrane is significantly permeable to Cl^-, as may be suggested by isotope flux data[29], equation (9) is not strictly valid. However, data from studies where the external Cl^- is largely replaced by isethionate or sulphate[11,14] suggest that the β-cell membrane is relatively impermeable to Cl^-. A problem arises also if a significant permeability to Ca^{2+} or other divalent ions is present. Then the solution corresponding to equation (9) is an implicit function of E. A further problem arises if the Na^+ pump is electrogenic[21] rather than a neutral ion-exchange mechanism. Then an electrogenic pump current must be added to equation (8). However, if the coupling ratio of the pump (n), the ratio active K^+ flux/active Na^+ flux, has a constant stoichiometry, independent of membrane potential, then equation (9) is valid but the apparent α is now nα. As the coupling ratio is decreased the membrane potential becomes closer to the K^+ Nernst (equilibrium) potential.

An advantage of slope conductance is that as, under steady-state
conditions, the total membrane current (I) is given by

$$I = \sum_{j=1}^{N} I_j \ ,$$

(14)

the differentials (slope conductances) are also additive. Thus the ion
slope conductances can be directly compared with the membrane slope
conductance determined from current-injection experiments.

It is important to appreciate the difference between the permeability
and conductance of a membrane to an ion. Permeability is interpreted as a
specific property of the membrane, whereas the conductance is determined
not only by this permeability property but also by the concentration of the
current-carrying ion on both sides of the membrane and the membrane
potential.

Investigations of the properties of single channels in small isolated
membrane patches have usually defined a single-channel conductance. For
small displacements of the membrane potential, the conductance defined can
be regarded as a slope conductance. Assuming that the behaviour of the
channel conforms to the Goldman model, the conductance can be corrected, if
necessary, using equation (11), to a value corresponding to physiological
ion concentrations and membrane potential[10]. However, as permeability is a
better description of membrane properties than conductance, it is probably
better to use equation (11) to define a single-channel permeability[10,24].
The relation between membrane permeability to ion j and single-channel
permeability (P'_{ji}) is given by

$$P_j = \sum_{i=1}^{N} P'_{ji}\, \rho_i\, P_{oi} \ ,$$

(15)

where N is the number of different types of channel present ρ_i is the
density of channel i in the membrane and P_{oi} is the probability of a
channel i being open.

Potassium Permeability and Conductance

Changes in the β-cell membrane potential in the presence of glucose have been interpreted, using equation (9), as due to a reduction in the ratio P_{Na}/P_K[2,16]. The value of α increases from 0.06 to 0.18 in the presence of 11.1 mM glucose. The effect of the sulphonylurea, glibenclamide, has been interpreted in a similar manner[16]. In both these cases, an increase in input resistance suggests that it must be a decrease in P_K, rather than an increase in permeability to another cation, that is involved.

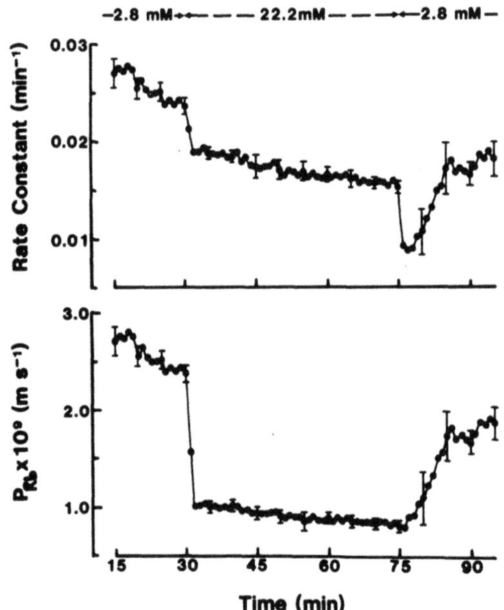

Fig. 1. Effect of glucose on efflux rate constant and Rb^+ permeability coefficient, P_{Rb} (37°C). The glucose concentration was raised to 22.2 mM at 30 min and returned to 2.8 mM at 75 min. The data are the mean of 4 experiments. Error bars represent SEM of the data normalised to 26-29 min as reference.

More specific information can be obtained from studies of K^+ efflux from β-cells. There have been many studies using $^{86}Rb^+$ as an "isotope" for K^+ and occasional studies using $^{42}K^+$ [20,26]. In general, the efflux is decreased in the presence of high external glucose concentrations.

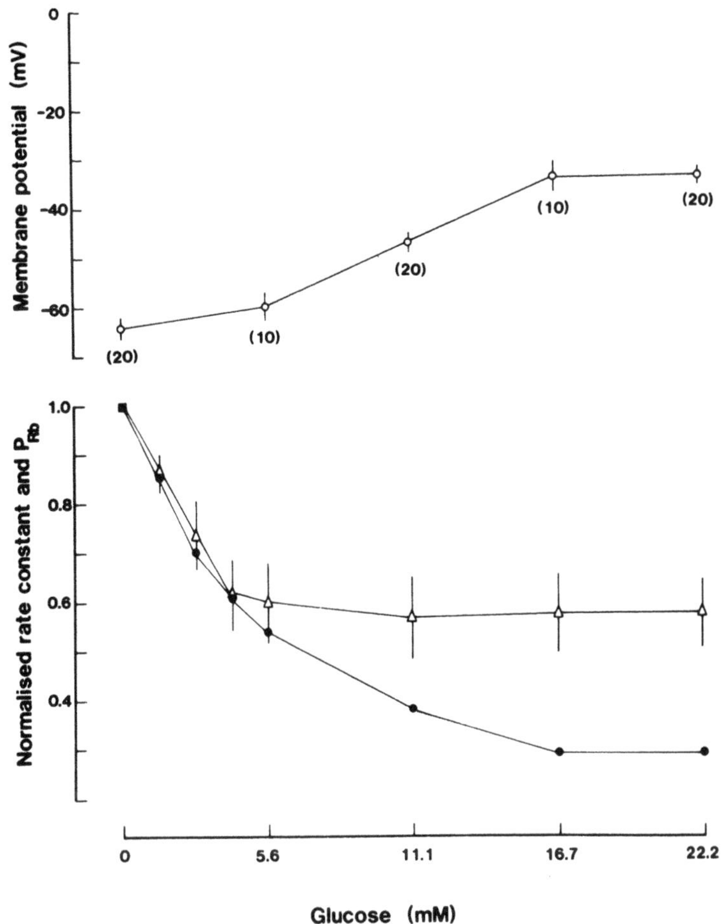

Fig. 2. Effect of glucose on the membrane potential (o), [86]Rb[+] efflux rate constant (Δ) and Rb[+] permeability coefficient (●) (37°C). The number of membrane potential records from different cells is indicated. The efflux rate constant was taken as the mean of the last 10 min of the 15 min exposure to the stated glucose concentration (n=4). The rate constants and permeabilities have been normalised to the mean value of the 10 min immediately preceding the first exposure to glucose (i.e. min 50-60 after the islets were introduced into the perifusion chamber). For the rate constant 1 = 0.0266 ± 0.0059 min^{-1} and for the permeability coefficient 1 = 3.08 ± 0.68 m s^{-1}. All results are given as the mean ± SEM.

However, as has been pointed out, measurements of efflux rate constants do not themselves characterise properties of the cell membrane. For this it is also necessary to use membrane potential data obtained from parallel experiments and to compute the permeability coefficients[8,9,10].

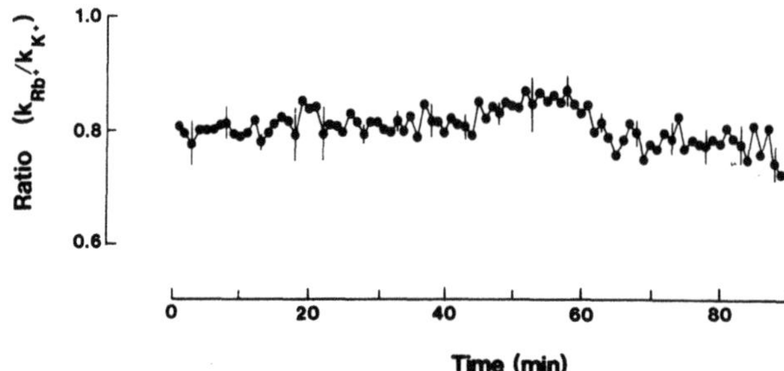

Fig. 3. Ratio of efflux rate constant k_{Rb}/k_K determined in a series of experiments (n=4) over a range of glucose concentrations (37°C).

In Fig. 1, modified from[8], the effect of a high external glucose concentration on $^{86}Rb^+$ efflux rate constant and Rb^+ permeability is given. It can be seen that the effect on the Rb^+ permeability coefficient is altogether more dramatic than the effect on efflux rate constant. This is a consequence of the depolarization of the membrane potential in high glucose that is itself a consequence of a decrease in K^+ permeability. Note also that the undershoot in the efflux rate constant, on return to low glucose, is not reflected in the permeability coefficient.

In Fig. 2[10], the effect of a range of glucose concentrations is shown on the experimental parameters (membrane potential and efflux rate constant) and on the derived quantity (permeability coefficient). It seems clear that the response to an increasing external glucose concentration is a decrease in the Rb^+ permeability of the cell membrane. The effect on permeability levels off above 16.7 mM glucose. It seems that the depolarization of the membrane potential from the resting value to the plateau value at the base of the action potential spikes could be interpreted as due to changes in P_K without any necessary change in the permeabilities of other cations.

In these flux experiments, the longer-life $^{86}Rb^+$ has usually been used rather than the short-life $^{42}K^+$. It is important to justify this use. A study of the simultaneous efflux of $^{86}Rb^+$ and $^{42}K^+$ has been reported in rat islets[20]. A more detailed study of the simultaneous efflux of $^{86}Rb^+$ and $^{42}K^+$ in mouse islets has been made[9,10]. The ratio of efflux rate constants (k_{Rb}/k_K), a ratio that is identical to that of the permeability coefficients (P_{Rb}/P_K), has been investigated in various external glucose

216

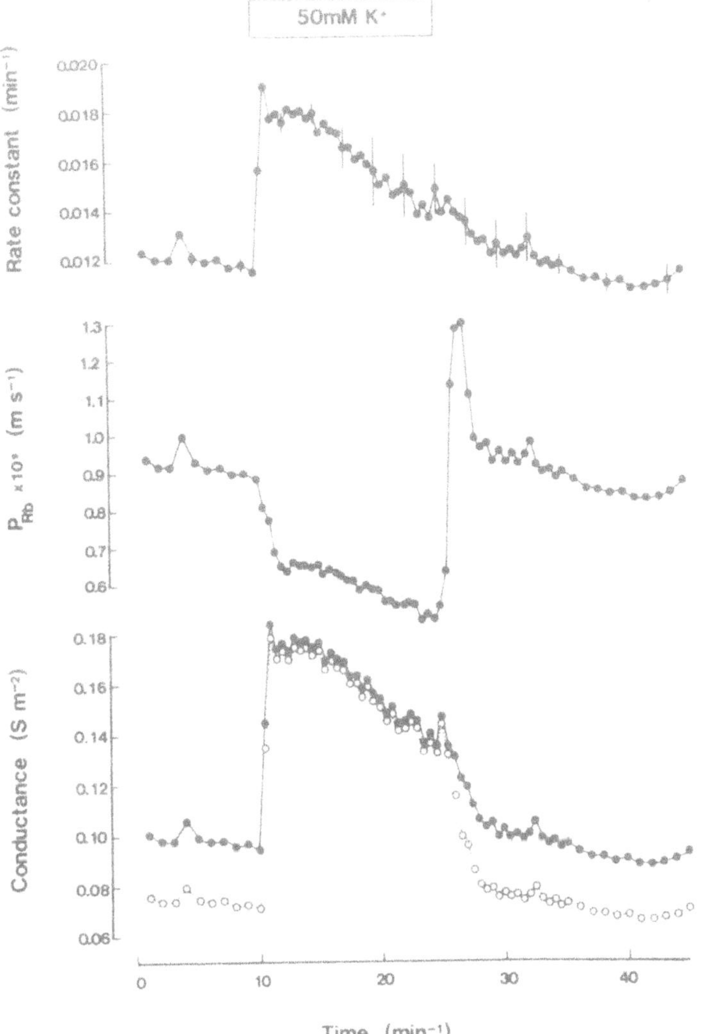

Fig. 4. The effect of 50 mM K^+ in the presence of 11.1 mM glucose on Rb^+ efflux rate constant, Rb^+ permeability coefficient and K^+ slope (•) and chord (o) conductance (37°C). During the exposure to 50 mM K^+ and for 10 min afterwards, samples were collected at 30 s intervals. The efflux rate constant values were normalised to the mean value for min 5-9 and expressed as mean ± SEM (n=4).

concentrations. It can be seen from Fig. 3, that the ratio is essentially constant at 0.80 over an external glucose concentration from 0 to 22.2 mM[10]. Thus $^{86}Rb^+$ seems an acceptable "isotope" for K^+, although strictly the small correction factor should be used. Using this correction factor, the mean value derived from $^{86}Rb^+$ efflux rate constant data (n=62) corresponds to P_K = 3.5 x 10^{-9} m s^{-1} for normal mouse islet cells in zero glucose[10]. This value of P_K is low compared to that of resting frog muscle[23] and toad lens fibres[13]. It is, however, higher than that of rat liver-cells[4] and much higher than that of human erythrocytes[3]. These values must be related to the functions of the cell membrane in the various specialised cell types.

Using the Goldman model, the K^+ conductance of the membrane can be calculated from the K^+ permeability, equation (11) and equation (13). As the value of the resting potential indicates that K^+ permeability and conductance are the dominant factors in the cell membrane, a change in K^+ conductance would be expected to have a marked effect on input resistance. The decrease in K^+ conductance predicted at high external glucose concentrations[8,10] agrees with the increase in input resistance found in current-injection experiments[2].

The small K-channel found in membrane patches, that is inhibited by glucose in cell-attached patches[1] and by ATP in isolated patches[5], must play a key role in controlling the changes in K^+ permeability that result in depolarization of the β-cell membrane. Using equation (11), the conductance of this channel is about 8 pS under physiological conditions[10]. If this conductance is compared to the K^+ conductance computed from isotope efflux data (0.28 S m^{-2}), it appears that there are about 10 small K^+ channels open per β-cell in the resting condition. The observed decrease in K^+ permeability in high external glucose concentrations is then attributable to a decreased probability of these channels being in the open state.

An interesting effect of high external K^+ concentration was found on Rb^+ and K^+ permeability and K^+ conductance[10]. The permeabilities decrease in high external K^+. An example is given in Fig. 4[10]. This finding was unexpected. In most cell types that have been investigated the reverse is true. However, a very similar effect was seen with cephalopod lens cells[12]. It is suggested that, in the β-cell, voltage-gated channels are not active over the potential range involved in these experiments and, further, that the K^+ channels involved have sites at the external surface that can be blocked by high concentrations of Rb^+ and K^+ ions. Interestingly, the K^+ conductances calculated from equation (11) and equation (13) increase in the presence of high external K^+. This is in agreement with the decrease in input resistance found in similar experiments[7]. Again this discrepancy in the direction in which K^+ permeability and conductance move is seen in the cephalopod lens cells in similar experiments[10,12]. The K^+ conductance is a product of K^+

permeability and K^+ concentration within the channel and, although the former may decrease in high external K^+, the latter would be expected to increase.

Sodium Permeability

Although there have been many measurements of Rb^+ or K^+ flux in islets, there have been relatively few studies of Na^+ flux[22,25,30]. A major problem in determining Na^+ flux across the cell membrane is that the concentration of Na^+ in the cytosol is much less than that of K^+. Also the amount of Na^+ in the extracellular space of the islet is considerable. Thus the analysis of efflux experiments to elucidate the membrane-limited component presents difficulties. In preliminary experiments, the efflux of Na^+ has been studied using a triple-labelled technique. The influence of the extracellular space has been studied using ^3H-sucrose, a species whose diffusion coefficient is not greatly different from Na^+. The efflux of $^{22}Na^+$ and $^{86}Rb^+$ was also studied simultaneously, as it is important to compare the efflux of these two ions under strictly comparable conditions. Unlike the situation with $^{86}Rb^+$, it was not possible to interpret the $^{22}Na^+$ efflux using a simple one-labelled compartment model. However, under steady-state conditions, the rate-of-loss method, equation (5), could be used. Thus it has been possible to compare the cell membrane-limited efflux of $^{22}Na^+$ and $^{86}Rb^+$ in the same groups of islets. Unlike the situation with Rb^+ or K^+, the efflux of Na^+ was unaffected by external glucose concentration from 0 to 22.2 mM. This confirms previous work on mouse islets[30] but differs from work on rat islets[25].

In experiments carried out under strictly steady-state conditions, the Na^+ permeability coefficient can be estimated using equation (7) and compared with that of Rb^+, using equation (6). The ratio P_{Na}/P_{Rb} changed from 0.05 in the absence of glucose to 0.21 in the presence of 22.2 mM glucose. These ratios are similar to the ratios obtained using the Goldman potential equation, equation (9).

Permeability of Other Ions

Studies of $^{36}Cl^-$ fluxes in ob/ob mouse islets have given unexpected results[29]. It was suggested that Cl^- is not passively distributed across the membrane and that glucose increases the Cl^- permeability of the β-cell

membrane. These conclusions seem incompatible with membrane potential
measurements that suggest that Cl^- is in equilibrium across the cell
membrane[14] and also that, as replacing most of the external Cl^- by
isethionate or sulphate had little effect on membrane potential[11,14], the
Cl^- permeability is low. As in the case with Na^+, the cytosolic
concentration is low and the extracellular space concentration is high and
this makes the interpretation of $^{36}Cl^-$ fluxes difficult. Clearly the Cl^-
permeability needs further investigation but, at the moment, it is
suggested that Cl^- permeability is unlikely to play any major role in the
electrical activity of the β-cell and that Cl^- flux is mainly electrically-
neutral coupled transport.

The investigation of Ca^{2+} fluxes using $^{45}Ca^{2+}$ has shown that, in the
presence of glucose, there is a transient decrease followed by a transient
peak in efflux and these affects have had various interpretations[19,33]. A
problem in investigating Ca^{2+} fluxes is that the cytosolic Ca^{2+} concentration is likely to be very low and variable and that the bulk of the intracellular Ca is in bound or in non-cytosolic compartments. This makes the
interpretation of $^{45}Ca^{2+}$ fluxes extremely difficult and it is possible that
information on the role of Ca^{2+} will have to be derived largely from
electrophysiological, including patch clamp, data.

Conclusions

In conclusion, it is suggested that it is correct to regard K^+ and Na^+
as the only significantly permeable ions in the region between the resting
potential and the plateau of the action-potential burst. The membrane
depolarization, that occurs in the presence of glucose, is caused by a
decrease in K^+ permeability. This is considered to be caused by a decrease
in the probability of the small K^+ channels being in the open state. It is
unnecessary to invoke changes in the permeability of the β-cell membrane to
other ions during this initial depolarization of the membrane that leads to
the initiation of the burst of action potentials and, ultimately, to
insulin release.

ACKNOWLEDGEMENTS

The authors wish to acknowledge the support of the British Diabetic
Association, the Medical Research Council and the Wellcome Trust. We thank
Mrs. Jill Gorton for patience in preparing the typescript.

REFERENCES

1. F.M. Ashcroft, D.E. Harrison and S.J.H. Ashcroft, Glucose induces
    closure of single potassium channels in isolated rat pancreatic
    β-cells, Nature 312:446 (1984).

2. I. Atwater, B. Ribalet and E. Rojas, Cyclic changes in potential and
    resistance of the β-cell membrane induced by glucose in islets of
    Langerhans from mouse, J. Physiol. 278:117 (1978).

3. L. Beauge & V.L. Lew, Passive fluxes of sodium and potassium across
    red cell membranes, in: "Membrane Transport in Red Cells", J.C.
    Ellory and V.L. Lew, eds., Academic Press, London, pp 39 (1977).

4. M. Claret and J.L. Mazet, Ionic fluxes and permeabilities of cell
    membranes in rat liver, J. Physiol. 223:279 (1972).

5. D.L. Cook and C.N. Hales, Intracellular ATP directly blocks K^+
    channels in pancreatic β-cells, Nature 311:271 (1984).

6. D.L. Cook, M. Ikeuchi and W.Y. Fujimoto, Lowering pH_i inhibits
    Ca^{2+}-activated K^+ channels in pancreatic β-cells, Nature 311:269
    (1984).

7. C.M. Dawson, I. Atwater and E. Rojas, The response of the pancreatic
    β-cell membrane potential to potassium-induced calcium influx in
    the presence of glucose, Quarterly J. Exp. Physiol. 69:819 (1984).

8. C.M. Dawson, P.C. Croghan, I. Atwater and E. Rojas, Estimation of
    potassium permeability in mouse islets of Langerhans, Biomed. Res.
    4:389 (1983).

9. C.M. Dawson, P.C. Croghan, A.M. Scott and J.A. Bangham, Direct
    comparison of K^+ and Rb^+ efflux in normal mouse islets,
    Diabetologia 27:267A (1984).

10. C.M. Dawson, P.C. Croghan, A.M. Scott and J.A. Bangham, Potassium and
     rubidium permeability and potassium conductance of the β-cell
     membrane in mouse islets of Langerhans, Quarterly J. Exp. Physiol.
     71:205 (1986).

11. P.M. Dean and E.K. Matthews, Electrical activity in pancreatic islet
     cells: effect of ions, J. Physiol. 210:265 (1970).

12. N.A. Delamere and G. Duncan, A comparison of ion concentrations, potentials and conductances of amphibian, bovine and cephalopod lenses, $\underline{J}$. $\underline{Physiol}$. 272:167 (1977).

13. G. Duncan and P.C. Croghan, Effect of changes in external ion concentrations and 2,4-dinitrophenol on the conductance of toad lens membranes, $\underline{Exp}$. $\underline{Eye}$ $\underline{Res}$. 10:192 (1970).

14. G.T. Eddlestone and P.M. Beigelman, Pancreatic β-cell electrical activity: the role of anions and the control of pH, $\underline{Amer}$. $\underline{J}$. $\underline{Physiol}$. 244:C188 (1983).

15. G.T. Eddlestone, A. Goncalves, J.A. Bangham and E. Rojas, Electrical coupling between cells in islets of Langerhans from mouse, $\underline{J}$. $\underline{Memb}$. $\underline{Biol}$. 77:1 (1984).

16. R. Ferrer, I. Atwater, E.M. Omer, A.A. Goncalves, P.C. Croghan and E. Rojas, Electrophysiological evidence for the inhibition of potassium permeability in pancreatic β-cells by glibenclamide, $\underline{Quarterly}$ $\underline{J}$. $\underline{Exp}$. $\underline{Physiol}$. 69:831 (1984).

17. I. Findlay, M.J. Dunne and O.H. Petersen, High-conductance K^+ channel in pancreatic islet cells can be activated and inactivated by internal calcium, $\underline{J}$. $\underline{Memb}$. $\underline{Biol}$. 83:169 (1985).

18. D.E. Goldman, Potential, impedance and rectification in membranes, $\underline{J}$. $\underline{Gen}$. $\underline{Physiol}$. 27:37 (1943).

19. B. Hellman, β-cell cytoplasmic Ca^{2+} balance as a determinant for glucose stimulated insulin release. $\underline{Diabetologia}$ 28:494 (1985).

20. J.C. Henquin, The potassium permeability of pancreatic islet cells: mechanisms of control and influence on insulin release, $\underline{Hormone}$ $\underline{and}$ $\underline{Metabolic}$ $\underline{Res}$. $\underline{Supp}$. $\underline{Series}$ 10:66 (1980).

21. J.C. Henquin and H.P. Meissner, The electrogenic sodium-potassium pump of mouse pancreatic β-cells, $\underline{J}$. $\underline{Physiol}$. 332:529 (1982).

22. J.C. Henquin and H.P. Meissner, Significance of ionic fluxes and changes in membrane potential for stimulus-secretion coupling in pancreatic β-cells, $\underline{Experientia}$ 40:1043 (1984).

23. A.L. Hodgkin and P. Horowicz, The influence of potassium and chloride ions on the membrane potential of single muscle fibres, $\underline{J}$. $\underline{Physiol}$. 148:127 (1959).

24. T.J.C. Jacob, J.A. Bangham and G. Duncan, Characterization of a cation channel on the apical surface of the frog lens epithelium, $\underline{Quarterly}$ $\underline{J}$. $\underline{Exp}$. $\underline{Physiol}$. 70:403 (1985).

25. S. Kawazu, A.C. Boschero, C. Delcroix and W.J. Malaisse, The stimulus-
    secretion coupling of glucose-induced insulin release, XXVIII,
    Effect of glucose on Na^+ fluxes in isolated islets, Pflügers Arch.
    375:197 (1978).

26. W.J. Malaisse, A.C. Boschero, S. Kawazu and J.C. Hutton, The stimulus-
    secretion coupling of glucose-induced insulin release, XXVII,
    Effect of glucose on K^+ fluxes in isolated islets, Pflügers Arch.
    373:237 (1978).

27. E.K. Matthews and D.A. Shotton, Efflux of ^{86}Rb from rat and mouse
    pancreatic islets: the role of membrane depolarization, Brit. J.
    Pharmacol. 83:831 (1984).

28. H.P. Meissner, J.C. Henquin and M. Preissler, Potassium-dependence of
    the membrane potential of pancreatic β-cells, FEBS Lett. 94:87
    (1978).

29. J. Sehlin, Interrelationship between chloride fluxes in pancreatic
    islets and insulin release, Amer. J. Physiol. 235:E501 (1978).

30. J. Sehlin and I.-B Taljedal, Transport of rubidium and sodium in
    pancreatic islets, J. Physiol. 242:505 (1974).

31. J.F. Thain, The analysis of radioisotopic tracer flux experiments in
    plant tissues, J. Exp. Botany 35:444 (1984).

32. H.H. Ussing, Active transport of inorganic ions, Symposia Soc. Exp.
    Biol. 8:407 (1954).

33. C.B. Wollheim and G.W.G. Sharp, The regulation of insulin release by
    calcium, Physiol. Rev. 61:914 (1981).

^{22}Na$^+$ EFFLUX FROM NORMAL AND ob/ob MOUSE ISLETS OF LANGERHANS

C.M. Dawson and P.C. Croghan

Dept. of Biophysics
School of Biological Sciences
University of Est Anglia
Norwich, NR4 7TJ, England

Previous studies have produced differing results for the effect of glucose on ^{22}Na$^+$ efflux from islets. Interpretation of ^{22}Na$^+$ efflux data is difficult due to compartmentalization and the high concentration of Na$^+$ in the extracellular space. An attempt has been made to resolve the problem using a triple-label method, with tracer loading in the perifusion chamber. This method was used to compare ^{22}Na$^+$ efflux rate in normal and ob/ob mouse islets. ^{3}H-sucrose was used as an extracellular space marker and ^{86}Rb$^+$ as a 'control' isotope, representing K$^+$, and present predominantly in the intracellular compartment, in addition to ^{22}Na$^+$. ^{3}H-sucrose efflux showed two rate constants: one due to the chamber (about 2 min^{-1}; confirmed by blank chamber experiments) and one due to the islet extracellular space (about 0.3 min^{-1}). The efflux of ^{22}Na$^+$ and ^{86}Rb$^+$ could each be described by three rate constants, the first two being equivalent to the sucrose rate constants and the third being due to efflux across the cell membrane. In the absence of glucose, the ^{22}Na$^+$ rate constant was 0.06 min^{-1} and the ^{86}Rb$^+$ rate constant was 0.03 min^{-1}. Glucose had no significant effect on ^{22}Na$^+$ efflux from normal or ob/ob islets but decreased ^{86}Rb$^+$ efflux as expected. Permeabilities (P_j) of Na$^+$ and Rb$^+$ were calculated from steady-state efflux data and membrane potentials. The value of P_{Na}/P_K was 0.05 in the absence of glucose and 0.21 in 22.2 mM glucose.

THE ROLE OF ANIONS IN THE REGULATION OF INSULIN SECRETION

Janove Sehlin

Dept. of Histology and Cell Biology
University of Umea
S-901 87 Umea 6, Sweden

INTRODUCTION

Previous studies have shown that glucose-induced insulin release
is inhibited by substitution of Cl^- with non-permeant mono-valent
anions such as isethionate,[9,10-13] para-aminohippurate,[11] or methyl
sulphate.[13] The precise mechanism behind this apparent Cl^- dependence
is unclear but the demonstration of a strong correlation between
radioactive Cl^- efflux from prelabelled islets and insulin release may
indicate that Cl^- transport in the β-cell is important for the
regulation of insulin release.[10,11] The present paper concerns the
basic question, whether glucose-induced insulin release is strictly
Cl^--dependent or whether it rather shows anion selectivity. To this
end, a study has been undertaken to measure the effects of a number of
anions on insulin release and to compare the secretory effects of
anions with their capacity to interfere with islet $^{45}Ca^{2+}$ uptake.

METHODS

Adult, noninbred ob/ob mice of the Umea colony (Umea ob/ob mice)
were used throughout. Prior to experiments, the animals were starved
overnight to normalize the blood sugar levels. These animals are
characterized by β-cell hyperplasia resulting in large islets with
more than 90% β-cells.[6] The present results are therefore probably
representative of this cell type. Although the animals are
metabolically abnormal, their β-cells respond normally to various
stimulators and inhibitors of insulin secretion.[2]

Pancreatic islets were microdissected freehand under
stereomicroscope.[5] Then, they were preincubated and incubated in
Krebs-Ringer-Hepes medium (KRH) as indicated in the legends to
Figures. The basal KRH medium had the following composition (mM):
130 NaCl, 4.7 KCl, 2.56 $CaCl_2$, 1.2 $MgSO_4$, 1.2 KH_2PO_4, 20 Hepes, as
well as 3 mM D-glucose and 1 mg/ml bovine serum albumin. The pH was
7.40 and the gas phase ambient air.

In studies of insulin secretion, after an initial preincubation
for 30 min in basal KRH medium and incubation for 10 min, batches of 3
islets were rapidly removed from the incubation vials, freeze-dried
and weighted on a quartz-fibre balance. The incubation media were
frozen for subsequent determination of insulin by radioimmunoassay.

In studies of $^{45}Ca^{2+}$ uptake, after 30 min of preincubation in
basal medium, batches of 4-5 isolated islets were incubated for 3 min
in medium supplemented with $^{45}Ca^{2+}$.[7] After incubation, the islets
were washed for 60 min in large volumes (5 ml) of non-radioactive
medium to remove extracellular and membrane-bound label without
significantly affecting the amount of intracellular $^{45}Ca^{2+}$.[7] Then,
the islets were freeze-dried and weighed and the radioactive contents
measured by liquid scintillation counting.

RESULTS AND DISCUSSION

Fig. 1 shows how the addition of low concentrations of various
anions (16-20 mM) affected glucose-induced insulin release (20 mM
D-glucose). Some of the anions, perchlorate, thiocyanate, and iodide,
clearly potentiated the secretion, whereas other anions, bromide,
nitrate, isethionate, and acetate, did not significantly influence the
glucose-stimulated insulin release when added at this low
concentration. Addition of an extra 20 mM NaCl to the medium did not
affect the secretion.

To further evaluate the effects of monovalent anions on insulin
release, a large fraction (120 mM) of the extracellular Cl⁻ was
replaced by different anions. As shown in Fig. 1, nitrate had the
capacity to functionally replace Cl⁻, whereas replacement of this
anion for isethionate or acetate substantially reduced acute insulin
release. Replacement with thiocyanate, however, led to a strong

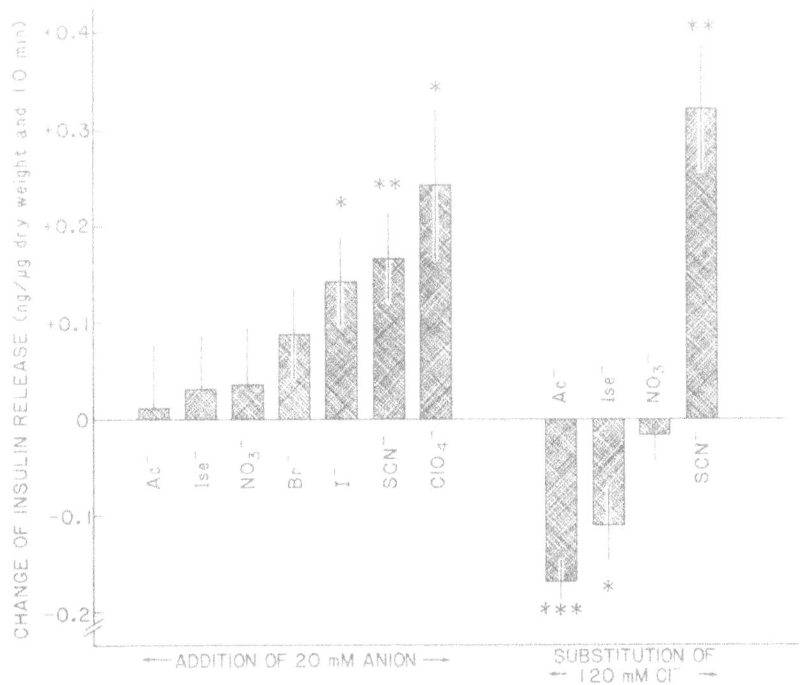

Fig. 1. Effects of anions on insulin release. Islets were incubated for 10 min in KRH medium. Test media were modified either by adding low concentrations of anions (Na$^+$ salts of 16 mM ClO_4^- and 20 mM of other anions) or by replacing 120 mM NaCl with equimolar sodium salts of other anions as indicated in the figure. The results are presented in terms of change in insulin release at 20 mM D-glucose. The control levels of insulin release were 0.28 ± 0.02 (n-90) at 20 mM glucose and 0.06 ± 0.01 ng/µg dry islet (n=90) at 3 mM glucose. The bars denote means ± S.E. for 7-14 separate experiments (animals). *P<0.05, **P<0.01, and ***P<0.001 for difference from 20 mM glucose alone (zero line in the figure).

potentiation of glucose-induced insulin release, suggesting that thiocyanate is more effective than Cl$^-$ in supporting the glucose-induced insulin secretion.

It has been suggested that ClO_4^- improves excitation-contraction coupling in skeletal muscle by interfering with the gating of Ca channels.[1,8] Since low concentrations of ClO_4^- also strongly potentiated the insulin secretion, it was of interest to see whether the different effects of anions on insulin release were paralleled by

similar effects on islet $^{45}Ca^{2+}$ uptake. Fig. 2 shows the initial
uptake of $^{45}Ca^{2+}$ (3 min) by the La^{3+}-nondisplaceable (intracellular)
pool. It may be seen that the general pattern for anion effects on
the $^{45}Ca^{2+}$ uptake is similar to that on insulin release. Thus, the
addition of 16-20 mM ClO_4^-, SCN^-, or I^- potentiated the $^{45}Ca^{2+}$ uptake
induced by 20 mM glucose, whereas the addition of Br^-, NO_3^-, or acetate showed no significant effect. Also, replacement of 120 mm
$Cl^-$with equimolar acetate or isethionate led to an inhibition of
glucose-induced $^{45}Ca^{2+}$ uptake, whereas replacement with SCN^-strongly
potentiated the uptake. A difference, however, was found when replacing Cl^- with NO_3^-. The clear potentiation of $^{45}Ca^{2+}$ uptake was not
paralleled by any potentiation of insulin release. Although we have
no ready explanation for this difference, it cannot be excluded that
the NO_3^- effect is short-lived and therefore is more readily seen in

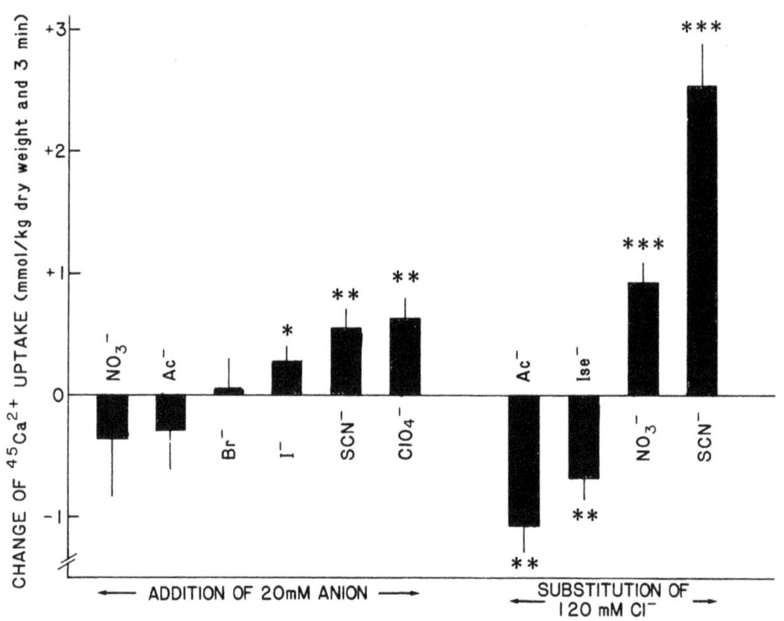

Fig. 2. Effects of anions on $^{45}Ca^{2+}$ uptake. Islets were incubated
for 3 min in KRH medium with $^{45}Ca^{2+}$ (144 GBq/mol) and with anionic
modifications as in Fig. 1. Results are presented as change in $^{45}Ca^{2+}$
uptake at 20 mM D-glucose. The control levels of $^{45}Ca^{2+}$ uptake were
3.78 ± 0.12 (n=97) at 20 mM glucose and 2.22 ± 0.11 mmol/kg dry islet

(n=81) at 3 mM glucose. The bars denote means ± S.E. for 7-16
separate experiments (animals). * P<0.05, **P<0.01, and ***P<0.001 for
difference from 20 mM glucose alone (zero line in the figure.)

the 3-min incubation ($^{45}Ca^{2+}$) than in the 10-min incubation (insulin
release).

Taken together, these results suggest that insulin release,
rather than being strictly Cl$^-$ dependent, shows anion selectivity.
Different types of anion selectivity have previously been observed in
many biological systems.[15] The selectivity sequence found in the
present studies of both insulin release and $^{45}Ca^{2+}$ uptake appears to
be: perchlorate>thiocyanate>iodide>(bromide, nitrate,
chloride)>(isethionate, acetate). This sequence resembles the
original Hofmeister series on the effects of anions on proteins and
biopolymers.[3,4] The resemblence may prove useful in attempts to
determine by which physico-chemical mechanisms the anions act in the
islets. It appears that the anions most distant from the salting-out
end of the Hofmeister series as well as the most "chaotropic" ones
tested, i.e. perchlorate and thiocyanate, were also the strongest
potentiators of glucose-induced insulin release and $^{45}Ca^{2+}$ uptake. It
is evident that change in anion composition may affect many different
cellular functions. However, it is known that glucose oxidation in
isolated pancreatic islets is not affected by either perchlorate or
thiocyanate (Sehlin, unpublished data) or isethionate.[12] This makes
it unlikely that with the possible exception of acetate (the conjugate
base of a permeant weak acid), the anions affect the β-cell activity
primarily by changing initial metabolic signals for insulin release.

The resemblence between the anion selectivity sequence for
insulin secretion and islet $^{45}Ca^{2+}$ uptake points at the possibility of
a causal relationship between the anion effects on the two processes.
As voltage-dependent calcium uptake by β-cells is generally implicated
as a key event in the stimulus-secretion coupling (for review, see
ref. 14), the present results and the previous observations on the
effects of perchlorate in skeletal muscle[1,8] makes it tempting to
suggest that one effect of anions in the β-cells is to modify the
gating of the voltage-dependent Ca channel. Naturally, this does
not exclude the possibility that mono-valent anions are also involved
in other regulatory steps in the β-cells.

ACKNOWLEDGEMENTS

This work was supported by grants from the Swedish Medical
Research Council (12x-4756), and the Swedish Diabetes Association.

REFERENCES

1. M. Gomolla, G. Gottschalk, and H. Ch. Luttgau, Perchlorate-in-
    duced alterations in electrical and mechanical parameters of
    frog skeletal muscle fibers, J. Physiol. 343:197 (1983).

2. H. J. Hahn, B. Hellman, A. Lernmark, J. Sehlin, and I.-B.
    Taljedal, The pancreatic β-cell recognition of insulin
    secretagogues. Influence of neuraminidase treatment on the
    release of insulin and the islet content of insulin, sialic
    acid, and cyclic adenosine 3':5'-monophosphate, J. Biol.
    Chem. 249:5275 (1974).

3. W. G. Hanstein, Chaotropic ions and their interactions with
    proteins, J. Solid-Phase Biochem. 4:189 (1979).

4. W. G. Hanstein, K. A. Davis, and Y. Hatefi, Water structure and
    the chaotropic properties of haloacetates, Arch. Biochem.
    Biophys. 147:534 (1971).

5. C. Hellerstrom, A method for the microdissection of pancreatic
    islets of mammals, Acta Endocrinol. (Kbh.) 45:122 (1964).

6. B. Hellman, Studies in obese-hyperglycemic mice, Ann. N. Y. Acad.
    Sci. 131:541 (1965).

7. B. Hellman, J. Sehlin, and I.-B. Taljedal, Effects of glucose on
    $^{45}Ca^{2+}$ uptake by pancreatic islets as studied with the
    lanthanum method, J. Physiol. 254:639 (1976).

8. H. C. Luttgau, G. Gottschalk, L. Kovacs, and M. Fuxreiter, How
    perchlorate improves excitation-contraction coupling in
    skeletal muscle fibers, Biophys. J. 43:247 (1983).

9. C. S. Pace and J. S. Smith, The role of chemiosmotic lysis in the
    exocytotic release of insulin, Endocrinol. 113:964 (1983).

10. J. Sehlin, Interrelationship between chloride fluxes in pancre-
    atic islets and insulin release, Am. J. Physiol. 235:E501
    (1978).

11. J. Sehlin, Univalent ions in islet cell function, Horm. Metab.
    Res. Suppl. Ser. 10:73 (1980).

12. G. Somers, A. Sener, G. Devis, and W. J. Malaisse, The stimu-
     lus-secretion coupling of glucose-induced insulin release.
     XLV. The anion-osmotic hypothesis for exocytosis. _Pflugers_
     Arch. 388:249 (1980).
13. T. Tamagawa and J. C. Henquin, Chloride modulation of insulin
     release, $^{86}Rb^+$ efflux and $^{45}Ca^{2+}$ fluxes in rat islets stimu-
     lated by various secretagogues. _Diabetes_ 32:416 (1983).
14. C. B. Wollheim and G. W. G. Sharp, Regulation of insulin release
     by calcium, _Physiol Rev._ 61:914 (1981).
15. E. M. Wright and J. M. Diamond, Anion selectivity in biological
     systems, _Physiol. Rev._ 57:109 (1977).

GRADED SPIKE ELECTROGENESIS IN MOUSE PANCREATIC β-CELLS

B. Soria and Rosa Ferrer

Departamento de Fisiologia
Facultad de Medicina
Universidad de Alicante
Alicante, Spain

INTRODUCTION

It is well established that the islet of Langerhans is responsible for the synthesis and secretion of insulin, glucagon and somatostatin. These hormones are produced in cells with distinct morphology and different topological distribution within the islet.[20] While 88% of the islet volume is comprised of centrally located insulin secreting β-cells, glucagon secreting α-cells appear to concentrate in the outer cell layers.[19] α-cells and β-cells constitute a morphological and functional unit.[9,16,19]

During the last decade, the β-cell membrane response to D-glucose has been characterized using radioactive tracer techniques to measure ionic fluxes and high resistance microelectrodes to measure membrane potentials from one[4,5,8,15,18,21] or two cells within the islet.[9]

Although the exact mechanisms to the glucose response are yet to be determined, there are several well established β-cell membrane properties. For example, the electrical response to stimulatory levels of glucose is known to depend mainly on the degree of activation of two membrane channels, namely, the glucose blockable and the $[Ca^{2+}]_i$-senstive K-channels (see this book). In the absence of glucose, the glucose blockable K-channel (also blocked by intracellular ATP; see this book), and the $[Ca^{2+}]_i$-activated K-channel (blocked by external application of quinine,[7] control the membrane potential. In the presence of stimulatory levels of glucose a burst pattern of electrical activity is obtained.[4,5,8,15,18] The membrane potential oscillates between two potentials: one at about -55 mV (referred to as the silent phase) and the other at about -35 mV (the active or plateau phase). During the plateau phase, Ca-channels are activated giving rise to Ca^{2+} action potentials. The repolarizing phase of the action potential is determined by the activation of the voltage-gated K-channel (blocked by tetraethylammonium, TEA).

Although, under steady-state glucose stimulation the burst pattern of electrical activity is regular, the amplitude of the spikes along the active phase of the burst is rather irregular. In fact, two distinct populations of spikes can be distinguished[11] under steady-state 11 mM glucose stimulation. For most cells, the amplitude of the spikes along the active phase of the bursts showed a bimodal distribution of amplitudes. Two possible interpretations for this were given. Firstly, the activation of two different voltage-gated Ca-channels in the β-cell membrane, as in other cell types[1,17] could give rise to spikes of variable amplitude and, secondly, the small spikes could result from the electrotonic spread of large action potentials generated in nearby β-cells via cell-to-cell electrical coupling.[9] However, yet another possible explanation is that the rapid and localized increase in the $[Ca^{2+}]_i$ along the active phase of the burst could stimulate $[Ca^{2+}]_i$-activated K-channels causing a random termination of the action potential (see Bangham, Smith & Croghan, this book).

To further examine these alternative explanations and, since different Ca-channels exhibit different pharmacological properties[14] we have compared the spike amplitude distribution along the active phase with the distributions of net inward and outward currents under different conditions including the presence of specific blockers of the channels. Two-microelectrode experiments were also carried out and the cross-correlation of the amplitude distributions was considered.

METHODS

The methods used for the intracellular injection of current and the simultaneous recording of the membrane potential with a single microelectrode have been described elsewhere.[4] Intracellular record-

ing from two electrically coupled cells was carried out as described
in detail elsewhere.[9]

The modified Krebs solution used had the following composition
(mM): 120 NaCl, 25 NaHCO$_3$, 5 KCl, 2.6 CaCl$_2$ and 1 MgCl$_2$ and was
equilibrated with a gas mixture containing 95% O$_2$ at 37°C.

RESULTS AND DISCUSSION

<u>Amplitude Distribution of the Net Membrane Inward Current During Spike</u>
<u>Activity</u>

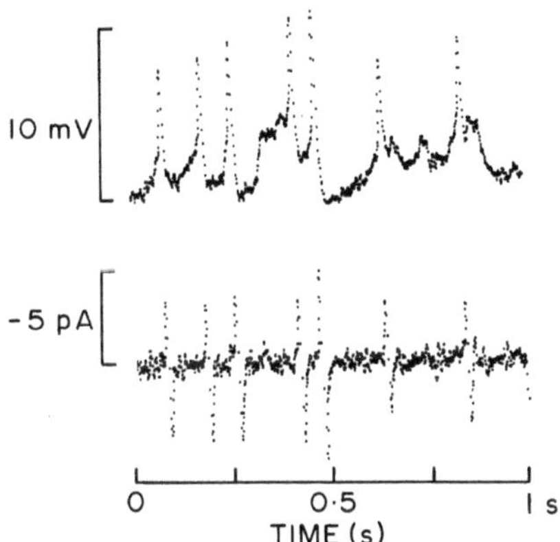

Fig. 1. Net membrane currents during spike electrogenesis.

It is possible to estimate the net ionic current during an action
potential recorded from a small spherical cell, by assuming that the
cell interior is isopotential, i.e., same electrical potential throughout the cytosol,

$$I_i = - C_m \, dV/dt \quad (1)$$

where C_m represents the membrane capacitance and dV/dt represents the
time derivative of the membrane potential. Taking C_m for a single
β-cell as 5 pF[4] and calculating the time derivative from the recorded
electrical activity as shown in Fig. 1, one can calculate I_i in the
range from −5 to 10 pA (see Fig. 1, lower part).

In order to compare the spike amplitude distribution with the distribution of the net inward current, which in pancreatic β-cells is carried by Ca ions, we calculated the maximum rate of depolarization during each spike (Fig. 2, upper right panel). It follows from equation (1) that this parameter is proportional to the size of the maximum inward current. Fig. 2 compares four frequency distributions, namely, spike amplitude (upper left), rate of depolarization (upper right), time-to-peak (middle left) and rate of repolarization (middle right). Also shown in Fig. 2 are examples of a burst of electrical activity (lower left) and action potentials (lower right). The frequency distribution of the spike amplitude and of the rate of repolarization (proportional to the peak value of the net outward K^+ current) are clearly bimodal. In the case of the rate of depolarization the distribution is complex and no attempt was made to separate the components.

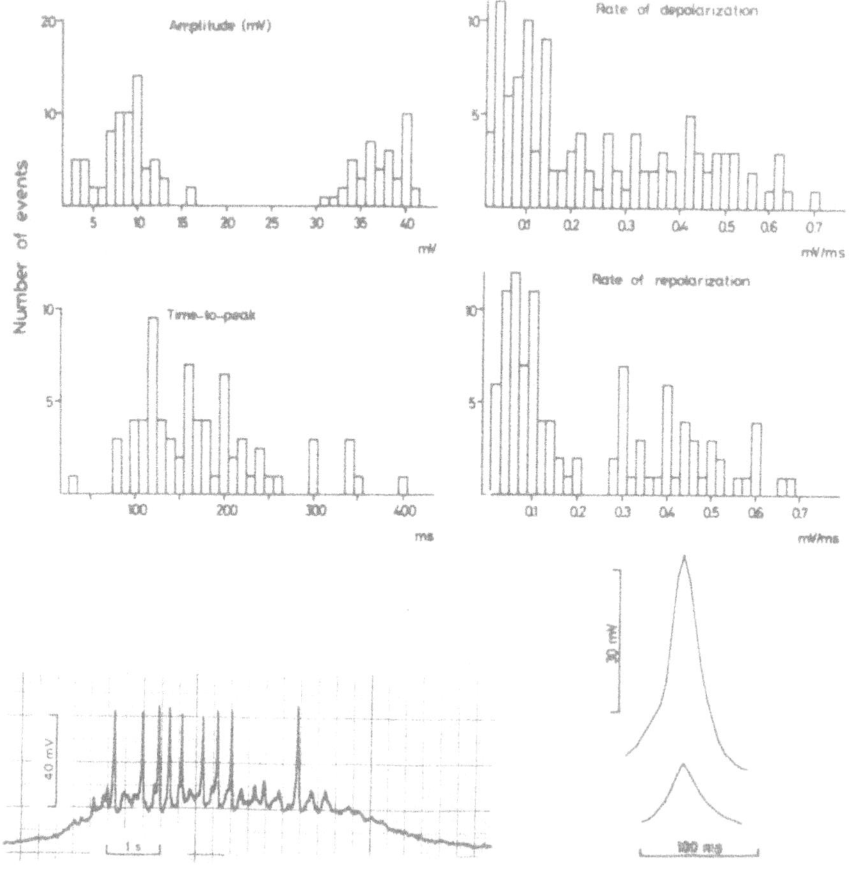

Fig. 2. Frequency distributions for spike amplitude (upper left), rate of depolarization (upper right), time-to-peak (middle left) and rate of repolarization (middle right). Lower left: burst of electrical activity on expanded time scale; lower right: examples of large and small action potentials.

Fig. 3 shows the effects of glucose on the spike amplitude distribution in the presence (upper) or absence (lower) of quinine, a
blocker of K-permeability.[7] In the presence of quinine (200 µM) and
in the absence of glucose (upper left) the distribution is bimodal,
and the distribution centered around 6 mV is more densely populated.
Addition of glucose (5.6 mM) causes a drastic change in the relative
proportion of the two populations, the distribution of the larger
spikes becoming more densely populated. In the absence of quinine
(lower panel), increasing the concentration of glucose from 11.1 to
22.2 mM causes qualitatively similar changes. However, the mean
amplitude of the large spikes also increased from about 30 mV measured
in 11.1 mM glucose to about 34 mV measured in the presence of 22.2 mM
glucose.

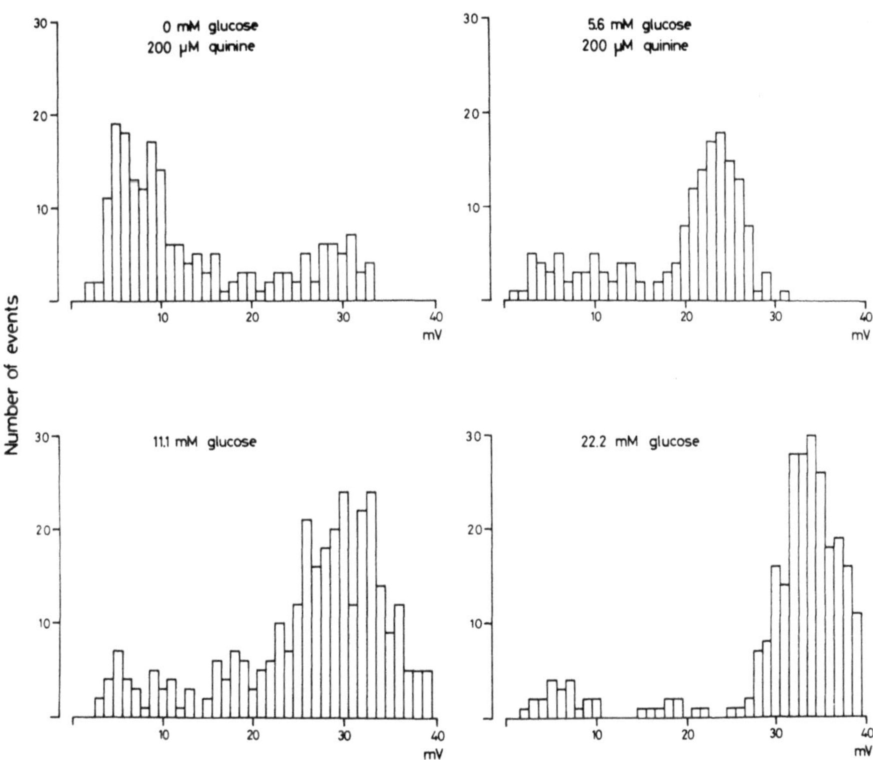

Fig. 3. Effects of glucose on the frequency distribution of spike
amplitudes. Upper (left and right): in the presence of quinine
(200 µM).

The simplest interpretation of these results is that, glucose, by
improving cell-to-cell coupling[9] and by blocking membrane K-channels

(this book), increases the effective space constant for electrotonic
spread of the action potentials within the cellular domains of the
islet.[19] While partial blockade of the β-cell membrane K^+-conductance
by glucose would tend to increase the size of the individual spikes,
the effects of glucose on cell-to-cell coupling would tend to increase
the size of the spikes generated in neighboring cells. The mean value
of the population of large spikes increased from 30 mV, measured in
the presence of 11.1 mM glucose, to 34 mV in the presence of 22.2 mM
glucose. The mean value of the other distribution was not affected.

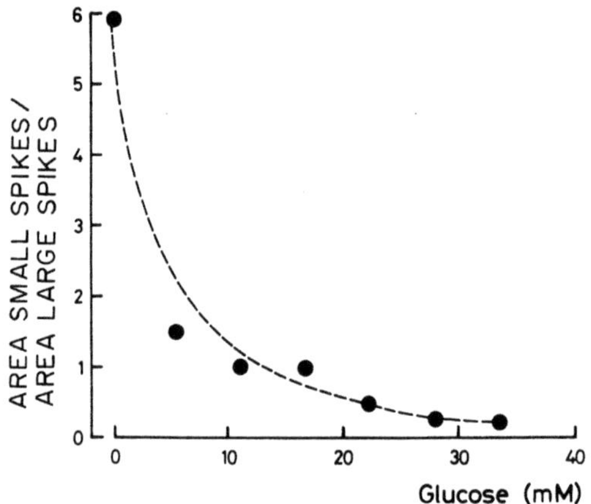

Fig. 4. Effects of glucose on the relative areas under the two amplitude distributions. Values at 0 and 56 mM glucose were obtained in
the presence of quinine (200 μM).

Another way of examining the data is to estimate the area under
each distribution of spike amplitudes and to calculate the ratio
Area Small Spikes/Area Large Spikes.

As shown in Fig. 4 the ratio decreases as the concentration of
glucose is increased.

240

<u>Effect of Membrane Potential on the Relative Density of the Two</u>
<u>Populations of Action Potentials</u>

To examine the effects of the membrane potential at which the
action potentials are generated on the relative densities of the two
populations, we injected current intracellularly during the active
phase of the bursts. The frequency distribution of the amplitudes of
the spikes during the application of current pulses of the same amplitude was calculated first and then the ratio of the areas. The results of one of such experiments are shown in Fig. 5.

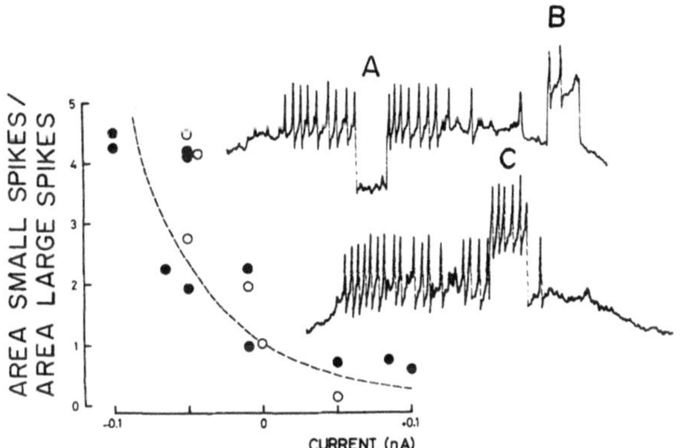

Fig. 5. Effect of membrane potential on the frequency distribution.
Inserts: upper, effects of injecting a pulse of hyperpolarizing
current (A) followed by a depolarizing pulse (B). Lower, effects of
injecting a depolarizing current pulse (C). Data from two different
islets in the presence of 11.1 mM glucose.

The amplitude of the small spikes decreases during the application of hyperpolarizing current. Blockers of K-channels such as
tetraethylammonium (TEA, 20 mM), glibenclamide (50 µM) and Ba^{2+} (10
mM) increased the relative proportion of spikes with larger amplitude.
On the other hand, blockers of Ca-channels such as verapamil (1 µM),
Zn^{2+} (<250 µM) and Co^{2+} induced changes in the opposite direction,
i.e., the relative proportion of the smaller spikes increased.

Frequency Distributions of the Amplitude of the Spikes in Electrically
Coupled Cells

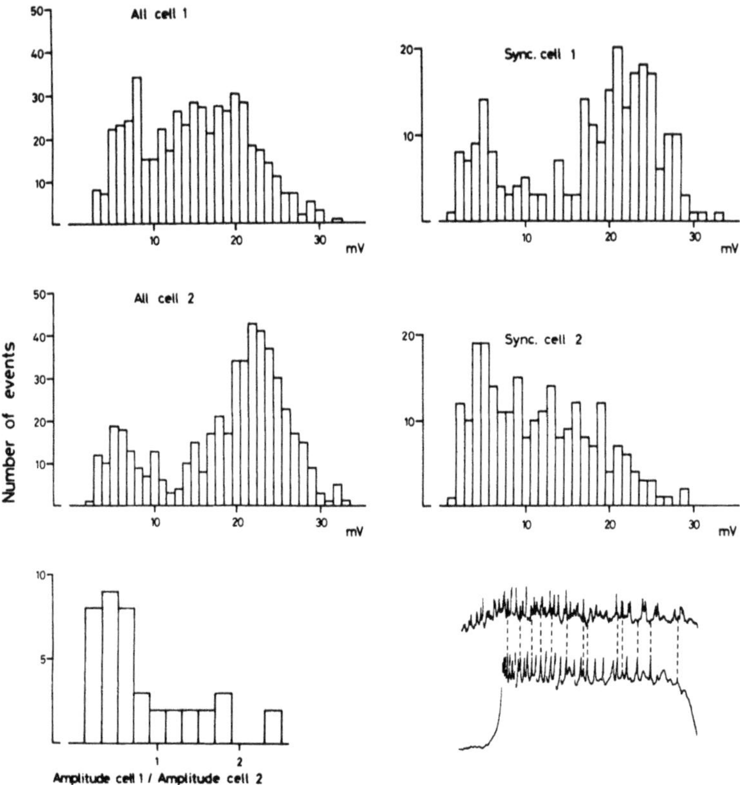

Fig. 6. Frequency distributions in electrically coupled cells.
Upper: cell held under current clamp. Lower: cell 30 µM apart.
Insert: Superimposed bursts to show coincident action potentials.

Fig. 6 shows the spike amplitude distributions for two electrically coupled cells. The distribution on the upper left panel reports
on the amplitude in the cell injected with current. The distribution
on the lower left panel, in the cell electrically coupled. It may be
seen that both distributions are bimodal. The distributions on the
right side were obtained by considering only the spikes appearing at
the same time in both records (see insert, lower right).

Effects of Various Agents on the Frequency Distribution

We have examined the effects of Ca^{2+}, Ca-channel blockers.[14]
K-channel blockers[2,12] and caffeine on the density of the distribution
and on the mean amplitude of the spikes. The data are summarized in
Table I.

Table I. Effects of Various Agents on Spike Amplitude Frequency
          Distribution

			Ratio of areas	A_1/A_2
Tetrodotoxin	(0.4	μM)	0.78	1.08
Tetrodotoxin	(1	μM)	0.80	1.04
$CaCl_2$	(1	mM)	1.12	1.10
$CaCl_2$	(5	mM)		
$ZnCl_2$	(0.05	mM)	0.87	0.79
$CaCl_2$	(7.5	mM)	0.97	0.82
Verapamil	(1	μM)	4.35	1.13
Co^{2+}	(250	μM)	4.06	1.26
Zn^{2+}	(10	μM)	2.51	0.95
Zn^{2+}	(50	μM)	10.22	0.62
Caffeine	(10	mM)	15.85	1.06
Tea	(20	mM)	0.52	0.45
Glibenclamide	(50	μM)	0.06	0.48
$BaCl_2$	(10	mM)	0.40	0.92
Cooling to 27°C[6]			3.52	0.98

A_1/A_2 = mean value of the amplitude of the small spikes (A_1)/mean
value of the amplitude of the large spikes (A_2). All experiments in
11.1 mM glucose. From ref. [13]

It is clear from the data in Table 1 that tetrodotoxin (TTX) does
not affect the amplitude of the spikes. Thus, activation of the
voltage-gated Na-channels does not appear to be the cause of the
graded spectrum of spike amplitudes. Increasing the concentration of
$CaCl_2$ in the external medium from 1 to 7.5 mM had no effect on these
two parameters of the frequency distribution of amplitudes. Only the
Ca-channel blockers verapamil, Co^{2+} and Zn^{2+} induced a clear increase
in the area under the frequency distribution of the small spikes. It
is interesting to notice that caffeine induced a 15-fold increase in
this ratio without affecting the mean values of the amplitudes in both
populations. Since caffeine is known to cause Ca^{2+} release from

internal stores,[10] this effect might be due to a decrease in cell-to-cell coupling.[9] It is also apparent from Table I that, Ba^{2+} increases the density of the distribution of large spikes. Atwater, Beigelman and Ribalet[3] proposed that Ba^{2+} blocks K-permeability and substitutes for Ca^{2+} as a current carrier through the Ca-channel in the pancreatic β-cell. We have seen that depolarization by current injection causes an increase in the density of the distribution of small spikes (see Fig. 5). We have also shown that blockers of the K-permeability, such as glucose and TEA, have the opposite effects. Thus, the added effects of Ba^{2+}, i.e., blockade of the K-channels, increase in the inward currents, and depolarization, give an overall increase in the density of the distribution of large spikes.

In conclusion, our results support the idea that the graded spike electrogenesis in the pancreatic β-cell results from electrotonic spread of the spike activity from neighboring cells in the islet.

ACKNOWLEDGEMENT

The authors are indebted to J.A. Bangham for computer programs used in spike analysis. The experiments involving two microelectrodes were done in collaboration with Dr. E. Rojas. This work has been supported in part by the Spanish CAYCIT (grant n 3167-84) and the U.S.-Spain Joint Committee for Scientific and Technological Cooperation.

REFERENCES

1. C. M. Armstrong and D. R. Matteson, Two distinct populations of calcium channels in a clonal line of pituitary cells, Science 227:65 (1985).
2. C. M. Armstrong, R. P. Swenson, and S. R. Taylor, Block of squid axon K channels by internally and externally applied barium ions, J. Gen. Physiol. 80:663 (1982).
3. I. Atwater, P. M. Beigelman, and B. Ribalet, Three actions of Ba^{2+} upon membrane potential in mouse pancreatic β-cells, J. Physiol. (London) 266:38 (1977).
4. I. Atwater, B. Ribalet, and E. Rojas, Cyclic changes in potential resistance of the β-cell membrane induced by glucose in islets of Langerhans from mouse, J. Physiol. (London) 278:117 (1978).

5. I. Atwater, C. M. Dawson, G. T. Eddlestone, E. Rojas, Voltage
       noise measurements across the pancreatic β-cell membrane:
       calcium channel characteristics, J. Physiol. (London)
       314:195 (1981).

6. I. Atwater, A. Goncalves, A. Herchuelz, P. Lebrun, W. J.
       Malaisse, E. Rojas, and A. Scott, Cooling dissociates glu-
       cose-induced insulin release from electrical activity and
       cation fluxes in rodent pancreatic islets, J. Physiol.
       (London) 348:615 (1984).

7. I. Atwater, C. M. Dawwon, B. Ribalet, and E. Rojas, Potassium
       permeability activated by intracellular calcium ion
       concentration in the pancreatic β-cell, J. Physiol 288:575
       (1979).

8. P. M. Dean and E. K. Matthews, Glucose induced electrical activ-
       ity in pancreatic islet cells, J. Physiol. (London) 210:255
       (1970).

9. G. T. Eddlestone, A. Goncalves, J. A. Bangham, and E. Rojas,
       Electrical coupling between cells in islet of Langerhans
       from mouse, J. Memb. Biol. 77:1 (1984).

10. M. Endo, Mechanism of action of caffeine on the sarcoplasmic
       reticulum of skeletal muscle, Proc. Japan Acad. Sci. 51:479
       (1975).

11. R. Ferrer, B. Soria, C. M. Dawson, I. Atwater, and E. Rojas,
       Effects of Zn^{2+} on glucose induced electrical activity and
       insulin release from mouse pancreatic islets, Amer. J.
       Physiol. 246:C520 (1984).

12. R. Ferrer, I. Atwater, E. M. Omer, A. A. Goncalves, P. C.
       Croghan, and E. Rojas, Electrophysiological evidence for the
       inhibition of potassium permeability in pancreatic β-cells
       by glibenclamide, Quart. J. Exp. Physiol. 69:831, (1984).

13. R. Ferrer, S. Sala, J. V. Sanchez-Andres, and B. Soria, Further
       evidence that Zn^{2+} blocks voltage dependent calcium channels
       in the mouse β-cell, Biochem. Soc. Trans. (in press 1986).

14. S. Hagiwara and L. Byerly, Calcium channel, Annu. Rev. Neurosci.
       4:69 (1981).

15. E. K. Matthews and Y. Sakamoto, Electrical characteristics of
       pancreatic islet cells, J. Physiol. (London) 246:421
       (1975).

16. P. Meda, R. L. Michaels, P. A. Halban, L. Orci, and J. D.
       Sheridan, In vivo modulation of gap junctions and dye cou-

pling between β-cells of the intact pancreatic islet, _Diabetes_ 32:858 (1983).

17. D. K. Meisheri, O. Hwang, and C. van Breemen, Evidence for two separate Ca^{2+} pathways in smooth muscle plasmalemma, _J. Memb. Biol._ 59:19 (1981).

18. H. P. Meissner and H. Schmelz, Membrane potential of beta-cells in pancreatic islets, _Pflugers Arch._ 351:195 (1974).

PREDICTION OF THE GLUCOSE-INDUCED CHANGES IN MEMBRANE IONIC PERMEA-
BILITY AND CYTOSOLIC Ca^{2+} BY MATHEMATICAL MODELING

J. Rinzel, T. R. Chay, D. Himmel and I. Atwater

LCBG and MRB, NIDDK
National Institutes of Health
Bethesda, MD

Dept. of Biological Sciences
University of Pittsburgh
Pittsburgh, PA

INTRODUCTION

Probably the most interesting property of pancreatic β-cells is
their ability to detect and respond to the concentration of glucose in
the plasma. Most experiments performed on islets are done in search
of understanding this enigmatic process. For example, electrophysiological studies on the β-cell have indicated that the oscillations in membrane potential, which involve repetitive bursts of
action potentials, are a necessary part of the glucose recognition
process.[4,10,28,29,33] This "burst pattern" is generated by the inter-
play between ionic channels in the β-cell membrane.[1,2,3,4,5,7,19,20,22]
One proposal is that the glucose sensitivity is due to changes in the
efflux rate of ionic Ca from the cytosol.[7] Recently, Chay and
Keizer,[14,15] and Chay[12] developed a mathematical model which reproduces the β-cell's unique glucose-sensitive bursting pattern of electrical activity. The model is based upon experimentally determined
membrane ionic permeabilities: voltage-gated Ca-channels,[2] voltage-
gated K-channels,[5] Ca-activated K-channels.[3,7,22] In the model,
glucose recognition is equated to the rate of Ca^{2+} uptake, k_{Ca} in the
β-cell. Numerical simulations of this model are in good agreement
with experimental measurements of the voltage oscillations recorded
from β-cells at different glucose concentrations in steady-state
conditions. The membrane-potential-independent lengthening of the

active phase of the burst pattern observed when increasing the glucose concentration[e10,28] is reproduced in the model by increasing k_{Ca}.

Another unique feature of β-cell behavior is that within an islet, most cells appear to be synchronous, probably due to cell-to-cell coupling.[21] The burst pattern observed in different β-cells within an islet is almost identical, whereas the pattern observed from different islets is enormously variable.[4] We have used the Chay-Keizer model to explain the observed islet-to-islet variabilities in the glucose-induced burst pattern and can thus predict that the size of the bound Ca compartment in the β-cell determines the burst frequency;[8,14] as the ratio of free to bound Ca in the cell increases, so the frequency of bursting increases, independently of the percent duration of the active phase. We have also compared the chaotic burst pattern observed in islets isolated from mice fed a higher lipid diet[26] to the chaotic electrical pattern generated by the model upon alteration of some of the channel gating properties.[16] Such concordance between measured and predicted behavior lends considerable credance to the validity of this model.

Patch-clamp experiments on a variety of β-cell types have uncovered another K-channel in the β-cell membrane.[19] This K-channel has been shown to be inhibited by glucose in whole cell experiments[1,34] and by ATP in isolated patches of membrane.[19,34] Thus, a second proposal is that the glucose sensitivity of the β-cell is due to changes in ATP concentration.[19] Chay has recently adapted the original β-cell model to include this new K-channel, K-Glu, and compared the effects of simulating the glucose sensitivity by varying k_{Ca} with those by varying K conductance, g_{Glu}, while keeping k_{Ca} constant.[13,17,24] Both approaches predict equally well the steady state burst pattern as a function of glucose. However, the two models have some subtle differences in the predicted electrical patterns, differences in the predicted ion fluxes and a marked difference in the predicted average cytosolic free Ca levels, $[Ca^{2+}]_i$. The purpose of this study is to analyze the two models, to illustrate these differences and to compare the experimental records with the theoretical calculations in order to shed some light on the underlying mechanism of glucose sensitivity in the β-cell. It is considered essential to establish the minimal restrictions which would have to be imposed on any theoretical model.

RESULTS

There are a number of examples of excitable cells which show
bursting electrical activity. However, there are several features of
the β-cell burst pattern which make it unique: first, an underlying
slow wave, without action potentials, has not been identified (the
plateau potential appears only with action potentials); second, the
time course of mean membrane potential (averaged over the time scale
of a few action potentials) is not a sinuosoidal like slow wave, but
rather like a square wave; third, the frequency of the burst activity
is not correlated to the absolute duration of the active phase,[4] and;
fourth, the modulation of the active phase duration is independent of

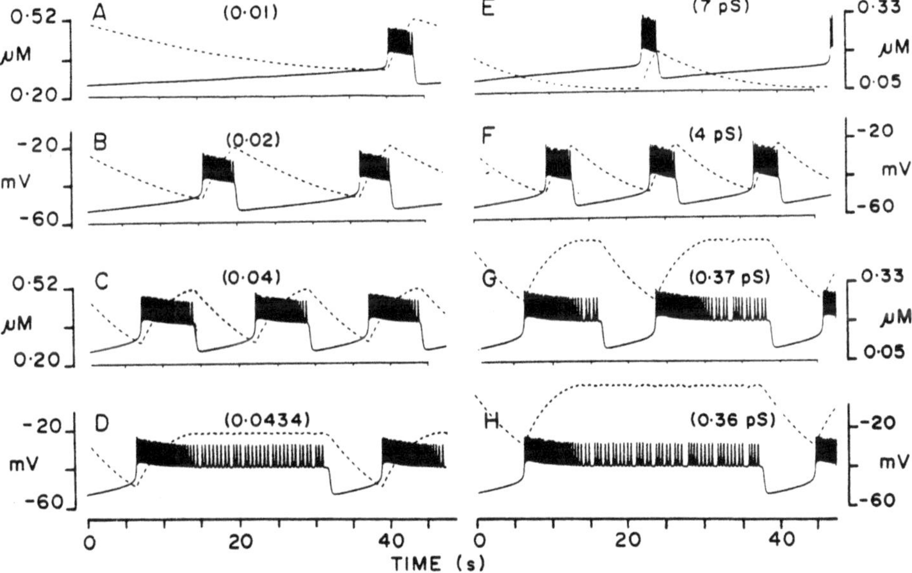

Fig. 1. Burst response for two different hypotheses of glucose-de-
pendence in theoretical model of β-cell electrical activity. Time
courses of membrane potential, V, (solid) and intracellular free
calcium concentration, $[Ca^{2+}]_i$ (dashed) determined by numerical integration of equations in Appendix. Left panels show effect of increasing glucose modelled as increasing calcium removal rate: active phase
duration increases; silent phase duration decreases; $[Ca^{2+}]_i$ range un-
altered. Here, $g_{Glu} = 0$. For right panels, increased glucose is
modelled as decreased g_{Glu}, conductance of ATP-sensitive channel:
active phase duration increases, silent phase duration decreases;
$[Ca^{2+}]_i$ range shifts upward. Here, $k_{Ca} = 0.045$ ms^{-1}.

changes in the mean membrane potential.[6] Both the original Chay-
Keizer model[14] and the later modification[5,12,13,15] reproduce well these
features. Figure 1 illustrates the mathematically predicted bursting
pattern as a function of glucose. The equations, and parameter values, for the model appear in the Appendix. The left panels (A-D) of
Fig. 1 express changes in glucose by varying k_{Ca} (the rate at which Ca
is removed from the cytosol) and the right panels (E-H) express
changes in glucose by varying g_{Glu} (the conductance of the glucose-
sensitive, ATP blockable K-channel). The solid traces represent
predicted membrane potential and the dashed lines represent the pre-

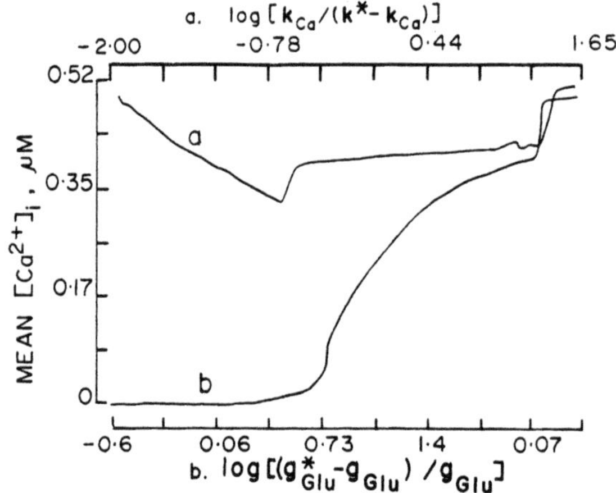

Fig. 2. Effect of glucose on mean (long time average) $[Ca^{2+}]_i$ as modelled
according to two different hypotheses. Curve "a" is (upper horizontal axis) is for $[Ca^{2+}]_i$ removal rate as glucose sensitive parameter (see
Appendix, eqn. (A.5), k^{*}_{Ca} = 0.045 ms^{-1}); curve "b" (lower horizontal
axis) is for g_{Glu} as glucose sensitive parameter (see Appendix, eqn.
(A.7), g^{*}_{Glu} = 0.05 ms).

dicted cytosolic Ca^{2+} concentration, $[Ca^{2+}]_i$. It is immediately
apparent that the bursting pattern is not significantly different as
predicted by both models, yet the underlying changes in $[Ca^{2+}]_i$, while

undergoing similar oscillations in time, are quite different in amplitude and in the mean concentration achieved at the different "glucose" concentrations. (Here, mean $[Ca^{2+}]_i$ is the time average over several bursts.)

The predicted mean $[Ca^{2+}]_i$ as a function of glucose for the two models is given in Fig. 2. Glucose concentration is equated to a function of either k_{Ca} or g_{Glu}. Both models show a fairly sharp transition in $[Ca^{2+}]_i$ when going from the resting state to the beginning of the bursting activity (threshold) and when going from the bursting activity to continuous activity (maximal stimulation). However, there is a dramatic difference in the resting level of predicted by $[Ca^{2+}]_i$ the two models. Even more importantly, there is a very different behavior with respect to glucose over the range of glucose concentrations between threshold and maximum electrical activity, the range over which one expects maximum difference in insulin release.[33] The absolute levels of $[Ca^{2+}]_i$ in the mathematical simulations are somewhat arbitrary as they depend upon an assumed value for the Ca-dependent K-permeability, and upon the $[Ca^{2+}]_i$ removal rate, and these have not been measured in an intact cell.

Since $[Ca^{2+}]_i$ is an integral part of the feed-back system which controls the burst pattern in these models, it is not immediately clear how this pattern can be so similar with such different levels of $[Ca^{2+}]_i$. One way to understand the underlying mechanisms which control the burst pattern is to interpret the models by graphical phase plane analysis.[8] This analysis exploits the slower time scale of $[Ca^{2+}]_i$ variations by first describing the membrane electrical characteristics with $[Ca^{2+}]_i$ treated as a parameter. Figure 3 illustrates each of the steady-state currents as a function of voltage, for fixed $[Ca^{2+}]_i$. (These currents might be called pseudo-steady-state currents since, in practice, $[Ca^{2+}]_i$ varies with changes in voltage). Over the voltage range shown in Fig. 3, the potassium currents are outward and the calcium current is inward; leakage current changes sign since the leakage reversal potential is set to -40 mv. Note, $I_{K,Ca}$, $I_{K,Glu}$, and I_L are linear since the conductances for these currents are not voltage dependent. Only $I_{K,Glu}$ is dependent upon glucose, and only for the second model; its slope decreases with increasing glucose. For the first model, $I_{K,Glu}=0$. In both cases,

the slope of $I_{K,Ca}$ increases with increasing $[Ca^{2+}]_i$. The total
steady-state current, i.e., sum of the individual currents, and its
dependence upon $[Ca^{2+}]_i$ is shown in Fig. 4. As $[Ca^{2+}]_i$ increases, the
current shifts upward as the outward contribution of $I_{K,Ca}$ increases;

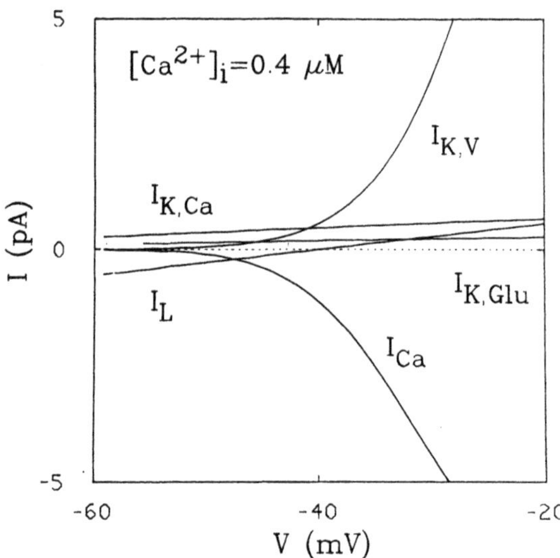

Fig. 3. Steady state current-voltage (I-V) relations for individual
channel types (denoted by subscripts) with $[Ca^{2+}]_i$ fixed at 0.4 µM.
Dotted line indicates zero current. The current $I_{K, Glu}$ is evaluated
here with g_{Glu} = 4 pS. Note, I_L crosses zero at -40 mV, the reversal
potential for "leakage." During silent phase, steady state currents
sum approximately to zero. For $[Ca^{2+}]_i$ = 0.4 µM, silent phase potential equals -51.6 mV.

for the second model (Fig. 4B), $I_{K,Glu}$ contributes additional outward
current which appears approximately as an upward translation of the
current curves. In each case, the current-voltage relation exhibits
three zero-crossings (e.g., open circles in Fig. 4A) for an intermediate range of $[Ca^{2+}]_i$. This range includes the values over which

$[Ca^{2+}]_i$ oscillates during a burst pattern and, for such values, the
left-most zero-crossing corresponds to the slowly increasing, pseudo-
steady-state, potential of the silent phase while the action potentials of the active phase oscillate around the voltage at the right-
most zero-crossing. The middle zero-crossing corresponds to an approximate threshold voltage for transitions between the active and
silent phases.

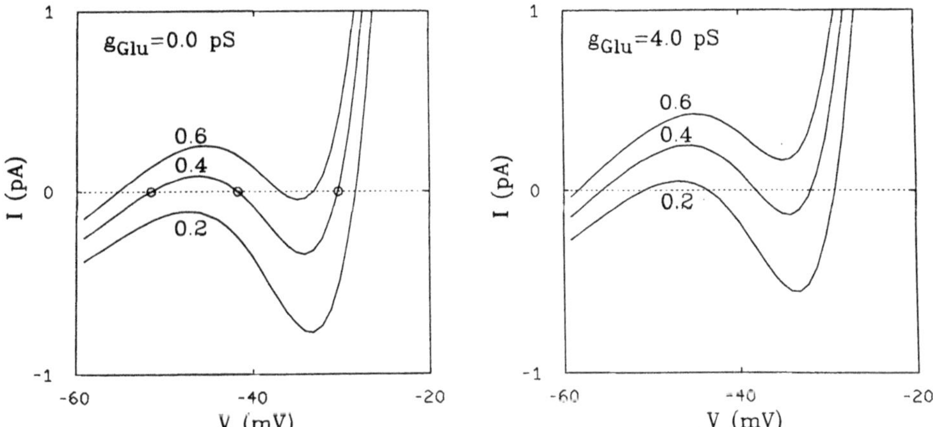

Fig. 4. Total steady state current-voltage relations for different
fixed values of $[Ca^{2+}]_i$ (μM); Left and right panels compare the effects of
g_{Glu}. Note I-V relation exhibits negative resistance in each case. If Ca is
either too small, or too large, the current is either all inward, or all
outward, over the negative resistance regime. For intermediate values
of $[Ca^{2+}]_i$, the I-V relation has three zero-crossings. Left-most zero-
crossing corresponds to silent phase potential; spikes of active phase
oscillate around right-most zero-crossing.

The Z-shaped curve of Fig. 5 summarizes the calcium dependence of
the steady state potentials (zero-crossings) from Fig. 4. It also
illustrates the dynamic properties of $[Ca^{2+}]_i$-handling in the model.
For V and $[Ca^{2+}]_i$ above the long-dashed curve, $[Ca^{2+}]_i$ is increasing,
i.e., d$[Ca^{2+}]_i$/dt > 0, while $[Ca^{2+}]_i$, decreases below the curve. This
curve is called the Ca-nullcline and it is obtained by equating to
zero the right-hand side of eqn. (A.3). The burst pattern, when
plotted as a curve (trajectory) of V vs. $[Ca^{2+}]_i$, is related to this
graphical (phase plane) representation in Fig. 5B. During the active
phase, the trajectory is above the Ca-nullcline so $[Ca^{2+}]_i$ is increas-

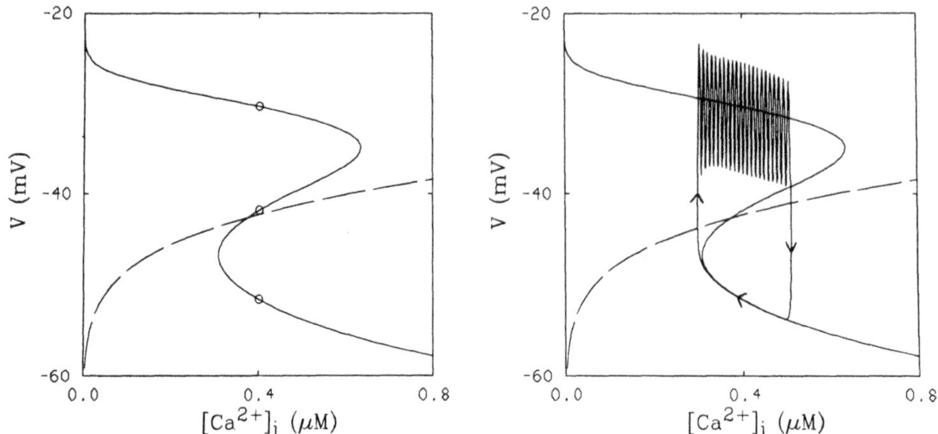

Fig. 5. Left, dependence of steady state potentials upon $[Ca^{2+}]_i$:
Z-shaped solid curve. Here, g_{Glu} = 0. Open circles correspond to
open circles in Fig. 4 left. Long-dashed curve represents Ca-
nullcline, above or below which $[Ca^{2+}]_i$ is increasing or decreasing,
respectively. Right, Burst pattern of Fig. 1B plotted as curve V vs.
$[Ca^{2+}]_i$ and superimposed on left panel. Silent phase is along lower
branch of Z-curve which corresponds to left-most zero-crossing of I-V
relation in Fig. 4 left. Active phase terminates when V falls below
threshold, represented by middle branch of Z-curve.

ing. The active phase ends when the trajectory falls below the
threshold voltage (middle branch of Z-curve). During the silent
phase, the trajectory tracks the low voltage steady state of the
Z-curve; $[Ca^{2+}]_i$ slowly decreases and V slowly increases. The silent
phase ends when V meets the threshold potential, at the Z-curve's left
knee.

Figure 6 illustrates the effect of glucose on the phase plane
characteristics. In Fig. 6A, the calcium removal rate, k_{Ca}, is varied, and g_{Glu}=0. For sub-threshold glucose concentrations (low k_{Ca}),
both $[Ca^{2+}]_i$ and membrane potential are at equilibrium where the
Ca-nullcline intersects the lower branch of the Z-curve; this corresponds to the rest state of the β-cell. For intermediate glucose
concentrations, the Ca-nullcline crosses the Z-curve in the unstable
portion (middle branch) and leads to the repetitive burst cycle (as in
Fig. 5B) and oscillations in membrane potential. At high glucose
concentrations (high k_{Ca}), the Ca-nullcline intersects the Z-curve in

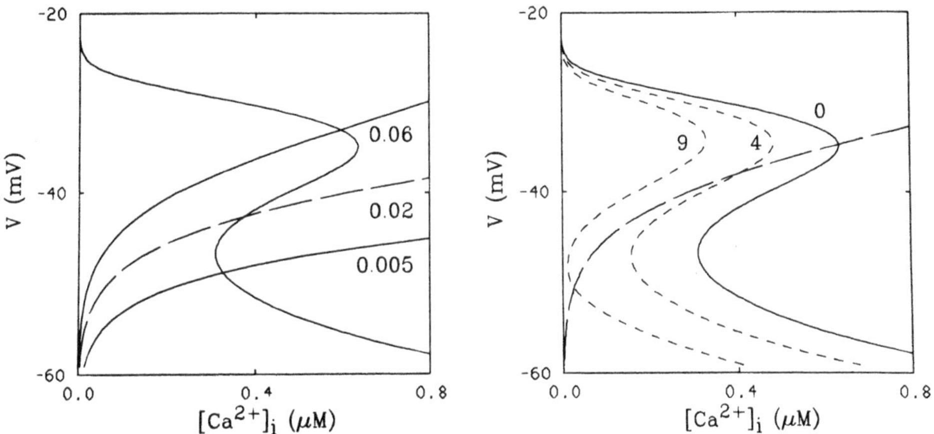

Fig. 6. Effect of glucose upon Z-curve and Ca-nullcline of Fig. 5.
In the left panel, increase of glucose is modelled as increase in
$[Ca^{2+}]_i$ removal rate, k_{Ca}. This affects only $[Ca^{2+}]_i$ handling dynamics; Ca-nullcline moves approximately upward on this scale (value of
k_{Ca} ms^{-1} indicated for each curve). Long dashed curve corresponds to
case of Fig. 5. In the right panel, increased glucose is modelled as
decreased g_{Glu} which effects I-V characteristics of membrane and
therefore alters Z-curve, but not Ca-nullcline. Value of g_{Glu} pS
indicated for each Z-curve. The case of solid curve, and Ca-
nullcline, are as in Fig. 5 left.

the spiking regime, and the membrane potential undergoes the fast
oscillations associated with continuous spiking. The cycle of Fig. 5B
(i.e., silent phase trajectory and envelope of active phase) is followed for all intermediate concentrations of glucose (all values of
k_{Ca}) which induce a burst pattern. The limits of the excursions in
$[Ca^{2+}]_i$ and in membrane potential are the same, only the velocity at
which the various parts of the cycle are passed is different. As k_{Ca}
increases through this range, the Ca-nullcline intersects the Z-curve
at higher voltage, so the accumulation of $[Ca^{2+}]_i$ during the active
phase is slower (i.e., $d[Ca^{2+}]_i/dt$ is less positive) and, consequently, the time spent in the depolarized state is longer. Similarly, during the silent phase, the rate of $[Ca^{2+}]_i$ disappearance from
the cytosol is faster and the time spent at the more negative membrane
potential is shorter.

The illustrations in Fig. 6B depict the phase plane properties at
three glucose concentrations for the adaptation of the model where a
decreasing g_{Glu} is correlated to an increasing glucose concentration

and k_{Ca} is kept constant. Similar to the examples in Fig. 6A, the
Ca-nullcline intersects the Z-curve on the lower branch for low concentrations of glucose (negative stable potential associated with the
resting state), on the central portion for intermediate glucose concentrations (unstable membrane potential, therefore bursting activity)
and well onto the upper portion at high glucose concentrations (spiking regime). Note that the burst patterns for both cases indicate
membrane potential oscillations between the same two potentials (about
-50 mV for the silent phase and about -30 mV for the active phase).
However, the glucose sensitivity in the first case is obtained by
sliding the Ca-nullcline up or down over a fixed Z-curve, whereas for
the second case, it is obtained by sliding the Z-curve left or right
across a fixed Ca-nullcline curve. For this reason, the absolute
range of $[Ca^{2+}]_i$ changes with glucose concentration in the second
case. At the lower glucose concentrations, the increased g_{Glu} is
compensated for by the resultant decrease of $g_{K,Ca}$.

DISCUSSION

The minimal model developed by Chay and Keizer[14] and later ver-
sions[12,13,15,17,18] reproduce the essential features of the β-cell electrical activity. The main purpose of the model was to reproduce the
burst pattern and to clarify the possible mechanisms of β-cell glucose
sensitivity. Because of experimental variability, the subtle differences in bursting behavior as predicted by the two models may not be a
reliable method to experimentally differentiate between them. A major
difference in the predictions from the two hypothesis for glucose
sensitivity is the dependence of mean $[Ca^{2+}]_i$ on glucose. Although
the oscillations in $[Ca^{2+}]_i$ cannot yet be experimentally monitored,
there are indirect evidences supporting the idea that $[Ca^{2+}]_i$ oscillates in phase with the bursts of electrical activity (Perez-
Armendariz and Atwater, this book). Up to now, average measurements of
$[Ca^{2+}]_i$ have been made in cell suspensions using the Ca^{2+} indicator,
quin-2.[11,23,31,32,35,36] However, the oscillatory electrical activity
appears to be a property of the β-cell in aggregates of coupled cells
and in any case, cells in suspension could not be expected to be in
phase. Furthermore, these measurements have mainly compared average
$[Ca^{2+}]_i$ at sub-threshold and supra-maximal concentrations of
glucose,[11,23] and therefore could not distinguish between the two
models. It may prove more informative to compare $[Ca^{2+}]_i$ in β-cells
exposed to two intermediate glucose concentrations (for example, 7 and
17 mM) where the two models predict quite different behavior.

As is apparent from the phase plane analysis, the two hypotheses result in similar qualitative behavior. Both rely on the juxtaposition of fast processes (i.e, Ca-channel gating) with slow processes (i.e., Ca^{2+} buffering) and the internal feedback mechanism between the two (i.e., Ca-activated K-channels). Addition of another voltage insensitive K-channel is merely balanced by a corresponding shift in cytosolic Ca (to reduce existing Ca-activated K currents).

Very recently, Chay[18] and Bangham et al.,[9] have consolidated the Ca-activated K-channel and the voltage-gated K-channel into one and the same, more in keeping with experimental observations.[30] Bangham has found that the shape of the action potentials then becomes more realistic. Chay reports on the sensitivity of the burst pattern, and $[Ca^{2+}]_i$ range, to parameters which describe this channel.

The beauty of the Chay-Keizer model is that it is based on experimentally determined mechanisms. The model can be used to predict variables we are as yet unable to measure, like variations in cytosolic Ca^{2+} during the bursting pattern. The usefulness of the model stems from its simplicity and its ability to suggest definitive (and reasonable) experiments which may help to differentiate between the proposed underlying mechanisms for glucose sensing in the pancreatic β-cell.

ACKNOWLEDGEMENTS

The data used to generate Figs. 1 and 2 constitute a part of the M.S. thesis of D. Himmel. This work was supported, in part, by NSF grant #PCM82 15583 to T. R. Chay.

APPENDIX

This version of the original Chay-Keizer model[14,15] was formulated and investigated by Himmel and Chay.[24] Inward current, carried by Ca^{2+}, is modelled similar to the sodium current in the Hodgkin-Huxley model[25]. Here, the activation and inactivation delays are ignored so that

$$I_{Ca} = g_{Ca}\, m^3 h\, (V - V_{Ca})$$

where $m = m_\infty(V)$, $h = h_\infty(V)$. There are three potassium currents in the model: the delayed rectifier, Hodgkin-Huxley-like current

$$I_{K,V} = g_{K,V}\, n^3\, (V-V_K).$$

in which n is V- and t-dependent (see eqn. (A.2);
the $[Ca^{2+}]_i$-activated, K-current,

$$I_{K,Ca} = g_{K,Ca} \; \frac{[Ca^{2+}]_i/K_{dis}}{1 + [Ca^{2+}]_i/K_{dis}} \; (V-V_K),$$

in which Ca^{2+} binding to the channel is with the apparent dissociation
constant K_{dis}; the current $I_{K,Glu}$ through the glucose-metabolite, ATP
sensitive channel,

$$I_{K,Glu} = g_{Glu}\, (V-V_K)$$

in which g_{Glu} decreases as glucose increases.

The model takes the following form:

$$(A.1) \qquad 4\pi r^2 C_m\, \frac{dV}{dt} = -\, I_{Ca} \;-\; I_{K,V} - I_{K,Ca} - I_{K,Glu} - I_L$$

$$(A.2) \qquad \frac{dn}{dt} = \lambda_n[n_\infty(V)-n]/\tau_n(V)$$

$$(A.3) \qquad \frac{d[Ca^{2+}]_i}{dt} = f\{-\alpha\, I_{Ca} - k_{Ca}[Ca^{2+}]_i]$$

where, the leakage current $I_L = g_L(V-V_L)$, and $1/\alpha = 4\pi r^3 F/3)$, in which
r is cell radius, F is Faraday's constant.

Voltage dependent functions for equations (A.1)-(A.3):

$$j\alpha(V)=\alpha_j(V)/[\alpha_j(V)+\beta_j(V)], \qquad\qquad j=m_\infty,\ h_\infty,\ or\ n_\infty$$
$$\tau_n(V)=1/[\alpha_n(V)+\beta_n(V)].$$
$$\alpha_m(V)=0.1\,(-V-25)/[\exp\{0.1(-V-25)\}-1], \quad \beta_m(V)=4\,\exp\{(-V-50)/18\},$$
$$\alpha_h(V)=0.7\,\exp\{(-V-50)/20\}, \qquad\qquad \beta_h(V)=1/[\exp\{0.1(-V-20)+1\}],$$
$$\alpha_n(V)=0.01(-V-20)/[\exp\{0.1(-V-20)\}-1], \quad \beta_n(V)=0.125\,\exp\{(-V-30)/80\}$$

Parameter values used here are:*

$$C_m = 1 \ [\mu F/cm^2], \quad r = .6 \ [\mu m].$$

$$V_{Ca} = \frac{RT}{2F} \ \ell n \ \frac{[Ca^{2+}]_o}{[Ca^{2+}]_i}, \quad [Ca^{2+}]_o = 2.0$$

$$V_K = \frac{RT}{F} \ \ell n \ \frac{[K^+]_o}{[K^+]_i} , \quad [K^+]_o = 5, \ [K^+]_i = 130$$

$$g_{Ca} = 7.0, \ g_{K,V} = 2.0, \ g_{K,Ca} = 0.06, \ g_L = 0.028,$$

$$g_{K,Glu} = 0.0, \ \text{or variable}, \ K_{Ca} = 0.045 \ [ms^{-1}], \ \text{or variable},$$

$$\lambda_n = 1/3, \ K_{dis} = 2.0 \ [\mu M], \ f = 0.002,$$

$$T = 310°K, \ R = 8.314 \ J/mole/K, \ F = 96487 \ C/mole.$$

*The units for t, V's, g's, and concentrations are ms, mV, nS, mM, respectively.

Glucose dependence of parameters.

A sigmoidal relationship is assumed in each case with the maximal parameter value denoted with superscript*. Glucose concentration is assumed to be scaled by a typical value and represented by the dimensionless variable G. For the case in which calcium removal rate increases with glucose:

$$(A.4) \qquad k_{Ca} = k_{Ca}^* \ \frac{G^n}{1+G^n}$$

which implies

$$(A.5) \qquad n \log G = \log \ [k_{Ca}/(k_{Ca}^* - k_{Ca})].$$

For the case in which g_{Glu} decreases with increasing glucose:

$$(A.6) \qquad g_{Glu} = g^*_{Glu} \; \frac{1}{1+G^n}$$

which implies

$$(A.7) \qquad n \log G = \log[(g^*_{Glu} - g_{Glu})/g_{Glu}].$$

Note, curves in Fig. 2 are plotted according to (A.5) and (A.7) so that no commitment to a value for n or to the scaling constant of glucose concentration is made.

REFERENCES

1. F. M. Ashcroft, D. E. Harrison, and S. J. H. Ashcroft, Glucose induces closure of single postassium channels in isolated rat pancreatic β-cells, Nature 312:446 (1984).

2. I. Atwater, C. M. Dawson, G. T. Eddlestone, and E. Rojas, Voltage noise measurements across the pancreatic β-cell membrane: calcium channel characteristics, J. Physiol. 314:195 (1981).

3. I. Atwater, C. M. Dawson, B. Ribalet, and E. Rojas, Potassium permeability activated by intracellular calcium ion concentration in the pancreatic β-cell, J. Physiol. (London) 288:575 (1979).

4. I. Atwater, C. M. Dawson, A. Scott, G. Eddlestone, and E. Rojas, The nature of the oscillatory behavior in electrical activity for pancreatic β-cells, J.Horm. Metabolic Res. Suppl. 10:101 (1980).

5. I. Atwater, B. Ribalet, and E. Rojas, Mouse pancreatic β-cells: tetraethylammonium blockage of the postassium permeability increase induced by depolarization, J. Physiol. (London) 288:561 (1979).

6. I. Atwater, A. Goncalves, and E. Rojas, Electrophysiological measurement of an oscillating potassium permeability during the glucose-stimulated burst activity in mouse pancreatic β-cell, Biomedical Research 3:645 (1982).

7. I. Atwater, L. Rosario, and E. Rojas, Properties of the Ca-activated K-channel in pancreatic β-cells, Cell Calcium 4:451 (1983).

8. I. Atwater and J. Rinzel, The β-cell bursting pattern and intra-
        cellular calcium, <u>in</u>: "Ionic Channels in Cells and Model
        Systems," R. Latorre, R., Ed., Plenum, New York (1986).

9. J. A. Bangham, P. A. Smith, and P. C. Groghan, Modelling the
        β-cell electrical activity, this book.

10. P. M. Beigelman, B. Ribalet, and I. Atwater, Electrical activity
        of mouse pancreatic β-cells. II. Effects of glucose and
        arginine, <u>J. Physiol. (Paris)</u> 73:201 (1977).

11. A. E. Boyd, III, R. S. Hill, T. Y. Nelson, J. M. Oberwetter, and
        M. Berg, The role of cytosolic calcium in insulin secretion
        from a hamster β-cell line, this book.

12. T. R. Chay, Chaos in a three-variable model of an excitable cell,
        <u>Physica. D.</u> 16:223 (1985).

13. T. R. Chay, Glucose response to bursting-spiking pancreatic
        β-cells by a barrier kinetic model, <u>Biol. Cybern.</u> 52:339
        (1985).

14. T. R. Chay and J. Keizer, Minimal model for membrane oscillations
        in the pancreatic β-cell, <u>Biophys. J.</u> 42:181 (1983).

15. T. R. Chay and J. Keizer, Theory of the effect of extracellular
        potassium on oscillations in the pancreatic β-cell,
        <u>Biophysical J.</u> 48:815 (1985).

16. T. R. Chay and J. Rinzel, Bursting, beating, and chaos in an
        excitable membrane model, <u>Biophysical J.</u> 47:357 (1985).

17. T. R. Chay, Oscillations and chaos in the pancreatic β-cell, <u>in</u>:
        "Biomathematics, Non Linear Oscillations in Biology and
        Chemistry," Lecture Notes in Biomathematics, Springer
        Verlag, NY (1986).

18. T. R. Chay, The role of intracellular calcium on the Ca^{2+} sensi-
        tive K^+-sensitive K^+-channel on the bursting pancreatic
        β-cell, <u>Biophys. J.</u>, in press (1986).

19. D. L. Cook and C. N. Hales, Intracellular ATP directly blocks K^+
        channels in pancreatic β-cells, <u>Nature</u> 311:271 (1984).

20. D. L. Cook, M. Ikeuchi, and W. Y. Fujimoto, Lowering of pH_i
        inhibits Ca^{2+}-activated K^+ channels in pancreatic β-cells,
        <u>Nature</u> 311:269 (1984).

21. G. T. Eddlestone, A. Goncalves, J. A. Bangham, and E. Rojas,
        Electrical coupling between cells in islets of Langerhans
        from mouse, <u>J. Membrane Biol.</u> 77:1 (1984).

22. I. Findlay, M. J. Dunne, and O. H. Petersen, High-conductance K^+
        channel in pancreatic islet cells can be activated and
        inactivated by internal calcium, <u>J. Membrane Biol.</u> 83:169
        (1985).

23. B. Hellman, E. R. Gylfe, and P. Bergsten, Mobilization of different pools of glucose-incorporated calcium in pancreatic
β-cells after muscarinic receptor activation, this book.

24. D. M. Himmel, T. R. Chay, Computer simulations of the electrical
activity of pancreatic β-cells as a function of glucose,
Biophys. J. (submitted).

25. A. Hodgkin and A. F. Huxley, A quantitative description of membrane current and its application to conduction and
excitation in nerve, *J. Physiol. (London)* 117:500 (1952).

26. P. Lebrun and I. Atwater, Chaotic and irregular bursting electrical activity in mouse pancreatic β-cells, *Biophys. J.* 48:529
(1985).

27. P. Meda, I. Atwater, A. Goncalves, A. Bangham, L. Orci, and E.
Rojas, The topography of electrical synchrony among β-cells
in the mouse islet of Langerhans, *Quart. J. Exp. Physiol.*
69:719 (1984).

28. H. P. Meissner and M. Preissler, Glucose-induced changes of the
membrane potential of pancreatic β-cells: their significance
for the regulation of insulin release, *in* "Treatment of
Early Diabetes," R. A. Camerini-Davalos and B. Hanover,
eds., Plenum Press, New York (1979).

29. H. P. Meissner and H. Schmelz, Membrane potential of beta-cells
in pancreatic islets, *Pflugers Arch.* 351:195 (1974).

30. O. H. Petersen and Y. Maruyama, Calcium-activated potassium
channels and their role in secretion, *Nature* 307:693
(1984).

31. M. Prentki and C. B. Wollheim, Cytosolic free Ca^{2+} in insulin
secreting cells and its regulation by isolated organelles,
Experientia 40:1052 (1984).

32. P. Rorsman, H. Abrahamsson, E. Gylfe, and B. O. Hellman, Dual
effects of glucose on the cytosolic Ca^{2+} activity of mouse
pancreatic β-cells, *FEBS Lett.* 170(1): 196 (1984).

33. A. M. Scott, I. Atwater, and E. Rojas, A method for the simultaneous measurement of insulin release and β-cell membrane
potential in single mouse islets of Langerhans, *Diabetologia*
21:470 (1981).

34. Trube and P. Rorsman, Calcium and potassium currents recorded
from pancreatic β-cells under voltage clamp control, this
book.

35. C. B. Wollheim and T. Pozzan, Correlation between cytosolic free
Ca^{2+} and insulin release in an insulin-secreting cell line,
J. Biol. Chem. 259:2262 (1984).

36. G. H. J. Wolters, M. Vank, and A. Pajma, Relationship between
        extracellular Na^+ and the total ionized Ca^{2+} content of rat
        pancreatic islets, this book.

MODELLING THE β-CELL ELECTRICAL ACTIVITY

J.A. Bangham, P.A. Smith and P.C Croghan

Department of Biophysics
School of Biological Sciences
University of East Anglia
Norwich, NR4 7TJ, England

A considerable body of quantitative information is emerging on the
processes involved in stimulus-secretion by the β-cells of the islet of
Langerhans. Many variables are involved and the relationships between them
are complex. However, it is possible to quantify these relationships by
producing mathematical models. Such models are based on simple physico-
chemical principles and provide a convenient quantitative catalogue of
relevant data. However in general they are non-linear and too complex for
intuition alone to give the insight necessary to predict their properties.
The alternative is to solve the equations numerically with the help of a
computer. This approach has contributed to the quantitative description of
for example purkinje fibres[27,32], cardiac muscle[13,3], epithelia[22], bursting
pacemaker neuron[34] and a start has been made with the β-cell[5,25,21,36].

METHODS

The models were written in FORTRAN 77 and executed on a VAX 780
(Digital Equipment Co.) computer system. The differential equations which
form the models were integrated numerically by using the Gear method
(suitable for stiff equations) with automatic step size control (subroutine
DO4EFF, Nag library, Oxford). Random numbers were obtained from subroutine
GO5CAF (Nag).

RESULTS AND DISCUSSION

Extending Existing β-cell Models

Matthews and O'Connor[25] produced an empirical model which allowed some aspects of the electrical bursting behaviour of β-cells to be described, and the model focussed attention onto parameters such as the voltage and calcium dependence of the calcium and potassium permeabilities. A more general approach based on well established principles drawn from nerve and muscle physiology was started by Chay and Keizer[5] (the CK model) and has subsequently been developed[21]. The CK model describes the β-cell action potential using the Hodgkin-Huxley equations with modifications to the parameters, calcium replacing sodium conductances, a calcium activated potassium conductance controlling the bursting and a glucose dependent calcium efflux. Several consequences of varying parameters of the CK model were explored such as the glucose concentration, ionic concentrations, blocking the voltage and calcium sensitive channels and the effect of altering sodium pump activity. It was shown that several aspects of experimentally observed electrical behaviour (cf. 17) could be explained. Further features of the model will be described in this paper before showing how a significant modification can improve the way it describes spikes.

Does the Intracellular Calcium Concentration go up or down in the CK Model?

A feature of the CK model is that electrical activity is controlled by the concentration of calcium inside the cytoplasm. It is supposed that glucose increases the rate at which the calcium is bailed out from the cell, this causes a lowering of the intracellular calcium reducing the potassium conductance and raising the potential. When the calcium channel voltage threshold is reached a train of action potentials (spikes) allow calcium to enter the cell. The calcium influx rate then exceeds the efflux rate and intracellular calcium concentration eventually becomes high enough to activate potassium conductance, to hyperpolarize the membrane and to terminate the burst. Since the potassium permeability is reduced by removing calcium from the cytoplasm a new burst is initiated as soon as the calcium concentration is low enough to bring the potential above the calcium channel threshold again. The result is that on stimulation by glucose the intracellular calcium concentration first goes down and then oscillates. If it is supposed that the initial depolarisation is governed

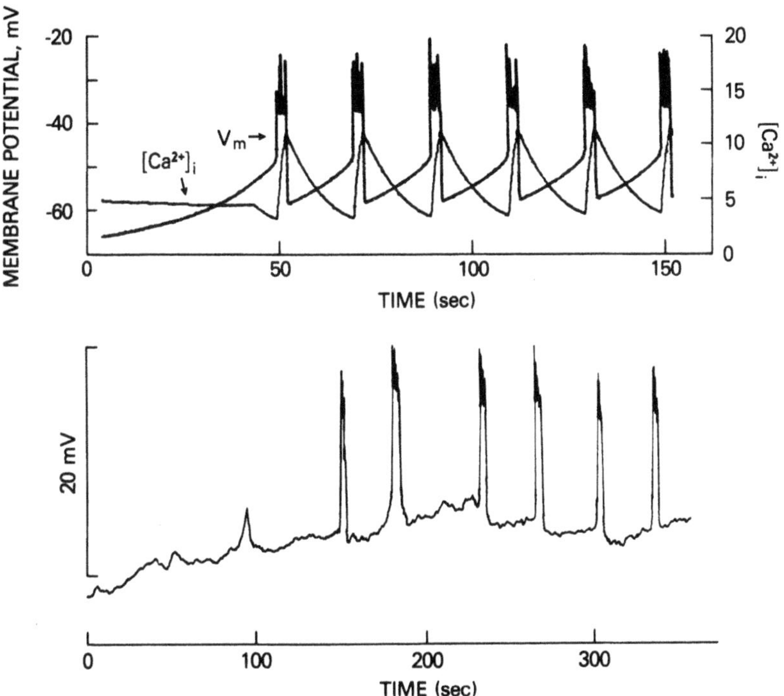

Fig. 1. Shows the effect of stimulating a CK model cell which
incorporates a glucose blocked potassium channel (top panel) and a
real β-cell (bottom panel) with 11.1 mM glucose. In this case the
parameters allowed the average calcium concentration in the cell
to rise, on average, after stimulation. Note that the real record
was chosen for the lack of biphasic before it started bursting in
the model the rate of glucose entry was made artificially high.
The CK model[5] was modified as follows: 1) Equation 9, (30000/2)
replaced value 3. This accounts for two charges on ionized
calcium and adjusts the units. 2) A temperature scaling factor
(TEMP = 4.5) was used to control the α and β parameters. 3) A
glucose blocked potassium conductance of 2×10^{-5} mScm^{-2} was added
to the total potassium conductance.

by ATP (generated from glucose[1], or from some other fuel[24]), blocking a
potassium channel[1,9,15] then the model provides no constraints as to the
initial change in the intracellular calcium concentration. Consequently
the cell calcium could go up, on average, or down during secretion. Fig. 1
shows the result of stimulating a CK model which includes a glucose blocked
potassium channel. In this case, the time-averaged (over several bursts)
intracellular calcium goes up despite the model depending on a glucose
stimulated calcium efflux.

Does the CK Model Account for Intercellular Coupling?

Cells of the islets of Langerhans have been shown to be
coupled[14,28,29,30] and this may be important [33]. Thus any difference in

267

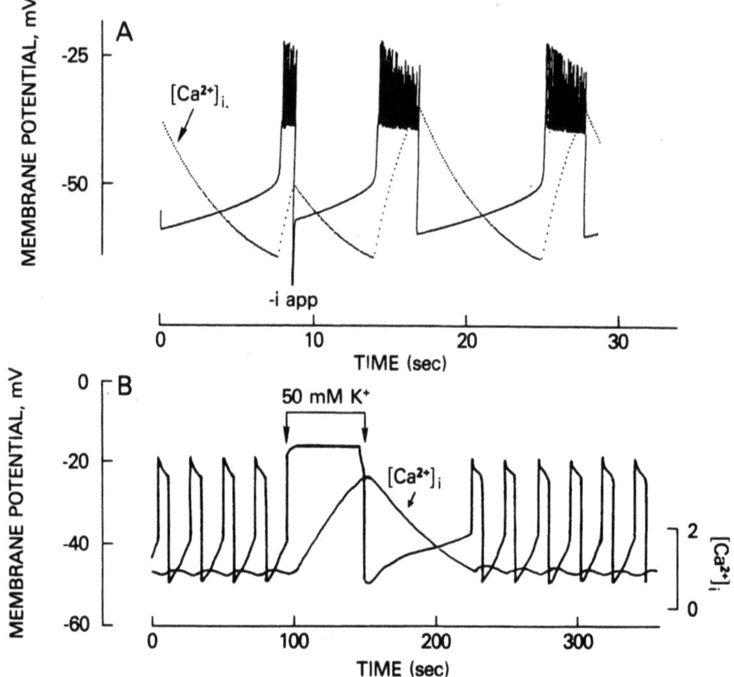

Fig. 2. Top panel A, shows how injecting negative current into a CK model
cell inhibits spiking and that the calcium concentration affects
the phase of the succeeding burst. Bottom panel B, shows that the
CK model responds to a 50 mM KCl depolarisation by allowing
calcium to enter and inhibit the bursting until the smoothed
(averaged) calcium concentration has returned to the normal
stimulated level. The hyperpolarizing current was added to the
total current and the spikes were suppressed by increasing TEMP
(see Fig. 1).

voltage between adjacent cells will result in current flowing between the
cells, this is equivalent to injecting current into a cell. In some
respects the CK model responds to the injection of current in a manner that
is consistent with published observations[2]. So for example, Fig. 2A shows
that spikes can be abolished by injecting negative current (-i app) because
the calcium channels, which initiate the spikes, are voltage sensitive.
The voltage dependent calcium channels also allow the model to emulate
several features of cell-cell coupling. If two of these model cells are
allowed to communicate through gap junctions, where the junction is
modelled as a 7.7×10^9 ohm cm^2 resistor[14], then they interact. The result
is illustrated in Fig. 3. Two separate cells are initially allowed to
burst at different rates. At point 'A' they are coupled together and any
difference in potential between them leads to current passing between them.
As a result, the cells entrain each other to the same frequency.

268

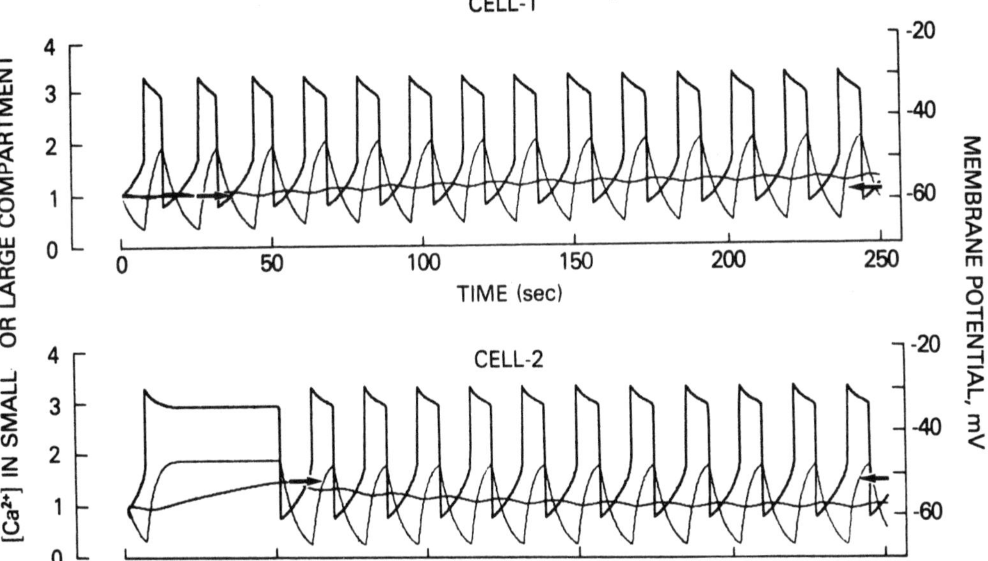

Fig. 3.   Shows the effect of coupling two CK modelled cells together.
Initially the cells are not connected and as cell 2 is made more
sensitive to glucose it is fully stimulated. However when they
are coupled (point A) the two become synchronous and the average
cell calcium in cell 2 (both the unsmoothed and the smoothed
calcium traces are shown) reduces whilst in cell 1 it increases.
The spikes were blocked by increasing TEMP (see Fig. 1) and the
glucose stimulation was varied by altering the calcium efflux rate
constant, k_{Ca}.

Notice that one of the cells was initially fully activated and that
the intracellular calcium level (and perhaps insulin secretion) tends to a
high value. However as soon as the cell becomes coupled, its activity
reduces and its co-respondent increases. Thus electrical coupling might
allow the cells to share the burden of secretion and permit the islet to
behave as a whole with a co-operative glucose sensitivity.

<u>Is Bursting Due to Voltage Dependent Channels?</u>

Injecting hyperpolarizing current into impaled β-cells can block
action potentials[40]. This is consistant with the view that the action
potentials are associated with voltage dependent channels. In contrast,
injecting the current does not appear to alter the burst frequency or
phase; however, as the cell is also likely to be electrically coupled to

adjacent cells, this observation does not rule out the possibility that the
bursts are due to voltage dependent channels. Alternative experimental
approaches (either passing current through the entire islet or briefly
exposing it to a high potassium concentration) produced results which
suggest that the phase of the bursts can be instantly reset[7,8]. These
results were interpreted as being consistant with a model in which the
bursts (in addition to the spikes) are due to voltage-gated channels
(permeabilities). Bursts in the CK model are controlled by changes in the
intracellular calcium concentration. The way it responds to hyperpolarizing
current is illustrated in Fig. 2A. Injecting current (-i app) stops
calcium entry and the next silent phase is truncated because the calcium
concentration at the beginning of the silent phase is lower than with other
cycles.

However the problem should be studied further as other results are
more consistent with there being a concentration term controlling the
bursts. For example the effect of longer potassium pulses has been
studied[11] and these results are strikingly similar to those shown predicted
by the CK model (see Fig. 2B) and their conclusion was that extra calcium
entering during the potassium depolarisation delayed the onset of the next
burst implying that the bursts are not voltage-gated but calcium dependent.
A second line of evidence for some other variable comes from the
observation (unpublished) that bursts can sometimes be seen at a series of
increasing voltages (-60 mV upwards) at low temperatures and when glucose
is added after a 50 mM potassium pre-pulse. As the bursts occur at a
variety of voltages either the bursts are not voltage dependent or that
dependency is modulated by another factor. Thirdly there is evidence that
some additional variable is controlling the bursting pattern from the way
11.1 mM glucose can generate bursts of spikes even when the membrane
potential between spikes is maintained in the range -45 to -40 mV by 8 mM
KCl[2].

As the evidence on voltage/concentration dependence of bursting is not
clear it is possible therefore that future models will have to accommodate
this. Meanwhile the models described in this paper will assume that
bursting is controlled in much the same way as in the CK model. This has
been found to be a satisfactory explanation for bursting in neurons[16].

One can envisage the 'burst-spiking apparatus' as a unit which
responds as a 'calcium input mechanism' to a reduction in intracellular
calcium by allowing a controlled calcium influx (see 12). By being
non-linear and switching between 'hard-on' (spikes) and 'hard-off' (silent
phase) it has digital characteristics. It may be that the local transient
high concentrations of calcium that this generates (compare 39) are
required for insulin release. By modelling a mechanism it is possible to
quantitatively predict the result that would be obtained in a given
experiment and so test the proposed mechanism. For example, it would be
interesting to use models to explore the effects of raising the external
calcium concentration. The CK model responds by reducing the time spent
actually spiking. Electrically this is very similar to what is observed.
The idea that the 'calcium input mechanism' is part of a feed back
mechanism which, once switched on by a depolarization, affects the
intracellular calcium concentration [12] might also explain a number of other
experimental observations. For example, the biphasic electrical response
to a glucose challenge may be due to a similar process. The initial flurry
of spikes which appear after a latency of about a minute, can be reduced if
the cell is preloaded with calcium, so perhaps the flurry is the 'calcium
input mechanism' responding to an initial lowering of intracellular
calcium.

<u>Modelling action potentials</u>. A significant discrepancy between the CK
model and experimental observations lies in the nature of the action
potentials themselves. The CK model uses the Hodgkin-Huxley equations[19]
modified so that the depolarising phase of the spike is due to calcium
instead of sodium permeability changes. The repolarisation is ascribed to
an increase in potassium permeability[35]. The spikes generated by the CK
model differ from experimental observations. The model exhibits smooth,
regular action potntials with a recovery phase which is slower than the
depolarisation phase. Figure 4 (insert) shows that the β-cell spikes can
be irregular, stochastic and with repolarisations that are often faster
than the depolarizations[26]. In the course of a typical burst the spikes
often run from being smooth and fast, through an irregular phase to less
frequent spikes which are again smooth. The parameters in the Hodgkin-
Huxley equations cannot be adjusted to reproduce these observations for if
the potassium permeability could indeed turn on quicker than the calcium
current then the spike would never manage to 'take off'.

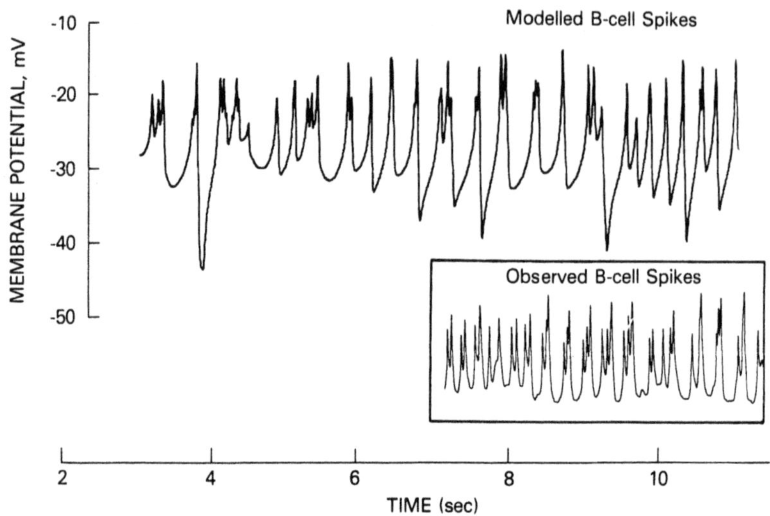

Fig. 4. Illustrates a train of action potentials modelled using a few large potassium channels which are allowed to open and shut at random, but with opening and closing probabilities based on experimental data. The insert shows a typical real cell voltage record. The model is outlined in Fig. 5.

A new starting point for a model of electrical activity. The initial depolarization phase of the action potential is similar to that in the CK model. It is due to calcium channels which are assumed to be so small and numerous that they can be modelled using continuous equations of the same form as those used before[4,38] in neurons (m^2h, where m is voltage and time dependent and h is a calcium dependent). The rapid repolarization is due to a small number of large voltage-gated (and perhaps calcium dependent) potassium channels[20]. Thus the random nature of the spikes would be due to the random potassium channel events. It is proposed that a recovery phase which is more rapid than the "take-off" is due to one or more potassium channel openings which chance to be open simultaneously and contribute sufficiently to the membrane permeability to repolarise the cell.

Figure 4 shows results from a model which is summarised in Figure 5. It incorporates calcium and voltage sensitive calcium channels, and calcium and voltage sensitive potassium channels. Fifteen large potassium channels are assumed to exist and have channel permeabilities (P')[10,20] of 3.4 nms^{-1} and mean open times of 10 ms.

Next steps. As it stands the model exhibits some of the other properties of the complete system. To this end parameters used to obtain

272

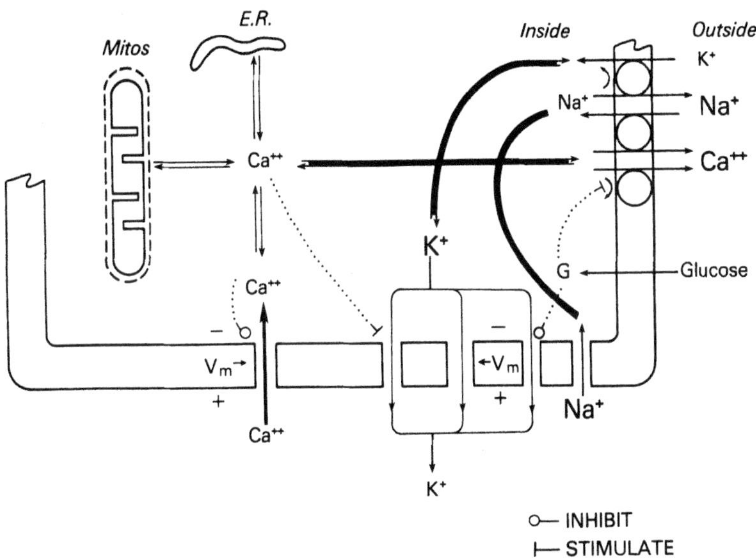

Fig. 5.   Shows the model that was used to generate Figs. 4 and 6 (at present the mitochondrial and endoplasmic reticulum compartments are not included). Ionic currents are calculated using function g (permeabilities, membrane potential, ionic gradients) which is the Goldman-Hodgkin-Katz equation. The total current flowing through the membrane is:

$$i_t = i_{pump} + i_{applied} + i_{Ca} + i_K$$

where

$$i_{pump} = -2/3 \text{ pump Na}_i \qquad (\text{pump} = \text{constant})$$

$i_{applied}$ = external current sources, normally zero

$$i_{Ca} = g[(P_{Ca}m^2h), V_m, Ca_o, Ca_i] \qquad (P_{Ca} = \text{constant})$$

$$i_{Na} = g(P_{Na}, V_m, Na_o, Na_i) \qquad (P_{Na} = \text{constant})$$

$$i_K = g[(P_{K,ATP} + P_{K,Ca} + P_{K,Vm}), V_m, K_o, K_i]$$

$P_{K,ATP}$ is ATP dependent K permeability, $P_{K,Ca}$ is Ca-activated K permeability, $P_{K,Vm}$ is the sum of the large K channels (P') that are open. The probability of opening is voltage dependent and the probablility of closing is constant.

the unstimulated resting potential are based on experimentally determined values for glucose (ATP) sensitive potassium channels (1.5 nm s^{-1}, also see[1,9,15]), calcium sensitive potassium channels (2.5 nm s^{-1}, also see[15]), and sodium currents (measured using tracer flux data, eg[23]). Similarly for the silent phase potential of −55 mV and the plateau potential of −40 to −35 mV. A sodium-potassium pump is included to inject current and depolarise the resting system by about 8 mV.[18,31] In keeping with recent flux measurements described elsewhere in this volume, this does not change with membrane potential. The equations used in the model include the Goldman equation for finding fluxes, Michaelis-Menten to describe

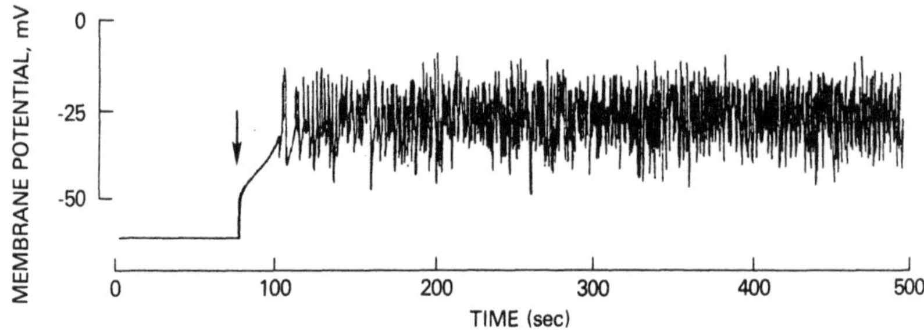

Fig. 6. Illustrates the response of the 'stochastic' model, as it now
stands, to stimulation by glucose (arrowed). The glucose influx
rate was artificially increased to reduce computation time and
this adversely affects the biphasic response.

reactions, stochastic systems for channel openings and diffusion equations
to cover the movement of molecules. The parameters are not yet optimised,
however a typical 'run' of the present model is illustrated in Fig. 6. The
glucose uptake rate constant is set much faster than it should be to reduce
the time taken by the computer to solve the equations and this may be
affecting the initial biphasic nature of the activity. Present
developments involve the inclusion of processes which allow the model to
simulate the way β-cells respond to glucose gradients[37] and generate slow
waves[6]. It is also hoped to couple the electrical events to a description
of the events which lead to insulin release.

It is worth recording that the parameters which are used in models
such as these must be chosen with care. It is found that even changes by
as little as a factor of 1.2 can prevent the model from displaying any
recognisible characteristic. Consequently it is a rigorous taskmaster. It
focuses attention onto each parameter, in turn, insisting that the very best
estimates are obtained by experiment. In return the model allows each
estimate to be tested for its contribution to the whole story. Finally it
is hoped that the model will eventually (years?) be a complete description
of the stimulation-secretion mechanism, providing a comprehensive catalogue
of all the information we have on the system.

ACKNOWLEDGEMENTS

P.A.S. is an SERC research student and the group is supported by the British Diabetic Association and the Wellcome Trust.

REFERENCES

1. F.M. Ashcroft, D.E. Harrison, and S.J.H. Ashcroft, Glucose induces closure of single potassium channels in isolated rat pancreatic β-cells, Nature (Lond.) 312:446 (1984).

2. I. Atwater, B, Ribalet, and E. Rojas, Cyclic changes in potential and resistance of the β-cell membrane induced by glucose in islets of Langerhans from mouse, J. Physiol. 278:117 (1978).

3. G.W. Beeler and H. Reuter, Reconstruction of the action potential of ventricular myocardial fibres, J. Physiol. Lond. 268:177 (1977).

4. R. Eckert and J.E. Chad, Inactivation of Ca channels, Prog. Biophys. molec. Biol. 44:215 (1984).

5. T.R. Chay and J. Keizer, A minimal model for membrane oscillations in the pancreatic β-cell, Biophys. J. 42:181 (1983).

6. D.L. Cooke, Isolated islets of Langerhans have slow oscillations of electrical activity, Metabolism 7:681 (1983).

7. D.L. Cook, W.E. Crill, and D. Porte, Jr., Plateau potentials in pancreatic islet cells are voltage-dependent action potentials, Nature Lond. 286:404 (1980).

8. D.L. Cook, D. Porte, Jr., and W.E. Crill, Voltage dependence of rhythmic plateau potentials of pancreatic islet cells, Am. J. Physiol. 240:E290 (1981).

9. D.L. Cook and C.N. Hales, Intracellular ATP directly blocks K^+ channels in pancreatic β-cells, Nature Lond. 311:271 (1984).

10. C.M. Dawson, P.C. Crogham, A.M. Scott, and J.A. Bangham, Potassium and rubidium permeability and potassium conductance of the β-cell membrane in mouse islets of Langerhans, Quart. J. Expt. Physiol. in press (1985).

11. C.M. Dawson, I. Atwater, and E. Rojas, The response of pancreatic β-cell membrane potential to potassium-induced calcium influx in the presence of glucose, Quart. J. Expt. Physiol. 69:819 (1984).

12. P.M. Dean and E.K. Matthews, Glucose-induced electrical activity in pancreatic islet cells, J. Physiol. 210:255 (1970).

13. D. DiFrancesco and D. Noble, A model of cardiac electrical activity
    incorporating ionic pumps and concentration changes, <u>Phil</u>. <u>Trans</u>.
    <u>R</u>. <u>Soc</u>. <u>Lond</u>. 307:353 (1985).

14. G.T. Eddlestone, A. Goncalves, J.A. Bangham, and E. Rojas, Electrical
    coupling between cells in islets of Langerhans from mouse, <u>J</u>.
    <u>Membr</u>. <u>Biol</u>. 77:1 (1984).

15. I. Findlay, M.J. Dunne, and O.H. Peterson, High-conductance K^+ channel
    in pancreatic islet cells can be activated and inactivated by
    internal calcium, <u>J</u>. <u>Membr</u>. <u>Biol</u>. 83:169 (1985).

16. A.L.F. Gorman, M.W. Thomas, and A. Hermann, Intracellular calcium and
    the control of neuronal pacemaker activity, <u>Fed</u>. <u>Proc</u>. 40:2233
    (1981).

17. J.C. Henquin and H.P. Meissner, Significance of ionic fluxes and
    changes in membrane potential for stimulus-secretion coupling in
    pancreatic β-cells, <u>Experientia</u> 40:1043 (1984).

18. J.C. Henquin and H.P. Meissner, The electrogenic sodium-potassium pump
    of mouse pancreatic β-cells, <u>J</u>. <u>Physiol</u>. 332:529 (1982).

19. A.L. Hodgkin and A.F. Huxley, A quantitative description of membrane
    current and its application to conduction and excitation in nerve,
    <u>J</u>. <u>Physiol</u>. <u>Lond</u>. 117:500 (1952).

20. T.J.C. Jacob, J.A. Bangham, and G. Duncan, Characterisation of a
    cation channel on the apical surface of the frog lens epithelium,
    <u>Quarterly</u> <u>J</u>. <u>Exp</u>. <u>Physiol</u>. 70:403 (1985).

21. Y.S. Lee, T.R. Chay, and T. Ree, On the mechanism of spiking and
    bursting in excitable cells, <u>Biophys</u>. <u>Chem</u>. 18:25 (1983).

22. V.L. Lew, H.G. Ferreira, and T. Moura, The behaviour of transporting
    epithelial cells. I. Computer analysis of a basic model, <u>Proc</u>. <u>R</u>.
    <u>Soc</u>. <u>Lond</u> <u>B</u> 206:53 (1979).

23. W.J. Malaisse, P. Lebrun, and A. Herchuelz, Ionic determinants of
    bioelectrical spiking activity in the pancreatic β-cell, Pfl<u>ügers</u>
    <u>Arch</u>. 395:201 (1982).

24. W.J. Malaisse, F. Malaisse-Lagae, and A. Sener, Coupling factors in
    nutrient-induced insulin release, <u>Experientia</u> 40:1035 (1984).

25. E.K. Matthews, and M.D.L. O'Connor, Dynamic oscillations in the
    membrane potential of pancreatic islet cells, <u>J</u>. <u>Exp</u>. <u>Biol</u>. 81:75
    (1979).

26. E.K. Matthews and Y. Sakamoto, Pancreatic islet cells: electrogenic
    and electrodiffusional control of membrane potential, <u>J</u>. <u>Physiol</u>.
    246:439 (1975).

27. R.E. McAllister, D. Noble, and R.W. Tsien, Reconstruction of the electrical activity of cardiac Purkinje fibres, _J. Physiol. Lond._ 251:1 (1975).

28. P. Meda, I. Atwater, A. Goncalves, A. Bangham, L. Orci, and E. Rojas, The topography of electrical synchrony among β-cells in the mouse islet of Langerhans, _Quart. J. Exp. Physiol._ 69:719 (1984).

29. P. Meda, A. Perrelet, and L. Orci, Increase of gap junctions between pancreatic β-cells during stimulation of insulin secretion, _J. Cell Biol._ 82:433 (1979).

30. H.P. Meissner, Electrophysiological evidence for coupling between β-cells of pancreatic islets, _Nature_ 262:502 (1976).

31. H.P. Meissner and J.C. Henquin, The sodium pump of mouse pancreatic β-cells: electrogenic properties and activation by intracellular sodium, _in_: "Electrogenic Transport: fundamental principles and physiological implications", M.P. Blaustein and M. Lieberman, eds., M. Raven Press, New York (1984).

32. D. Noble and R.W. Tsien, Reconstruction of the repolarization process in cardiac Purkinje fibres based on voltage clamp measurements of the membrane current, _J. Physiol. Lond._ 200:205 (1969).

33. D. Pipeleers, Islet cell interactions with pancreatic β-cells, _Experientia_ 40:1114 (1984).

34. R.E. Plant and M. Kim, Mathematical description of a bursting pacemaker neuron by a modification of the Hodgkin-Huxley equations, _Biophys. J._ 16:227 (1976).

35. B. Ribalet and P.M. Beigelman, Calcium action potentials and potassium permeability activation in pancreatic β-cells, _Am. J. Physiol._ 239:C124 (1980).

36. J. Rinzel, Bursting oscillations in an excitable membrane model, _in_: "Proc. 8th Dundee Conf. on the Theory of Ordinary and Partial Differential Equations", B.D. Sleeman, R.J. Jarvis and D.S. Jones, eds., Springer Press, (in press).

37. A.M. Scott, J.A. Bangham, C.M. Dawson, and P.C. Crogham, Insulin output, ion flux and electrical measurements show that mouse islets distinguish fast from slow changes of glucose concentration, _Diabetologia_ 27:267A (1984).

38. N.B. Stanton and P.R. Stanfield, A binding-site model for calcium channel inactivation that depends on calcium entry, _Proc. Roy. Soc. Lond._ B217:101 (1982).

39. R.S. Zucker and N. Stockbridge, Presynaptic calcium diffusion and the
     time course of transmitter release and synaptic facilitation at the
     squid giant synapse, J. Neurosci. 3:1263 (1983).

40. I. Atwater, A.A. Goncalves and E. Rojas, Electrical Measurement of an
     Oscillating Potassium Permeability during the Glucose-Stimulated
     Burst Activity in Mouse Pancreatic β-cell, Biomed. Res., 3:645
     (1982).

INSULIN SECRETION STUDIED IN ISLETS PERMEABILISED BY HIGH VOLTAGE DISCHARGE

P.M. Jones and S.L. Howell

Department of Physiology
King's College London (KQC)
University of London
Campden Hill Road
Kensington, London W8 7AH
England

Studies of the cellular mechanisms of insulin secretion from intact β-cells are hampered, to some extent, by the existence of an intact plasma membrane which limits the degree to which the intracellular environment can be experimentally manipulated. One means of circumventing this problem is to permeabilise the plasma membrane of the β-cells by a high voltage discharge technique.

Exposure to a high intensity electric field induces the formation of stable pores in the plasma membranes of a variety of cell types[8,10,11,16,27], allowing access to the interior of the cell and direct manipulation of the cytosolic environment. Although chemical methods, such as digitonin treatment, have recently been used to permeabilise insulin-secreting cells[5,20,25], electrical permeabilisation is particularly appropriate when studying membrane associated events such as exocytosis, since the membrane disruption produced by high voltage discharge is confined to the locality of the pores and there is no generalised chemical alteration of the plasma membrane. A further advantage of electrical permeabilisation over chemical methods is that, by careful choice of the electric field employed, pore formation can be limited to the plasma membrane without any danger of inadvertantly damaging intra-cellular organelles[8].

We have therefore used isolated islets of Langerhans permeabilised by
high voltage discharge to study the exocytotic release of insulin. In
particular, we have used the tumour-promoting phorbol ester, 12-0-tetra-
decanoylphorbol 13-acetate (TPA), to activate the phospholipid- and Ca^{2+}-
dependent protein kinase C(PK-C), and studied the relationship between
cytosolic concentrations of Ca^{2+} and activation of PK-C in the regulation
of insulin secretion.

MATERIALS AND METHODS

Islets of Langerhans were isolated from rat pancreata by collagenase
digestion[6], and incubated for 60 minutes at 37°C in a bicarbonate-buffered
physiological salt solution containing 2 mM glucose and 2 mM $CaCl_2$. The
islets were then washed 5 times at 4°C with a glutamate-based Ca^{2+}/EGTA
buffer (permeation buffer), containing 140 mM K^+ glutamate, 15 mM HEPES, 7
mM $MgSO_4$, 5 mM adenosine 5'triphosphate (ATP, Sigma), 5 mM glucose, 1 mM
EGTA, 0.5 mg/ml bovine serum albumin (fraction V, Sigma), pH 6.6, with
$CaCl_2$ added to produce a Ca^{2+} concentration of 10 nM. Ca^{2+} concentrations
in the permeation buffer were calculated using the dissociation constants
of Portzehl et al.[17], and confirmed by direct measurements with a Ca^{2+}
electrode using ionophore 1001 (Fluka), as described elsewhere[12].

Groups of 50-100 islets were re-suspended in permeation buffer (4°C)
of various Ca^{2+} concentrations (10^{-8} M to 10^{-4} M) and permeabilised by 5
exposures (200 μs) to an electric field of 3.4 kV/cm. The permeabilised
islets were washed twice with permeation buffer at 4°C, and groups of 10
islets transferred by micropipette to incubation vials containing 1.0 ml or
permeation buffer of various Ca^{2+} concentrations.

In experiments designed to study the role of protein kinase C on
insulin secretion, TPA was added to the incubation medium dissolved in
dimethyl sulphoxide (DMSO). Controls (no TPA) contained DMSO alone.

Islets were incubated for 30 minutes at 37°C, the incubation vials
centrifuged at 9000 g for 30 seconds, and insulin measured in the
supernatant by radioimmunoassay[27]. Differences between treatments were
assessed by analysis of variance or Student's unpaired t test, as
appropriate.

Incubation of permeabilised islets in buffers containing increasing concentrations of Ca^{2+} (10^{-8} to 10^{-4} M) produced dose-related increases in insulin secretion (Fig. 1). The threshold for stimulation of secretion was approximately 100 nM Ca^{2+}, with maximum secretion being achieved at Ca^{2+} concentrations of about 10 μM.

The Ca^{2+}-induced release of insulin from permeabilized islets was temperature dependent, with a maximum secretory response at 37°C (Fig. 2). Incubation at 22°C produced a much reduced, although still significant, response to Ca^{2+}, whilst the Ca^{2+}-induced insulin secretion was totally abolished at 4°C. Insulin secretion was also dependent on the presence of

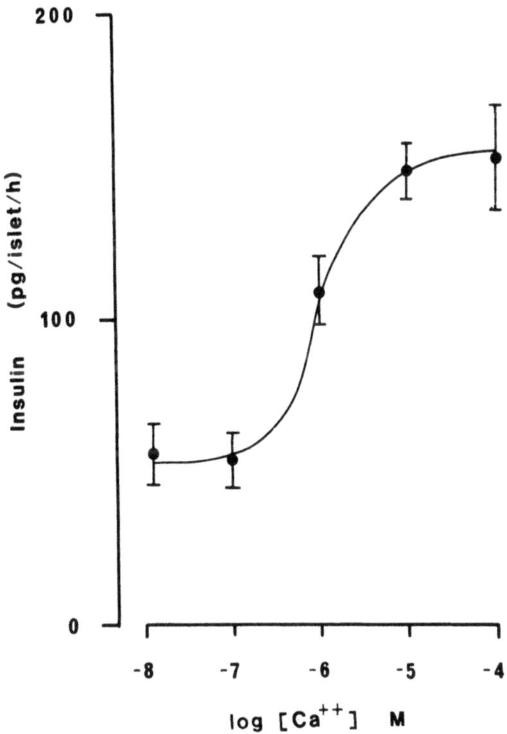

Fig. 1. Ca^{2+}-dependent exocytosis in permeabilised rat islets. Increasing concentrations of Ca^{2+} produced dose-related increases in insulin secretion by electrically permeabilised islets Points show means ± SEM for 8 or 9 observations.

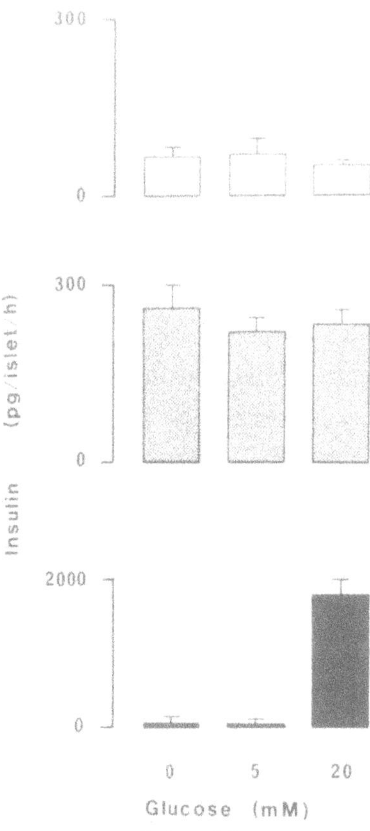

Fig. 2.   Temperature dependence of insulin secretion by permeabilised
islets.   Permeabilised islets were incubated in buffers containing
10 nM (open bars) or 10 μM Ca^{2+} (stippled bars) at 4, 22 or 37°C.
Islets incubated at 4°C showed no secretory response to Ca^{2+}.
Incubation at 22°C or 37°C produced significant Ca^{2+}-dependent
secretion of insulin (p < 0.05 and < 0.01, respectively).   Bars
show mean ± SEM, n=4.

MgATP.   Omitting ATP from the permeation buffer caused a reduction of
78 ± 13% (± SEM, n=4) in the response to 10 μM Ca^{2+}.

The secretion of insulin by the permeabilised islets in response to
elevated Ca^{2+} was not linear with time.   Figure 3 shows the results of an
experiment in which insulin release was measured after different incubation
periods, demonstrating that the secretory response is maximal in the first
15-30 minutes of the Ca^{2+} challenge.

Permeabilised islets did not respond to increases in glucose concentration (Fig. 4).   Neither the basal insulin secretion in 10 nM Ca^{2+} (upper

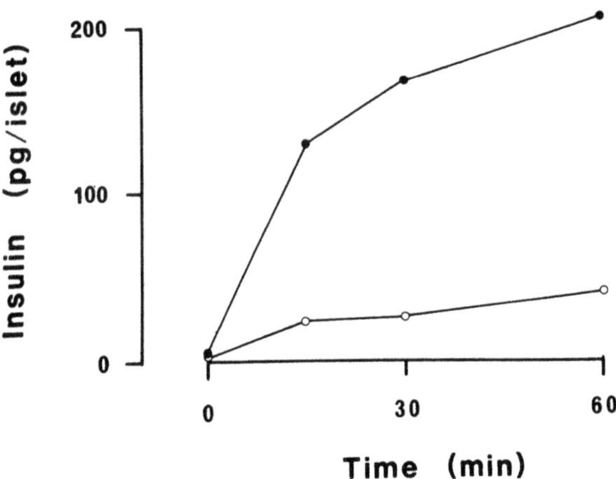

Fig. 3. Time-course of Ca^{2+}-induced insulin secretion. Permeabilised islets were incubated for various times in buffers containing 10 nM (o) or 10 μM (●) Ca^{2+}. The secretory response to 10 μM Ca^{2+} was greatest in the first 15-30 minutes.

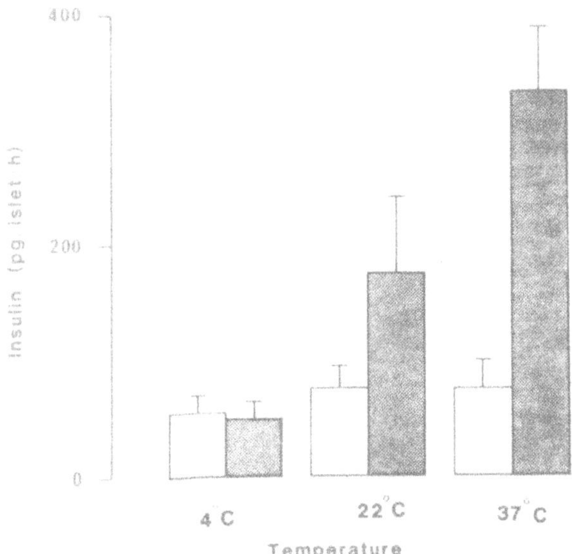

Fig. 4. Effects of glucose on insulin secretion by permeabilised islets. The presence of glucose (0.2 or 20 mM) did not affect insulin secretion by permeabilised islets incubated in 10 nM Ca^{2+} (open bars, top panel) or 10 μM Ca^{2+} (stippled bars, middle panel). The bottom panel (solid bars) shows glucose-induced insulin secretion by non-permeabilised islets incubated in a physiological salt solution containing 1 mM $CaCl_2$. Note that the maximum Ca^{2+}-stimulated secretion by permeabilised islets (10 μM Ca^{2+}, middle panel) is only about 15% of the glucose-induced secretion in intact islets.

panel), nor the stimulated secretion in 10 μM Ca^{2+} (middle panel) were
affected by the presence of 0, 5, or 20 mM glucose in the incubation
medium. Note also that the maximum secretory response of the permeabilised
islets to Ca^{2+} (10 μM, middle panel), is considerably less that that of
non-permeabilised islets to 20 mM glucose (Fig. 4, lower panel).

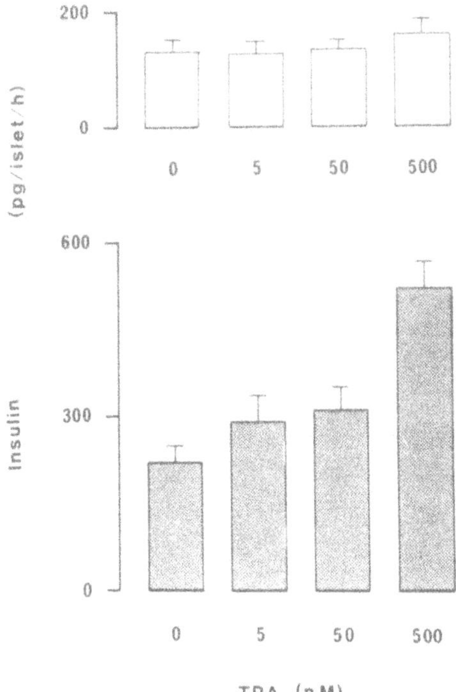

Fig. 5. Effects of TPA on insulin secretion by permeabilised islets. In
the absence of Ca^{2+}, TPA had no effect on insulin release (upper
panel, open bars). However, in the presence of 1 μM Ca^{2+}, TPA
produced a dose-related stimulation of insulin secretion (lower
panel, stippled bars). Bars show mean ± SEM, n=5.

Addition of the phorbol ester TPA, to the incubation medium produced a
dose-related stimulation of the Ca^{2+}-induced insulin secretion by the
permeabilised islets (Fig. 5, lower panel). However, in the absence of
Ca^{2+} (and the presence of 1 mM EGTA) TPA had no significant effects on
insulin secretion (Fig. 5, upper panel).

The stimulatory effect of TPA on insulin secretion in the presence of
Ca^{2+} appeared to be a sensitisation of the exocytotic mechanism to Ca^{2+},

promoting a greater secretory response to lower concentrations of Ca^{2+}
(Table 1). TPA also promoted insulin secretion above the maximum Ca^{2+}-
activated response (Fig. 6). At concentrations of 50 nM, 500 nM and 5 µM
TPA produced increases of 14 ± 5%, 82 ± 20%, and 124 ± 16%, respectively,
over the insulin release evoked by 10 µM Ca^{2+} alone.

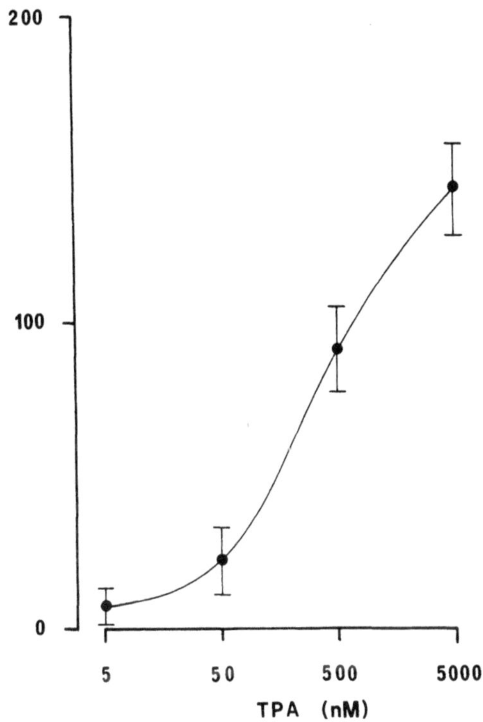

Fig. 6. Stimulation of the maximum Ca^{2+}-induced exocytosis by TPA.
Permeabilised islets were incubated in buffers containing 10 µM
Ca^{2+} and increasing concentrations of TPA. The phorbol ester
stimulated the maximum Ca^{2+}-induced exocytosis of insulin from
permeabilised islets. Mean ± SEM, n=5 or 6.

The effects of TPA on insulin release from permeabilised islets were
temperature and ATP dependent. The stimulatory effects of TPA at both 10
nM and 10 µM Ca^{2+} were totally abolished by incubation at 4°C, while
omission of ATP from the permeation buffer caused a reduction of 64 ± 7% (±
SEM, n=6) in the response to 0.5 µM TPA in the presence of 10 µM Ca^{2+}.

Table 1. Insulin secretion (pg/islet/h, mean ± SEM, n=5) by permeabilised islets incubated in buffers containing various concentrations of Ca^{2+} and TPA.

		Ca^{2+} (M)		
TPA (M)	0	10^{-8}	10^{-7}	10^{-5}
0	128 ± 22	112 ± 16	198 ± 40	396 ± 44
5×10^{-8}	131 ± 23	142 ± 24	172 ± 38	430 ± 28
5×10^{-7}	127 ± 19	270 ± 26	442 ± 86	722 ± 88
5×10^{-6}		364 ± 38	456 ± 70	886 ± 72

DISCUSSION

There is considerable evidence that elevated cytosolic Ca^{2+} is an important initiator of insulin secretion[24], and in the present studies increasing the concentration of free Ca^{2+} produced dose-related increases in insulin secretion from electrically-permeabilised islets.

The threshold of about 100 nM Ca^{2+} for stimulation of insulin secretion from the permeabilised islets is in good agreement with estimates of basal cytosolic Ca^{2+} in unstimulated insulin-secreting cells, whether measured by a fluorescent probe[18,23], or by a Ca^{2+} electrode in suspensions of digitonin permeabilised cells[2]. The maximum secretory response of electrically permeabilised islets to Ca^{2+} were seen at concentrations of about 10 μM, somewhat higher than those measured in simulated intact tumour cells, but similar to those giving secretory response in digitonin permeabilised islets[5,20].

The Ca^{2+}-induced insulin release from the permeabilised islets was temperature dependent, as might be expected of an exocytotic event, and also demonstrated an absolute requirement for MgATP, in accordance with similar findings in a number of other types of permeabilised cells[8,10,11].

In the present study, the permeabilised islets no longer showed secretory response to glucose. Since the pores produced by the permeabilisation procedure are of about 2 nm in diameter[8], many small soluble cytosolic components will be lost, and the lack of responsiveness to glucose could simply reflect an inability of the permeabilised cells to metabolise the substrate. The maximum Ca^{2+}-induced insulin secretion by the permeabilised islets was considerably less than glucose-induced secretion by non-permeabilised islets. This was not due to differences in the insulin content of permeabilised and non-permeabilised islets, but may reflect differences in the duration of the secretory response of permeabilised and intact islets. The kinetics of Ca^{2+}-induced insulin secretion from permeabilised islets requires further study.

There is a growing body of evidence to suggest that insulin secretion may be regulated in part by the hydrolysis of the plasma membrane phospho-lipid, phosphatidylinositol 4,5-bisphosphate, to produce myo-inositol-1,4,5 trisphosphate (IP3) and 1,2 diacylglycerol (DAG)[1,14]. Since IP3 can mobilise intracellular Ca^{2+} in islets[25] and insulin secreting tumour cells[2], it may be directly involved in the regulation of insulin secretion by regulating cytosolic Ca^{2+}. DAG may also play an important role in the regulation of insulin secretion by activating protein kinase C, an enzyme which has been identified in islets[21] and insulin-secreting cell lines[7]. The tumour-promoting phorbol esters, such as TPA, which mimic the DAG activation of PK-C[4] are potent insulin secretagogues[7,19,22,26], suggesting a physiological role for PK-C in insulin secretion.

In electrically permeabilised islets, TPA produced significant increases in Ca^{2+}-dependent insulin secretion. Since the experiments were performed in permeabilised islets, it is unlikely that the responses to TPA were due to its reported effects on ion fluxes across the β-cell plasma membrane[13,15]. While high concentrations of TPA may cause non-specific disruption of membranes[15], the temperature sensitivity and ATP-dependence of the exocytotic response to TPA in the present experiments suggest more than a simple solvent effect on the plasma or secretory granule membranes. Furthermore, in the absence of Ca^{2+}, TPA had no stimulatory effect on insulin secretion, as might be expected if the response was due to activation of a Ca^{2+}-dependent kinase.

One effect of TPA on insulin secretion was a shifting of the Ca^{2+}-activation curve to the left, thus inducing exocytosis at lower levels of intracellular Ca^{2+}. Similar findings have been reported in a number of other secretory tissues[3,9,10,11], although the effective doses of TPA varies between tissues. The concentrations of TPA which stimulate secretion from electrically permeabilised islets are similar to those which enhance insulin release from digitonin treated islets[20] and promote secretion[7,19,22,26] and affect the membrane potential[15] in intact islets. In addition to sensitising the secretory mechanism to Ca^{2+}, TPA also produced considerable increases in the maximum secretory response to Ca^{2+}. A similar, though somewhat smaller effect has been reported in permeabilised pancreatic acinar cells[10], but TPA does not increase the maximum Ca^{2+}-dependent exocytosis in either permeabilised chromaffin cells[9] or platelets[11], perhaps indicating that PK-C activation is quantitatively more important in regulating secretion from β-cells than from some other tissues.

These studies in permeabilised islets suggest that PK-C may play an important role in modulating the magnitude of the secretory response of β-cells, and raise the intriguing possibility that the physiological activation of PK-C by DAG could stimulate insulin secretion without the need for elevations in cytosolic Ca^{2+} above the reported basal levels of 100-200 nM[2,18,23]. Whether a physiological release of insulin can occur in intact islets without any increase in cytosolic Ca^{2+} remains to be seen.

In conclusion, high voltage discharge is a rapid and simple method of permeabilising isolated islets of Langerhans, and these permeabilised islets offer a useful model system in which to study the relationships between cytosolic Ca^{2+} and other intracellular control mechanisms in the regulation of insulin secretion.

ACKNOWLEDGEMENTS

Financial assistance from the Medical Research Council and the British Diabetic Association is gratefully acknowledged.

REFERENCES

1. L. Best, M. Dunlop, and W.J. Malaisse, Phospholipid metabolism in pancreatic islets, _Experientia_ 40:1085 (1984).

2. T.J. Biden, M. Prentki, R.F. Irvine, M.J. Berridge, and C.B. Wollheim, Inositol 1,4,5-trisphosphate mobilizes intracellular Ca^{2+} from permeabilised insulin-secreting cells, _Biochem. J._ 223:467 (1984).

3. K.W. Brocklehurst and H.B. Pollard, Enhancement of Ca^{2+}-induced catecholamine release by the phorbol ester TPA in digitonin-permeabilised cultured bovine adrenal chromaffin cells, _FEBS Lett._ 183:107 (1985).

4. M. Castanaga, Y. Takai, K. Kaibuchi, K. Sano, A. Kikkawa, and Y. Nishizuka, Direct activation of calcium-activated phospholipid-dependent protein kinase by tumor-promoting phorbol esters, _J. Biol. Chem._ 257:7847 (1982).

5. J.R. Colca, B.A. Wolf, P.G. Comens, and M.L. McDaniel, Protein phosphorylation in permeabilised pancreatic islet cells, _Biochem. J._ 228:529 (1985).

6. S.L. Howell and K.W. Taylor, K^{+} ions and the secretion of insulin by islets of Langerhans incubated in vitro, _Biochem. J._ 108:17 (1968).

7. J.C. Hutton, M. Peshavaria, and K.W. Brocklehurst, Phorbol ester stimulation of insulin release and secretory granule protein phosphorylation in a transplantable rat insulinoma, _Biochem. J._ 224:483 (1984).

8. D.E. Knight and P.F. Baker, Calcium dependence of catecholamine release from bovine adrenal medullary cells after exposure to intense electric fields, _J. Membr. Biol._ 68:107 (1982).

9. D.E. Knight and P.F. Baker, The phorbol ester TPA increases the affinity of exocytosis for calcium in leaky adrenal medullary cells, _FEBS Lett._ 160:98 (1983).

10. D.E. Knight and E. Koh, Ca^{++} and cyclic nucleotide dependence of amylase release from isolated rat pancreatic acinar cells rendered permeable by intense electric field, _Cell Calcium_ 5:401 (1984).

11. D.E. Knight and M.C. Scrutton, Cyclic nucleotides control a system which regulates Ca^{2+} sensitivity of platelet secretion, _Nature_ 309:66 (1984).

12. F. Lanter, R.A. Steiner, D. Ammann, and W. Simon, Critical evaluation of the applicability of neutral carrier Ca^{2+}-selective microelectrodes, _Anal. Chim. Acta_ 135:51 (1982).

13. W.J. Malaisse, P. Lebrun, A. Herchuelz, A. Sener, and A. Malaisse-Lagae, Synergistic effect of a tumor-promoting phorbol ester and a hypoglycemic sulfonylurea upon insulin release, Endocrinology 113:1870 (1983).

14. W. Montague, N.G. Morgan, G.M. Rumford, and C.A. Prince, Effect of glucose on polyphosphoinositide metabolism in isolated rat islets of Langerhans, Biochem. J. 227:483 (1985).

15. C.E. Pace and K.T. Goldsmith, Action of a phorbol ester on β-cells: potentiation of stimulant-induced electrical activity, Am. J. Physiol. 248:C527 (1985).

16. C.E. Pace, J.T. Tarvin, A.S. Neighbors, J.A. Pirkle, and M.H. Greider, Use of a high voltage technique to determine the molecular requirements for exocytosis in islet cells, Diabetes 29:911 (1980).

17. H. Portzehl, P.C. Caldwell, and J.C. Ruegg, The dependence of contraction and relaxation of muscle fibres from the crab Maia squindo on the internal concentration of free calcium ions, Biochim. Biophys Acta 70:581 (1964).

18. P. Rorsman, H. Abrahamsson, E. Gylfe, and B. Hellman, Dual effects of glucose on the cytosolic Ca^{2+} activity of mouse pancreatic β-cells, FEBS Lett. 170:196 (1984).

19. J. Stutchfield, F. Sullivan, K.C. Pedley, and S.L. Howell, Effects of polymyxin B, a protein kinase C inhibitor, on insulin release, Diabetes Res. Clin. Practice Supp. 1:S539 (1985).

20. T. Tamagawa, H. Niki, and A. Niki, Insulin release independent of a rise of cytosolic free Ca^{2+} by forskolin and phorbol ester, FEBS Lett. 183:430 (1985).

21. K. Tanigawa, H. Kuzuya, H. Imura, H. Taniguchi, S. Baba, Y. Takai, and Y. Nishizuka, Calcium-activated phospholipid-dependent protein kinase in rat pancreas islets of Langerhans, FEBS Lett. 138:183 (1982).

22. M.A.G. Virji, M.W. Steffes, and R.D. Estensen, Phorbol myristate acetate: effects of a tumor promoter on insulin release from isolated rat islets of Langerhans, Endocrinology 102:706 (1978).

23. C.B. Wollheim and T. Pozzan, Correlation between cytosolic free Ca^{2+} and insulin release in an insulin-secreting cell line, J. Biol. Chem. 259:2262 (1984).

24. C.B. Wollheim and G.W.G. Sharp, Regulation of insulin release by calcium, Physiol. Rev. 61:914 (1981).

25. B.A. Wolf, P.G. Comens, K.E. Ackerman, W.R. Sherman, and M.L.
     McDaniel, The digitonin-permeabilised pancreatic islet model.
     Biochem. J. 227:965 (1985).

26. S. Yamamoto, T. Nakadate, T. Nakaki, K. Ishii,and R. Kato, Tumor
     promoter 12-o-tetradecanoylphorbol-13-acetate induced insulin
     secretion, Biochem. Biophys. Res. Comml 105:759 (1982).

27. M.A. Yaseen, K.C. Pedley, and S.L. Howell, Regulation of insulin
     secretion from islets of Langerhans rendered permeable by electric
     discharge, Biochem. J. 206:81 (1982).

REGULATION OF INSULIN RELEASE INDEPENDENT OF CHANGES OF CYTOSOLIC Ca^{2+}
CONCENTRATION

T. Tamagawa, H. Niki, A. Niki and I. Niki

Department of Internal Medicine
Aichigakuin University
School of Dentistry
Chikusa-ku, Nagoya, 464
 Third Department of Internal Medicine
University of Nagoya
School of Medicine
Showa-ku, Nagoya 466
Japan

Although the central role of cytosolic Ca^{2+} in insulin release has been widely accepted mainly based on the $^{45}Ca^{2+}$ flux studies[20], it is not conclusive because of inability to measure changes of cytosolic Ca^{2+} concentration ($[Ca^{2+}]_i$) directly in the pancreatic β-cell. Recently, a method to estimate $[Ca^{2+}]_i$ using a fluorescent Ca^{2+} indicator, quin 2, has been developed and applied to the pancreatic β-cell[5,14,21,22]. Some of the findings by this method do not accord with those observed in the $^{45}Ca^{2+}$ flux studies[22]. Since quin 2 chelates calcium[12] and may affect insulin release[22], the data obtained using quin 2 should be interpreted with caution. Another approach to analyze the relationship between $[Ca^{2+}]_i$ in the pancreatic β-cell and insulin release is to clamp $[Ca^{2+}]_i$ at arbitrary levels. For this purpose, three methods have been developed in other cells[2]; i) microinjection of Ca or a suitable Ca-buffer into a cell, ii) intracellular dialysis or perfusion, and iii) permeabilization of the plasma membrane to permit extracellular Ca or Ca-EGTA buffers easy access to the intracellular environment. The first two methods are not suitable for small cells such as the pancreatic β-cell. The last method has been applied to pancreatic islets using high voltage electric discharge[11,23] or

digitonin treatment[4,16]. We describe here our findings obtained using
digitonin to permeabilize the cells in the islet.

MATERIALS AND METHODS

Islets were isolated from pancreases of fed male Wistar rats by
collagenase digestion. Details of the treatment of the islets with
digitonin have been published[16]. Briefly, the isolated islets were treated
with 20 µM digitonin for 5 min and washed 3 times with the basal medium
described below. For static incubations, groups of 5 islets were incubated
at 37°C for 30 min in the media containing test substances as indicated.
For dynamic studies, about 200 islets were placed in the perifusion
chambers (kindly provided by Dr. J.-C. Henquin, University of Louvain,
Belgium) and perifused with the media at a constant rate of 1 ml/min as
previously reported[15]. The dead time of the system is 2 min, and
correction was made in the presentation of the results. The composition of
the basal medium for digitonin treatment and static incubation or
perifusion of the treated islets (mM): K-glutamate, 100; Na-glutamate, 42;
Mg-ATP, 1; Hepes, 16; glucose, 3; EGTA, 1; and bovine serum albumin (5
mg/ml). $CaCl_2$ was added appropriately and the pH was adjusted to 7.0 with
NaOH. Insulin release in the media was determined by radioimmunoassay
using rat insulin as standard. Statistical significance was assessed by
Student's _t_-test for unpaired data.

The abbreviations used are: Hepes, N-2-hydroxyethylpiperazine-N'-2-
ethanesulfonic acid; EGTA, ethylenglycol-bis-(β-aminoethyl ether)-N,N'-
tetraacetic acid; TPA, 12-0-tetradecanoyl-phorbol-13-acetate; IBMX,
3-isobutyl-1-methylxanthine.

RESULTS AND DISCUSSION

Digitonin is known to permeabilize cholesterol-rich plasma membranes,
but not the membranes of intracellular organelles[7]. The islets thus
permeabilized with digitonin were incubated with Ca-EGTA buffers to fix
concentrations of free Ca^{2+} ($[Ca^{2+}]$) to arbitrary levels. Since the total
calcium content of an islet was reported to be 4.7 pmol on average[20], even
the maximal contribution of stored calcium in intracellular organelles to

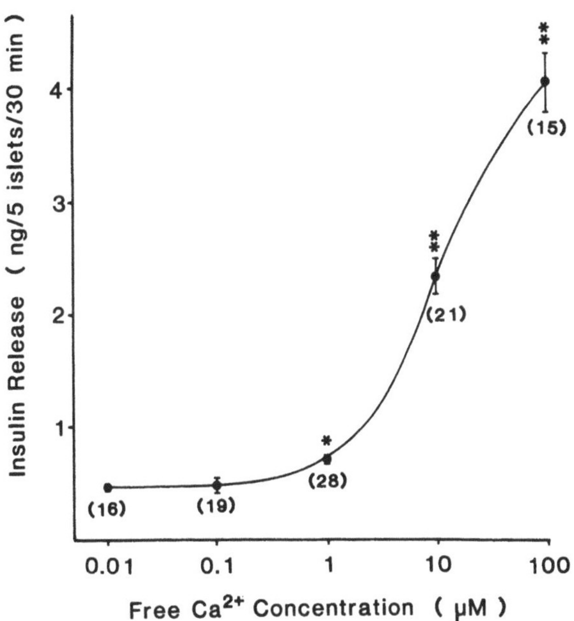

Fig. 1. Dependence of insulin release from digitonin-treated pancreatic islet upon Ca^{2+}.[16] Isolated islets were treated with digitonin as described in the text. Groups of 5 treated islets were incubated for 30 min at 37°C in the media containing indicated concentrations of Ca^{2+}. Values are means ± SE (represented by bars), for the numbers of experiments shown in parentheses. Significance levels: *p <0.01, **p <0.001, vs the value at 0.01 µM Ca^{2+}.

$[Ca^{2+}]_i$ should be nullified with the amount of EGTA (1 mM) used in the buffer.

Ca^{2+} Modulation of Insulin Release from Digitonin-treated Pancreatic Islets

At 10 and 100 nM Ca^{2+}, insulin level in the media remained basal, indicating that the digitonin-treated islets do not leak insulin. Significant insulin release was observed at 1 µM Ca^{2+}, and increased depending on $[Ca^{2+}]$ up to 100 µM (Fig. 1). The value of insulin release induced by 10 µM Ca^{2+} from digitonin-treated islets is about 70% of that induced by 10 mM glucose from intact sham-treated islets in Krebs Ringer Bicarbonate buffer. (data unshown). These data indicate that insulin release can be evoked only by a rise in $[Ca^{2+}]_i$ without a concomitant presence of any other initiator[16].

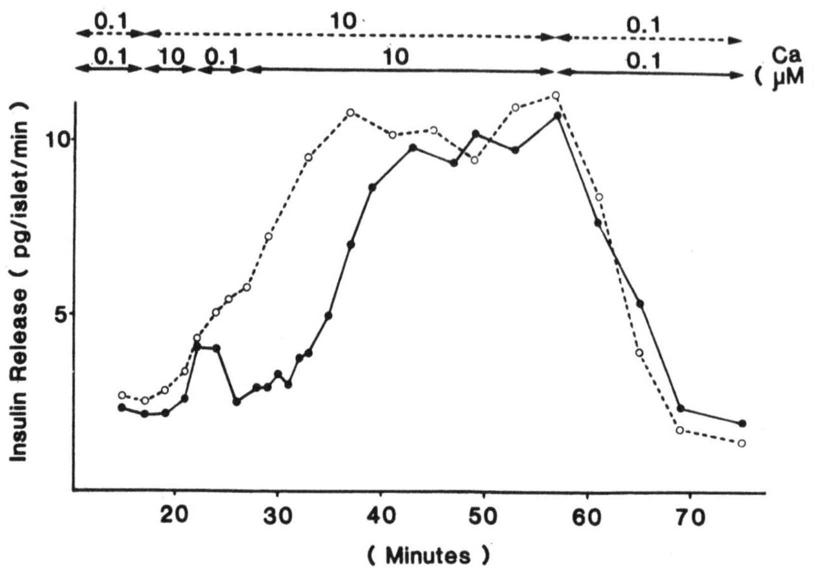

Fig. 2. Insulin release from perifused islets treated with digitonin. After the digitonin treatment, about 200 islets were transferred to the perifusion chambers and perifused at 37°C. [Ca^{2+}] was changed as indicated at the top of the figure. Effluent fractions were collected every min. The dead time (2 min) of the system was corrected in the presentation of the results. This experiment is representative of three other experiments that gave qualitatively similar results.

The threshold concentration of Ca^{2+} for insulin or catecholamine release from permeabilized cells seems to be between 0.1 and 1 µM[1,4,6,16,19,23]. However, Ca^{2+}-activation curves for hormone release are not necessarily identical among reports. It may be due to differences in the methods of permeabilization, the composition of buffers and temperature for incubation. Under our experimental condition, insulin release increased as Ca^{2+} concentration was raised and did not reach a plateau at any Ca^{2+} concentration tested, as in adrenal medullary chromaffin cells permeabilized with digitonin[6,19]. In three reports using permeabilized pancreatic islets[4,11,23], the maximal release of insulin seems to be observed at a Ca^{2+} concentration between 0.1 and 1 µM; however, the effects of higher concentrations of Ca^{2+} were not fully examined in these reports. It should be mentioned that insulin release from permeabilized islets is very much influenced by the composition of the buffers. Colca et al.[4] reported a failure to show Ca^{2+}-stimulated insulin release in their Tris/Pipes buffer at 37°C because of the large non-specific background release of insulin. We have tested several buffers other than the

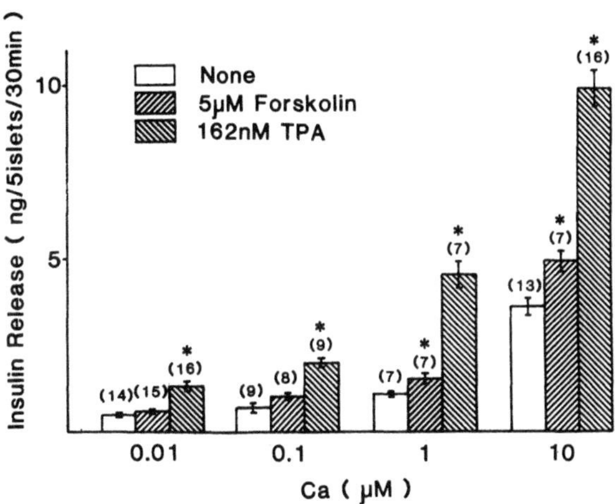

Fig. 3. Effects of TPA and forskolin on insulin release from digitonin-treated islets at various concentrations of Ca^{2+}. Experimental procedures as for Fig. 1. Groups of 5 treated islets were incubated for 30 min at 37°C in the media containing the indicated concentrations of Ca^{2+} in the figure with or without 162 nM TPA or 5 μM forskolin. *p <0.01 vs control without addition.

glutamate buffer used in the present study, and observed the high level of Ca^{2+}-independent insulin release in the buffer containing mainly Cl^- or sucrose. On the other hand, we failed to observe Ca^{2+}-stimulation of insulin release in the buffer containing isethionate which was used by Yaseen et al.[23] for incubation of permeabilized islets with high-voltage electric discharge.

Estimation of $[Ca^{2+}]_i$ with the use of quin 2 revealed that $[Ca^{2+}]_i$ increased from 0.1 to 0.2 μM on stimulation of the pancreatic β-cell by 17 or 20 mM glucose[5,14]. The range of $[Ca^{2+}]_i$ in which insulin release was stimulated in our experiments was higher than that observed in the experiments using quin 2. One of the possibilities may be that the permeabilized islets leak some factors which lower the threshold for Ca^{2+}-induced insulin release. Addition of some appropriate factors could shift the Ca^{2+}-activation curve for insulin release to the left. In fact, administration of TPA and forskolin did so, as described below (See Fig. 3). In addition, as shown in the perifusion study (Fig. 2), the regulatory mechanism of insulin release in the treated islets is considered to be functioning, since these islets did not leak insulin and promptly responded to the fluctuation of $[Ca^{2+}]$ in a completely reversible manner. Fig. 2 also shows that changing $[Ca^{2+}]$ evoked a biphasic pattern similar to that induced by glucose in intact islets.

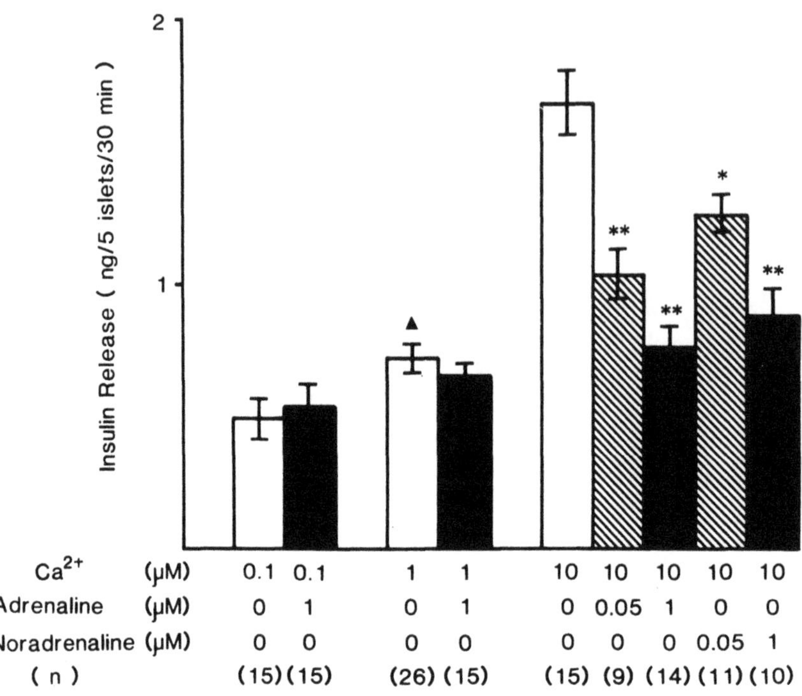

Fig. 4.   Effects of adrenaline and noradrenaline on insulin release at
various concentrations of Ca^{2+} from digitonin-treated islets.
Experimental procedures as for Fig. 1. Groups of 5 treated islets
were incubated for 30 min at 37°C in the media containing
indicated concentrations of Ca^{2+} and adrenaline or noradrenaline.
Values are the means ± SE (represented by bars), for the numbers
of experiments shown in parentheses. Significance levels: ▲ p
<0.01 vs the value by 0.1 µM Ca^{2+} alone; *p <0.05, **p <0.01 vs
the value by 10 µM Ca^{2+} alone.

Effect of Glucose and Tolbutamide on Insulin Release from Digitonin-treated Islets

Addition of 10 or 20 mM glucose to the media failed to influence
insulin release in the range of 0.1 - 10 µM Ca^{2+} (data not shown), in
agreement with the result reported by Pace et al.[11]. The finding suggests
that glucose-stimulation of insulin release is mainly dependent on changes
in [Ca^{2+}]$_i$. This failure may also be due to cytosolic dilution of
important factors such as glycolytic enzymes. Addition of tolbutamide
(0.15 mg/ml) also failed to influence insulin release at 1 µM Ca^{2+}, an
observation which is consistent with the view that the sulfonylurea evokes
insulin release by increasing Ca^{2+} influx.

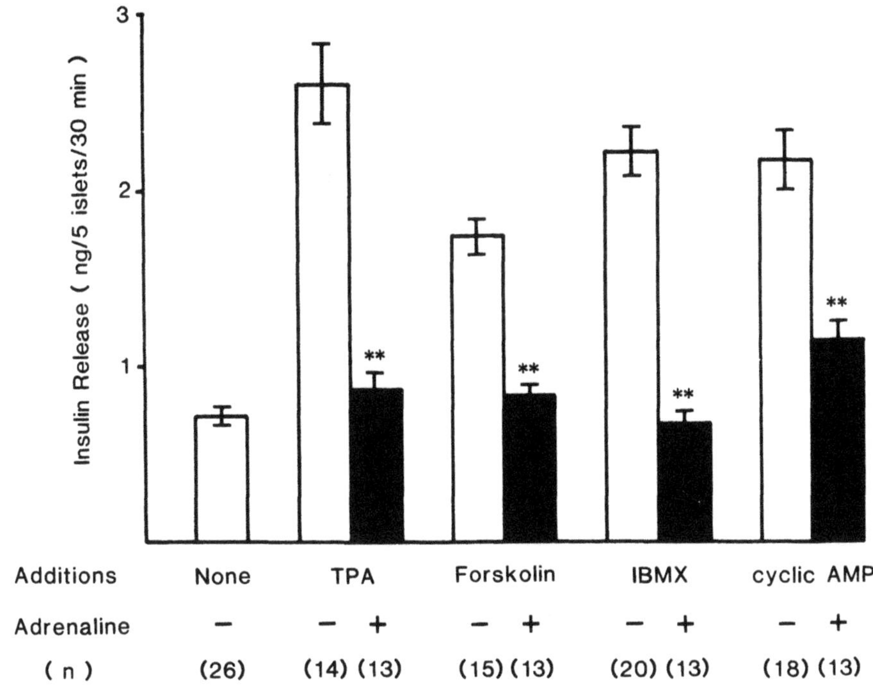

Fig. 5. Effects of adrenaline on insulin release from digitonin-treated
islets in the presence of TPA or substances which raise cyclic AMP
levels.[17] Experimental procedures as for Fig. 1. Groups of 5
treated islets were incubated for 30 min at 37°C in the media
containing 1 μM Ca^{2+} and test substances, concentrations of which
were as follows: TPC, 162 nM: forskolin, 5 μM; IBMX, 1 mM: cyclic
AMP, 5 mM: and adrenaline, 1 μM. Values are the means ± SE
(represented by bars), for the numbers of experiments shown in
parentheses. Significance levels: *p <0.001 vs each control
without adrenaline.

<u>Effects of TPA and Substances Which Raise Cyclic AMP Levels on Insulin</u>
<u>Release from Digitonin-treated Islets</u>

We further examined the actions of two other major intracellular
messenger systems; i.e., protein kinase C and elevation of cyclic AMP
levels[16].

TPA is known to substitute for diacylglycerol and increase the
affinity of protein kinase C for Ca^{2+}, resulting in its activation without
elevating $[Ca^{2+}]_i$ [10]. As shown in Fig. 3, addition of TPA (162 nM)
increased insulin release at all the concentrations of Ca^{2+} tested (0.01 –
10 μM). The findings indicate that the phorbol ester modulates insulin
release by a mechanism independent of a rise in $[Ca^{2+}]_i$, as in adrenal

medullary cells[9] and in platelets[13]. Thus it is likely that activation of
protein kinase C can increase insulin release without changes in $[Ca^{2+}]_i$.

Based on the $^{45}Ca^{2+}$ efflux, the action of cyclic AMP to enhance
insulin release in the presence of secretagogues has been thought to be due
to a mobilizing effect of intracellular stored calcium, leading to an
increase in $[Ca^{2+}]_i$[20]. However, a recent finding using a fluorescent Ca^{2+}
indicator revealed that cyclic AMP stimulation of insulin release was not
accompanied by increased $[Ca^{2+}]_i$[22]. Thus the interpretation of the results
obtained by $^{45}Ca^{2+}$ flux studies should be reconsidered. Our present data,
in support of the results using the Ca^{2+} indicator, showed that 5 µM
forskolin significantly augmented insulin release at 1 or 10 µM Ca^{2+}, as
shown in Fig. 3. Further, addition of IBMX (1 mM), a phosphodiesterase
inhibitor, or cyclic AMP (5 mM) to the media augmented insulin release at 1
µM Ca^{2+} (Fig. 5). The underlying mechanism(s) of stimulation of insulin
release at fixed $[Ca^{2+}]_i$ by TPA and cyclic AMP is not yet clear. One
possibility may be that these agents change the affinity of the secretory
process for Ca^{2+}, as suggested by Knight and Baker[9].

Effect of Adrenaline and Noradrenaline on Insulin Release from Digitonin-treated Islets

Although catecholamines are known to inhibit insulin release induced
by a wide variety of secretagogues, the mechanism by which the amines
inhibit insulin release remains obscure[20]. It has been suggested that the
α-adrenergic inhibitory action on insulin release involves a lowering of
$[Ca^{2+}]_i$ by redistributing Ca^{2+} into intracellular organelles and increasing
Ca^{2+} efflux to extracellular space[3].

We tested the effects of catecholamines on insulin release from the
digitonin-treated islets[17]. Fig. 4 shows that adrenaline as well as
noradrenaline dose-dependently inhibited insulin release at 10 µM Ca^{2+},
though the inhibition by adrenaline (1 µM) was not significant at 1 µM
Ca^{2+}. The inhibitory effect of adrenaline (1 µM) at 10 µM Ca^{2+} was
antagonized by 10 µM yohimbine, an α_2-adrenergic blocker, but not by 10 µM
prazosin, an α_1-adrenergic blocker, as was observed in intact islets[15].
The catecholamine (1 µM) also inhibited the potentiated insulin release by
TPA, forskolin, IBMX or cyclic AMP at 1 µM Ca^{2+}, as shown in Fig. 5. Thus,
adrenaline-inhibition of insulin release occurs even in the Ca^{2+}-clamped

islets. The recent observation using quin 2 also revealed that adrenaline
inhibited insulin release without lowering $[Ca^{2+}]_i$ [18].

CONCLUSIONS

The present data indicate that insulin release is regulated by two
modalities; one is dependent on, and the other is independent of changes in
$[Ca^{2+}]_i$.

ACKNOWLEDGEMENTS

We thank Miss H. Andoh and Miss F. Itoh, Aichigakuin University for
their technical assistance and secretarial help. We extend our thanks to
Prof. T. Koide, Aichigakuin University, and Prof. N. Sakamoto, University
of Nagoya, for their encouragement during the study. This work was
supported in part by Research Grant (No. 60770929) from the Ministry of
Education, Science and Culture, Japan.

REFERENCES

1. P.F. Baker and D.E. Knight, Calcium-dependent exocytosis in bovine
       adrenal medullary cells with leaky plasma membranes, _Nature_ 276:620
       (1978).

2. P.F. Baker, D.E. Knight, and J.A. Umbach, Calcium clamp of the
       intracellular environment, _Cell Calcium_ 6:5 (1985).

3. G.R. Brisson and W.J. Malaisse, The stimulus-secretion coupling of
       glucose-induced insulin release. XI. Effects of theophylline and
       epinephrine on ^{45}Ca efflux from perifused islets, _Metabolism_ 22:455
       (1973).

4. J.R. Colca, B.A. Wolf, P.G. Comens, and M.L. McDaniel, Protein
       phosphorylation in permeabilized pancreatic islet cells, _Biochem._
       _J._ 228:529 (1985).

5. M. Deleers, M. Mahy, and W.J. Malaisse, Glucose increases cytosolic
       Ca^{2+} activity in pancreatic islet cells, _Biochem Int._ 10:97 (1985).

6. L.A. Dunn and R.W. Holz, Catecholamine secretion from
       digitonin-treated adrenal medullary chromaffin cells, _J Biol.Chem._
       258:4989 (1983).

7. G. Fiskum, S.W. Craig, G.L. Decker and A.L. Lehninger, The
    cytoskeleton of digitonin-treaed rat hepatocytes, _Proc. Natl. Acad.
    Sci. USA_ 77:3430 (1980).

8. J.-C. Henquin, Tolbutamide stimulation and inhibition of insulin
    release: studies of the underlying ionic mechanisms in isolated rat
    islets, _Diabetologia_ 18:151 (1980).

9. D.E. Knight and P.F. Baker, The phorbol ester TPA increases the
    affinity of exocytosis for calcium in 'leaky' adrenal medullary
    cells, _FEBS Lett._ 160:98 (1983).

10. Y. Nishizuka, Y. Takai, A. Kishimoto, U. Kikkawa, and K. Kaibuchi,
    Phospholipid turnover in hormone action, _Rec. Prog. Horm. Res._
    40:301 (1984).

11. C.S. Pace, J.T. Tarvin, A.S. Neighbors, J.A. Pirkle, and M.H. Greider,
    Use of high voltage technique to determine the molecular
    requirements for exocytosis in islet cells, _Diabetes_ 29:911 (1980).

12. H. Rasmussen and P.Q. Barrett, Calcium messenger system: An integrated
    view, _Physiol. Rev._ 64:938 (1984).

13. T.J. Rink, A. Sanchez, and T.J. Hallam, Diacylglycerol and phorbol
    ester stimulate secretion without raising cytoplasmic free calcium
    in human platelets, Nature 305:317 (1983).

14. P. Rorsman, H. Abrahamsson, E. Gylfe, and B. Hellman, Dual effects of
    glucose on the cotysolic Ca^{2+} activity of mouse pancreatic β-cells,
    FEBS Lett 170:196 (1984).

15. T. Tamagawa and J.-C. Henquin, Epinephrine modifications of insulin
    release and of $^{86}Rb^{+}$ or $^{45}Ca^{2+}$ fluxes in rat islets, _Amer. J.
    Physiol._ 244:E245 (1983).

16. T. Tamagawa, H. Niki, and A. Niki, Insulin release independent of a
    rise in cytosolic free Ca^{2+} by forskolin and phorbol ester, _FEBS
    Lett_ 183:430 (1985).

17. T. Tamagawa, I. Niki, H. Niki, and A. Niki, Catecholamines inhibit
    insulin release independently of changes in cytosolic free Ca^{2+},
    Biomed. _Res._ 6:429 (1985).

18. S.J. Ullrich and C.B. Wollheim, Adrenaline inhibition of insulin
    release is accompanied by a paradoxical rise in cytosolic Ca^{2+} in
    RINm5F cells, _Diabetologia_ 27:340A (1984).

19. S.P. Wilson and N. Kirshner, Calcium-evoked secretion from digitonin-
    permeabilized adrenal medullary chromaffin cells, _J. Biol. Chem._
    258:4994 (1983).

20. C.B. Wollheim and G.W.G. Sharp, Regulation of insulin release by
    calcium Physiol. Rev. 61:914 (1981).

21. C.B. Wollheim and T. Pozzan, Correlation between cytosolic free Ca^{2+}
    and insulin release in an isulin-secreting cell line, J. Biol.
    Chem. 259:2262 (1985).

22. C.B. Wollheim, S. Ullrich, and T. Pozzan, Glyceraldehyde, but not
    cyclic AMP-stimulated insulin release is preceded by a rise in
    cytosolic free Ca^{2+}, FEBS Lett 177:17 (1984).

23. M.A. Yaseen, K.C. Pedley, and S.L. Howell, Regulation of insulin
    secretion from islets of Langerhans rendered permeable by electric
    discharge, Biochem. J. 206:81 (1982).

THE ROLE OF CYTOSOLIC CALCIUM IN INSULIN SECRETION FROM A HAMSTER BETA CELL
LINE

A.E. Boyd III, R.S. Hill, T.Y. Nelson
J.M. Oberwetter and M. Berg

Departments of Medicine and Cell Biology
Baylor College of Medicine
Houston, Texas 77030

Determining the molecular mechanisms of insulin secretion requires
detailed biochemical understanding of: a) the specific unique effects of
each individual secretagogue at the beta cell membrane, b) the
intracellular events by which each signal is mediated (second messengers)
and c) the modes and sites of action of second messengers within the
cytoskeletal system and the membrane-associated exocytotic process.

Because of the diversity of cell types and the limited amount of
tissue available for biochemical studies of pancreatic islets, a number of
laboratories have used insulin-secreting cell lines as model systems to
elucidate the biochemical mechanisms of insulin secretion. In the last few
years two cell lines have been studied extensively. The RIN cell, a
radiation-induced insulinoma developed by Chick et al.[4], releases both
insulin and somatostatin. It was cloned into cell lines which secrete
predominantly either insulin or somatostatin[7]. Santerre and colleagues
treated dispersed hamster beta cells with Simian virus 40, isolating the
transformants and cloned the cells into lines which are high insulin
secretors[19]. This line is called the HIT cells. Our laboratory has used
both cell lines in our studies[3,13,9].

A problem with the use of any hormone-secreting cell line is that with
time in culture hormone secretion gradually decreases as the more rapidly-
growing, less-differentiated cells gain a selective growth advantage over
their hormone-secreting siblings. This has been a major problem with the
RIN cell line. In early investigations of RIN cells insulin secretion was
stimulated by glucose; however, recent studies from a number of
laboratories (reviewed in 13) show that the RIN cells no longer release
insulin acutely in response to the major physiologic secretagogue, glucose.
Another problem with the RIN cell line is that the quantity of insulin
released acutely to any secretagogue is very small.

It was clear from a comparison of the two cell lines that HIT cells
offer many advantages as a model system to study the mechanism of insulin
secretion. The cells release insulin acutely in response to all the known
secretagogues we have tested including glucose. Furthermore, the amount of
hormone released by the HIT cells is dramatic and it is possible to study
insulin secretion in a standard perifusion system using 4×10^5 HIT cells[9].
The major difficulty with using the HIT cell line is that after about 80
passages the acute release of insulin following a glucose challenge is
lost. Thus, it is necessary to periodically thaw out early passage cells
which retain glucose responsiveness. Our recent studies have used quin 2,
a new fluroescent Ca^{2+} indicator, to directly measure the cytosolic free
Ca^{2+} concentration, $[Ca^{2+}]_i$, and test the hypothesis that $[Ca^{2+}]_i$ is a
primary intracellular signal or "second messenger" which triggers acute
insulin release. The nonpolar methyl-ester derivative of quin 2, quin/AM,
readily enters cells where it is hydrolyzed by esterases to quin 2 which is
highly charged and is trapped within the cell. Upon binding of Ca^{2+} to quin
2, the intensity of fluroescence increases. Flourescence can be monitored
in cell suspensions to continuously measure $[Ca^{2+}]_i$.

METHODS

Insulin was measured using a double antibody radioimmunoassay modified
from that of Hales and Randle[9] as previously described[2]. HIT cells were
cultured and insulin secretion monitored in cell perifusions or static
incubations as described[9,3]. The cytosolic free calcium concentration,
$[Ca^{2+}]_i$, was measured in suspensions of quin 2-loaded cells using a Perkin
Elmer LS-5 fluorescence spectrofluorimeter as outlined in detail in[3]. All

of the agents used as secretagogues were tested for fluorescence. Only
glyburide was found to fluoresce at an excitation wavelength of 340 nm.
Glyburide fluorescence peaked at an emission wavelength of 410 nm and
endogenous glyburide fluorescence was negligible at 520 nm. To eliminate
the contribution of intrinsic glyburide fluorescence, an emission
wavelength of 520 nm was used to monitor quin 2 fluorescence in the
glyburide experiments. Excitation and emission wavelengths for all other
studies were 340 and 490 nm respectively. The calibration of quin 2
fluorescence as a function of $[Ca^{2+}]_i$ was similar to that used in RIN cells
by Wollheim and Pozzan[21] as described in[3]. Statistics were also performed
as detailed in[9,3].

RESULTS

We first established the time course and pattern of insulin secretion
in perifusions of HIT cells following the addition of three different types
of secretagogues – high concentrations of potassium (40 mM), the
sulfonylureas, glyburide and tolbutamide, and the major physiologic
secretagogue glucose. As seen in Fig. 1, depolarization of the cells with
40 mM potassium or changing the glucose concentration of the buffer to 19.7
mM glucose after a 30 min glucose-free perifusion stimulated a monophasic
pattern of insulin release. Sulfonylureas also trigger a monophasic
pattern of insulin release (data not shown)[13]. In static incubations
glucagon and the amino acids, leucine, arginine and alanine, also stimulate
insulin release from the HIT cells. In perifusion the pattern of insulin
release with leucine and arginine is also monophasic. Insulin release
triggered by any secretagogue is dependent upon the presence of
extracellular calcium in the buffer and is completely blocked if the cells
are perifused with an EGTA containing, Ca^{2+} free buffer[3].

Since we are interested in the signals which trigger acute insulin
release, we have focused on the $[Ca^{2+}]_i$ during the first 5 min after
addition of each secretagogue. In control experiments quin 2-loaded cells
release about 75-80% as much insulin as cells treated with the DMSO, the
agent used to solubilize quin 2. Thus, despite the addition of a compound
which buffers Ca^{2+} transients in the beta cell, insulin release is still
sufficient to test the hypothesis that an increase in the $[Ca^{2+}]_i$ is the
primary signal for acute insulin release. As shown in Fig. 2, although the

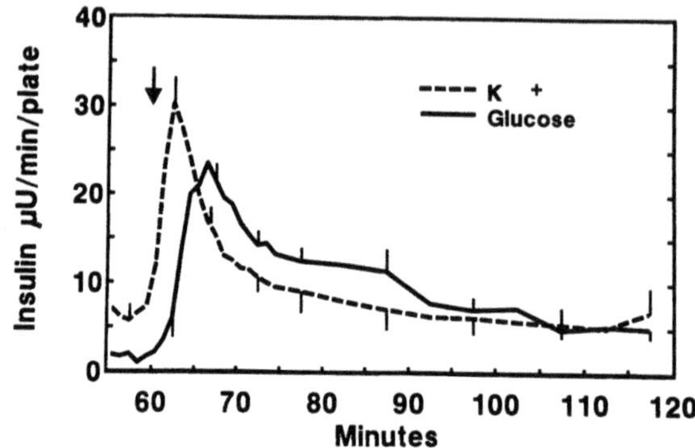

Fig. 1. Glucose and K^+-stimulated insulin secretion. HIT cells were perifused for a 30 min stabilization period in glucose-free buffer. At 30 min the glucose in the buffer was increased to 19.7 mM glucose (n = 6). In another group of perifusions HIT cells were exposed to 40 mM K^+ at 30 min.

basal $[Ca^{2+}]_i$ is not changed by the amount of quin loaded into the HIT cells, the rate of rise of $[Ca^{2+}]_i$ and the ultimate level of $[Ca^{2+}]_i$ is dependent upon the amount of quin 2 loaded into the HIT cells. The sensitivity of the method can be increased by decreasing the amount of quin 2 in the cells. This becomes important when one is trying to determine the lowest concentration of a secretagogue which results in a significant change in $[Ca^{2+}]_i$ or if an agent stimulates or potentiates insulin release by an effect on $[Ca^{2+}]_i$. However, it is not possible to load HIT cells

Table 1. Basal and Stimulated $[Ca^{2+}]_i$ in HIT Cell Line

Glucose	Basal	40 mM K^+	19.7 mM Glucose	10 mM Alanine	Fold Stimulation
		$[Ca^{2+}]_i$, nM			
1.67 mM	50 ± 2 (61)	371 ± 27 (61) *			7.4
1.67 mM	44 ± 3 (8)			71 ± 6 (8) *	1.6
0	37 ± 4 (11)		31 ± 4 (11)		
1.67 mM	37 ± 5 (9)		37 ± 5 (9)		

*p<0.001 from basal

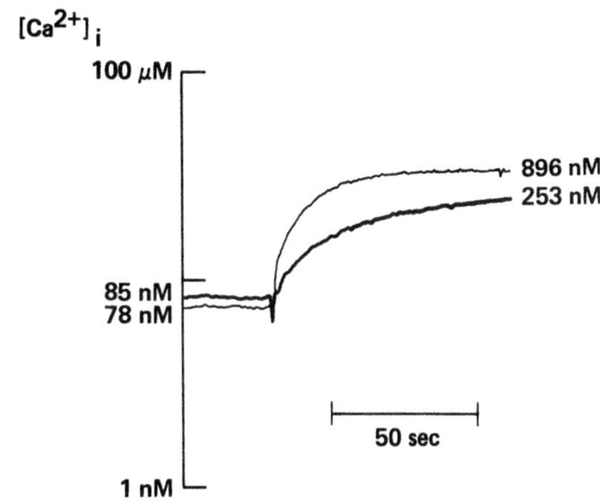

Fig. 2. Effect of varying quin 2/AM loading concentrations on $[Ca^{2+}]_i$. Shown are representative tracings from 6 different experiments. The amount of quin 2 in the HIT cells did not alter the basal $[Ca^{2+}]_i$. In cells loaded with 50 μM the rapidity of the rise was greater and the peak $[Ca^{2+}]_i$ higher than in cells loaded with 100 μM quin 2/AM.

with less than 50 μM quin 2/AM because at loading concentrations of quin 2/AM below 50 μM, the hydrolysis of the quin 2/AM to quin 2 is incomplete. Thus, the sensitivity of the $[Ca^{2+}]_i$ method is greatest at a loading concentration of 50 μM quin 2/AM. In most of the studies reported in this paper, we have used 100 μM quin 2/AM to load the cells. However, in all studies where sensitivity is important, the $[Ca^{2+}]_i$ measurements have been repeated at 50 μM quin 2/AM loading concentration.

Table 2. Basal and Stimulated $[Ca^{2+}]_i$ in HIT Cell Line

		$[Ca^{2+}]_i$, nM		
Basal	Glyburide 200 μM	Tolbutamide 370 μM	Carboxy-Tolbutamide 330 μM	Fold Stimulation
40 ± 2 (24)	56 ± 3 (19) *			1.4
33 ± 3 (19)		48 ± 5 (19)		1.5
36 ± 5 (7)			36 ± 4 (7)	

* p<0.001 from basal

Depolarization of the cells with 40 mM potassium increased $[Ca^{2+}]_i$ 7.4-fold from 50 ± 2 to 371 ± 27 nM(Table I, n = 61). The variability of the $[Ca^{2+}]_i$ rise to 40 mM potassium ranges from as low as 110 nM or 2.2-fold to values of over 1 μM (1,109 nM), a 27-fold increase. Alanine (10 mM) also increases the $[Ca^{2+}]_i$ (Table 1). Preliminary experiments with 10 mM leucine show that this amino acid also stimulates insulin release by increasing $[Ca^{2+}]_i$. In contrast, despite stimulating a significant burst of insulin release (Fig. 1), increasing the concentration of glucose from 0 to 19.7 mM glucose resulted in a slight, but significant drop in $[Ca^{2+}]_i$. If glucose levels were increased from 1.67 mM to 19.7 mM, there was no change in $[Ca^{2+}]_i$ (Table 1).

The mechanisms by which the sulfonylureas signal insulin secretion were examined. Dose response relationships of glyburide and tolbutamide were established. In static incubations, glyburide was approximately 300 times more potent as an insulin secretagogue than tolbutamide. Basal $[Ca^{2+}]_i$ increased 1.4-fold within 90 seconds of glyburide exposure, whereas tolbutamide increased $[Ca^{2+}]_i$ 1.5-fold above the basal level (Table 2). At similar concentrations the less active metabolite, carboxytolbutamide, had no effect on $[Ca^{2+}]_i$ or insulin secretion. Chelation of extracellular calcium with 4 mM EGTA completely inhibited the sulfonylurea-induced changes in $[Ca^{2+}]_i$ and blocked insulin release triggered by the sulfonylureas[14].

We have used the organic calcium channel blockers, verapamil and nifedipine, to test if the secretagogues that trigger a rise in $[Ca^{2+}]_i$ act on the "calcium channel". These drugs interfere with the entry of calcium into cells through voltage-sensitive channels, whereas processes that involve the release of intracellular calcium stores are relatively insensitive to these compounds[12]. In static incubations of HIT cells there was remarkable agreement between the concentrations of verapamil or nifedipine that blocked the rise in $[Ca^{2+}]_i$ and inhibited the release of insulin triggered by either 15 mM potassium or 100 μM glyburide. We used 15 mM potassium in these experiments because it stimulates the same amount of insulin release as 100 μM glyburide. The IC_{50} of verapamil on glyburide-induced insulin release and the effects on $[Ca^{2+}]_i$ were quite similar being 6.7 and 5.3×10^{-6} M respectively. The inhibitory effects of nifedipine on K^+ or glyburide-induced insulin release occurred at lower doses than verapamil with an IC_{50} of 9×10^{-8} M and 3.6×10^{-7},

respectively. Dose response relationships of the calcium channel blockers
have not yet been performed on cells stimulated with amino acids but
preliminary experiments indicate that amino acids also act at the voltage-
dependent calcium channel.

DISCUSSION

In a large number of experiments the resting $[Ca^{2+}]_i$ in this beta cell
line was 50 ± 2, n = 61. This is somewhat lower than the mean level of 105
nM reported in RIN cells[17], but agrees with the $[Ca^{2+}]_i$ concentration of 44
nM in single pituitary cells, the GH_3 line[9]. After loading of both RIN[19]
and HIT cells and extensive washing, significant amounts of extracellular
quin 2 are detected. This is probably secondary to leakage of
intracellular quin 2 from unhealthy or dead cells[21]. It is necessary to
correct for extracellular quin 2 or the basal levels of $[Ca^{2+}]_i$ will be
falsely elevated by interactions of the fluorescent probe with the mM
concentrations of Ca^{2+} in the buffer. In HIT cells quin 2 leakage is
decreased by working at room temperature. When calculations are made of
the amount of quin 2 in the HIT cells, using a loading concentration of 100
µM of quin 2/AM, the HIT cells contain approximately 1 mM quin 2[3]. As
shown in these experiments, although the quin 2 content of the cell does
not alter the basal value of $[Ca^{2+}]_i$ the rapidity of the increase in
$[Ca^{2+}]_i$ and the level to which $[Ca^{2+}]_i$ rises is affected by intracellular
quin 2 content. Since the loading efficiency varies from day to day, the
$[Ca^{2+}]_i$ measurements are considered as relative numbers.

In these studies four classes of insulin secretagogues were studied.
Three of them, potassium, amino acids and the sulfonylureas all stimulated
insulin secretion by elevating the $[Ca^{2+}]_i$. High K^+ (40 mM), which depolarizes the cells, increased the $[Ca^{2+}]_i$ rapidly with peak $[Ca^{2+}]_i$ levels
occurring within 30 to 60 sec of the stimulus. This rise in $[Ca^{2+}]_i$
precedes the peak of insulin secretion by one to two minutes in HIT cells
perifused with 40 mM K^+. Although the $[Ca^{2+}]_i$ levels then fall, they never
reached the basal level over the 4-5 minutes period of these continuous
measurements. By chelating extracellular Ca^{2+} with EGTA or adding
Ca-channel blockers, verapamil or nifedipine, prior to the K^+-stimulus, it
is clear that this secretagogue causes extracellular Ca^{2+} to enter the beta
cell through verapamil- or nifedipine-sensitive Ca-channels.

The amino acids also increase $[Ca^{2+}]_i$ by an action on the calcium Ca-channel. The rise in $[Ca^{2+}]_i$ with alanine was approximately 1.6-fold over the basal level. Verapamil or EGTA also blocked the effect of amino acids on $[Ca^{2+}]_i$ suggesting that these amino acids also initiate insulin secretion by an effect on the voltage-dependent Ca-channels. Our conclusions using HIT cells are similar to those in the RIN cells, which show that K^+- or alanine-induced insulin secretion is accompanied by a rise in $[Ca^{2+}]_i$ [14,16,21]. Using another fluorescent probe, bisoxonal, to monitor changes in membrane potential, Wollheim and Pozzan also established that K^+ and alanine depolarize the cell in parallel with their effect on $[Ca^{2+}]_i$ [21]. Thus, both K^+ and amino acids signal insulin release by depolarizing the beta cell, opening voltage-dependent Ca-channels and increasing $[Ca^{2+}]_i$. These studies also show that the sulfonylureas, glyburide and tolbutamide, also trigger insulin release by an effect on verapamil- or nifedipine-sensitive, voltage-dependent Ca-channels. The influx of extracellular Ca^{2+} increases the $[Ca^{2+}]_i$ approximately 1.5 to 2.0-fold and the rise in $[Ca^{2+}]_i$ triggered by the sulfonylureas stimulated a monophasic pattern of insulin release. Verapamil of nifedipine inhibited the glyburide-induced insulin release and the rise in $[Ca^{2+}]_i$ in parallel.

The mechanism by which a rise in $[Ca^{2+}]_i$ in the beta cell is coupled to the distal steps of exocytosis is unknown but is the subject of intense investigation. We and others have used the Ca^{2+}-dependent generation of tension in smooth muscle cells, which has been elegantly characterized (reviewed in 10), to develop hypotheses to identify Ca^{2+}-dependent reactions which regulate insulin release. In smooth muscle cells the Ca^{2+}-calmodulin complex activates the myosin light chain kinase which phosphorylates the M_r 20 kD myosin light chain (LC_{20}). This phosphorylation then allows actin to interact with the head of the myosin molecule. The resultant hydrolysis of ATP generates the tension necessary for contraction[10]. In the beta cell the rise in free Ca^{2+} triggered by K^+, amino acids or the sulfonylureas would presumably increase the amount of Ca^{2+} bound to the ubiquitious Ca^{2+}-binding protein, calmodulin, changing the configuration of the molecule to an active state. In the beta cell we have shown that calmodulin can interact with a number of calmodulin-binding proteins which can be identified by a gel overlay technique[13]. Many of these calmodulin-binding proteins are enzymes like the Ca^{2+}-calmodulin-dependent kinase or cytoskeletal elements like microtubules. We have recently shown that 40 mM K^+ rapidly phosphorylates three proteins in a

Ca^{2+}-dependent manner[15]. The 20 kD phospoprotein has a pI that is
consistent with its tentative identification as the myosin light chain.
The coordinate inhibition of insulin secretion, the rise in $[Ca^{2+}]_i$ and
phosphorylation of the 20 kD protein by verapamil are also consistent with
the hypothesis that, as in smooth muscle cells, depolarization of the cell
with K^+ causes Ca^{2+} to enter the beta cell through voltage-dependent,
verapamil-sensitive Ca-channels which regulates insulin release by
activating the myosin light kinase. However, direct proof of this
hypothesis is not yet available.

In these studies a rise in the total $[Ca^{2+}]_i$ was not linked to the
burst of insulin release triggered by glucose. Caution is necessary in
hypothesizing a non-Ca^{2+} signal for glucose-stimulated insulin release.
The quin 2 technique only measures the average $[Ca^{2+}]_i$ in a population of
cells. Furthermore, quin 2 is insensitive to Ca^{2+} changes in the μM range.
Thus, this technique could fail to detect a small, but physiologically
significant, localized rise in $[Ca^{2+}]_i$ in HIT cells exposed to glucose.
Furthermore, several laboratories have reported that glucose increases
$[Ca^{2+}]_i$ in suspensions of beta cells dispersed from isolated islets[16,6].
In addition, although the RIN cells do not respond to glucose, they release
insulin when exposed to the triose, glyceraldehyde which enters glycolysis
at a step later than glucose. In Wollheim's studies glyceraldehyde appears
to increase $[Ca^{2+}]_i$ by two mechanisms opening voltage-sensitive Ca-channels
and altering intracellular Ca-handling at another step[21].

In contrast, several lines of evidence support the findings that a
rise in $[Ca^{2+}]_i$ is not the signal that triggers acute insulin release
elicited by glucose. First, in quin 2 studies on the effect of glucose on
$[Ca^{2+}]_i$ in suspensions of beta cells of obese hyperglycemic mice, Rorsman
et al[18] concluded that initially glucose exposure lowers the $[Ca^{2+}]_i$. This
effect is later masked by the subsequent increase in $[Ca^{2+}]_i$ secondary to
depolarization of the cells and influx of Ca^{2+} through the voltage-
dependent channels. Second, in RIN cells exposure to 4 mM glucose
decreases[17] $[Ca^{2+}]_i$ and it has been hypothesized that defective depolarization of the RIN cell accounts for the lack of insulin secretion to glucose
exposure. Third, we have not found an increase in phosphorylation of the
Ca^{2+}-dependent proteins during glucose-stimulated insulin release from the
HIT cells[15] giving independent biochemical support for a non-Ca^{2+} signal
for the acute release of insulin stimulated by glucose. Finally, the

K-selective channel identified by Cook, et al[5], and by Ashcroft,et al[1], is
insensitive to varying concentrations of intracellular Ca^{2+} and pH, but, in
excised patches of beta-cell membranes, is inhibited by ATP or, in cell-
attached patches, by the metabolism of glucose. Since the HIT cells
release insulin in response to both glyceraldehyde and glucose, direct experiments comparing the two cell lines should clarify if the signal for the
immediate release of insulin occurs as a result of an action of glucose or
glyceraldehyde on the voltage-dependent Ca-channel or on intracellular
Ca^{2+}-redistribution.

REFERENCES

1. F.M. Ashcroft, D.E. Harrison, and S.J.H. Ashcroft, Glucose induces
      closure of single potassium channels in isolated rat pancreatic
      β-cells. Nature 312:446 (1984).
2. A.E. Boyd III, W.E. Bolton, and B.R. Brinkley, Microtubules and beta
      cell function: Effect of colchicine on microtubules and insulin
      secretion in vitro by mouse beta cells, J. Cell Biol. 92:425
      (1982).
3. A.E. Boyd III, R.S. Hill, J.M. Oberwetter, and M. Zabelshansky,
      Calcium dependency and free calcium concentrations during insulin
      secretion in a hamster beta cell line (HIT cells), J. Clin. Invest.
      77:774 (1986).
4. W.L. Chick, S. Warren, R.N. Chute, A.A. Like, V. Lauris, and K.C.
      Kitchen, A transplantable insulinoma in the rat, Proc. Natl. Acad.
      Sci. USA 74:628 (1977).
5. D.L. Cook, M. Ikeuchi, and W.Y. Fujimoto, Lowering of pH_i inhibits
      Ca^{2+}-activated K^+ channels in pancreatic β-cells, Nature 311:269
      (1984).
6. M. Deleers, M. Mahy, and W.J. Malaisse, Glucose increases cytosolic
      Ca^{2+} activity in pancreatic islet cells, Biochemistry International
      10:97 (1985).
7. A.F. Gazdar, W.L. Chick, H.K. Oie, H.L. Sims, D.L. King, G.C. Weir and
      V. Lauris, Continuous, clonal, insulin- and somatostatin-secreting
      cell lines established from a transplantable rat islet cell tumor,
      Proc. Natl. Acad. Sci. USA 77:3519 (1980).
8. C.N. Hales and P.J. Randle, Immunoassay of insulin with insulin
      antibody precipitate, Biochem. J. 88:137 (1963).

9. R.S. Hill and A.E. Boyd III, Perifusion of a clonal cell line of simian virus 40-transformed beta cells. Insulin secretory dynamics in response to glucose, 3-isobutyl-1-methylxanthine, and potassium, Diabetes 34:115 (1983).

10. K.E. Kamm and J.T. Stull, The function of myosin and myosin light chain kinase phosphorylation in smooth muscle, Ann. Rev. Pharmacol. Toxicol. 25:593 (1985).

11. B.A. Kruskal, C.H. Kerth, and F.R. Maxfield, Thyrotropin-releasing hormone-induced changes in intracellular $[Ca^{2+}]_i$ measured by microspectrofluorometry of individual quin 2-loaded cells, J. Cell Biol. 99:1167 (1984).

12. R.J. Miller and S.B. Freedman, Are hydropyridine binding sites voltage-sensitive to calcium channels? Life Science 34:1205 (1984).

13. T.Y. Nelson, J.M. Oberwetter, J.G. Chafouleas, and A.E. Boyd III, Calmodulin-binding proteins in a cloned rat insulinoma cell line, Diabetes 32:1126 (1983).

14. T.Y. Nelson, M. Berg and A.E. Boyd III, Cytosolic calcium: A signal for sulfonylurea-stimulated insulin release from beta cells, (in review).

15. J.M. Oberwetter and A.E. Boyd III, Insulin secretion and protein phosphorylation, J. Cell Biol. 101:293A (1985).

16. M. Prentki and C.B. Wollheim, Cytosolic free calcium in insulin secreting cells and its regulation by isolated organelles, Experientia 40:1052 (1984).

17. P. Rorsman, P.-O. Berggren, E. Gylfe, and B. Hellman, Reduction of the cytosolic calcium activity in clonal insulin-releasing cells exposed to glucose, BioScience Reports 3:939 (1983).

18. P. Rorsman, H. Abrahamsson, E. Gylfe, and B. Hellman, Dual effects of glucose on the cytosolic Ca^{2+} activity of mouse pancreatic β-cells, FEBS LETT. 170:1434 (1984).

19. R.F. Santerre, R.A. Cooke, R.M.D. Crisel, J.D. Shar, R.J. Schmidt, D.C. Williams, and C.P. Wilson, Insulin synthesis in a clonal cell line of simian virus 40-transformed hamster pancreatic beta cells, Proc. Natl. Acad. Sci. USA 78:4339 (1981).

20. R.Y. Tsien, T. Pozzan, and T.J. Rink, Calcium homeostasis in intact lymphocytes: Cytoplasmic free calcium monitored with a new, intracellularly trapped fluorescent indicator, J. Cell.Biol. 94:325 (1982).

21. C.B. Wollheim and T. Pozzan, Correlation between cytosolic free
     calcium and insulin release in an insulin secreting cell line, _J.
     Biol. Chem._ 259:2262 (1984).

DIFFERENTIAL EFFECT OF NUTRIENT AND NON-NUTRIENT SECRETAGOGUES ON CYTOSOLIC
FREE Ca^{2+} IN PANCREATIC ISLET CELLS

A. Herchuelz, M. Juvent, E. Van Ganse, and P. Gobbe

Laboratory of Pharmacology
Brussels University
School of Medicine
Brussels, Belgium

Glucose may increase cytosolic free Ca^{2+} ($[Ca^{2+}]i$) in the pancreatic
β-cell by increasing Ca^{2+}-inflow. Whether glucose may also increase
$[Ca^{2+}]i$ by inhibiting Ca^{2+}-extrusion by Na^+/Ca^{2+} exchange, remains a matter
of debate. Cytosolic free Ca^{2+} was measured in rat pancreatic islets using
the Ca^{2+} indicator Quin 2. The mean $[Ca^{2+}]i$, in the presence of 1.7 mM
glucose, was 109 ± 9 nM (n = 21). A rise in the glucose concentration from
1.7 mM to 16.7 mM induced a biphasic increase in $[Ca^{2+}]i$ which reached a
value of about 640 nM after 15 min. This increase was doubled in the
presence of tetraethylammonium (20 mM), reduced by nifedipine (1 µM) and
abolished by mannoheptulose (20 mM). Glucose (16.7 mM) also increased,
although to a lower extent, $[Ca^{2+}]i$ in the absence of extracellular Ca^{2+} or
at a non-insulinotropic concentration (4.2 mM) in the presence of
extracellular Ca^{2+}. However, the effect of glucose (16.7 mM) was abolished
in the absence of both extracellular Ca^{2+} and Na^+. Other nutrient
secretagogues (e.g. mannose and alanine) also induced a biphasic increase
in $[Ca^{2+}]i$ which persisted in the absence of extracellular Ca^{2+}. In
contrast, non-nutrient secretagogues [e.g. 20 mM K^+ or the sulfonylurea
gliclazide (15 µg/ml)] induced a monophasic increase in $[Ca^{2+}]i$ which was
completely suppressed in the absence of extracellular Ca^{2+}.

In conclusion, our data indicate that glucose increases $[Ca^{2+}]i$ in the pancreatic β-cell by both increasing Ca^{2+} inflow and by inhibiting Ca^{2+}-extrusion by Na^{+}/Ca^{2+} exchange.

RELATIONSHIP BETWEEN EXTRACELLULAR Na^+ AND THE TOTAL IONIZED Ca^{2+} CONTENT
OF RAT PANCREATIC ISLETS

G.H.J. Wolters, M. Vonk and A. Pasma

Dept. of Experimental Endocrinology
University of Groningen
1 Bloemsingel, 9713 BZ Groningen
The Netherlands

It is assumed that sodium ions, Na^+, are involved in a Na-Ca exchange
process at the β-cell membrane and in the mobilization of calcium ions,
Ca^{2+}, from intracellular stores[4,5]. Na-Ca exchange, which exchanges
extracellular Na^+ for intracellular Ca^{2+}, has been observed in several cell
types and has been characterized in detail in cardiac cells and membrane
vesicles[6]. Na-Ca exchange depends on the electrochemical gradient of Na^+
(and of Ca^{2+}) across the cell membrane. This process is sensitive to
changes in the membrane potential and the extracellular Na^+ concentration,
$[Na^+]_o$[6]. If this process exists in β-cells, depolarization of the membrane
will inhibit the exchange and enhance the intracellular Ca^{2+} concentration,
$[Ca^{2+}]_i$, in the β-cells.

The aim of the present study was to investigate the effects of gradual
lowering of $[Na^+]_o$ on the Ca^{2+} content of β-cells.

Batches of 30 rat pancreatic islets were incubated at 2.5 mM glucose
at various $[Na^+]_o$ for 0 to 90 min. Extracellular Na^+ was replaced with
choline chloride, KCl or LiCl. $NaHCO_3$ was replaced by the respective
bicarbonates. Na^+ was also replaced with sucrose. After incubation the
islets were subjected to freeze-substitution, sectioned and stained with
the metallochromic indicator glyoxal-bis-(2-hydroxyanil), (=GBHA). The
ionized Ca^{2+} content of β-cells was determined by measuring the

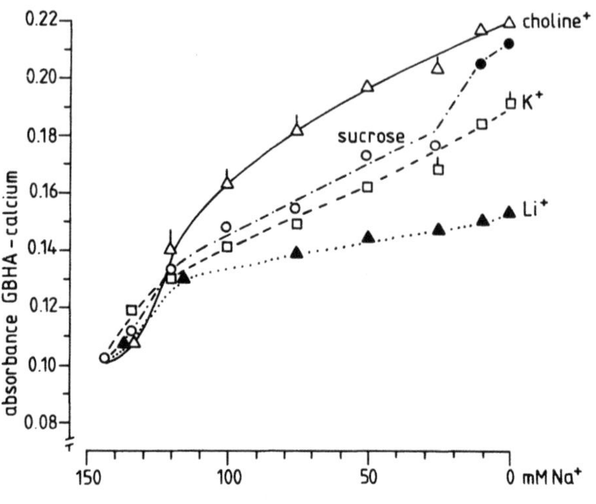

Fig. 1. Effects of Na$^+$ on Ca^{2+} content.

GBHA-calcium complex with a cytophotometer[7]. This study is based on the determination of the Ca^{2+} in 8730 islets in 873 batches from 4 independent experiments for each substitute used. Figure 1 shows mean ± SEM of 4 experiments. When no bar is shown it is too small to be indicated.

The ionized Ca^{2+} content of β-cells as measured with GBHA is very sensitive to changes in $[Na^+]_o$. Lowering $[Na^+]_o$ results in a rapid rise of GBHA-Ca^{2+}. The rate of the GBHA-Ca^{2+} rise during the first 5 min after complete removal of extracellular Na$^+$ is shown in table 1. The values are expressed as percentage of the GBHA-Ca^{2+} content at 144 mM $[Na^+]_o$. The highest rate of increase (9.7% per min) is observed when choline is used as substitute and the lowest rate (5.7% per min) when Li$^+$ is used. After 30 min more or less stable levels are attained (data not shown). The GBHA-Ca^{2+} levels attained between 60 and 90 min at the various $[Na^+]_o$ on using different substitutes for Na$^+$ are plotted in Figure 1. The concentration response relationship demonstrates that the size of the increase of GBHA-Ca^{2+} on lowering $[Na^+]_o$ depends on the substitute used to replace Na^+_o. Moreover, all curves demonstrate a break between 120 and 100 mM $[Na^+]_o$.

The increase of GBHA-Ca^{2+} on lowering $[Na^+]_o$ from 144 to 120 mM using Li$^+$, sucrose or K$^+$ as substitute is comparable, 1.14 - 1.27% increase per mM Na$^+$ decrease. When choline is used as substitute a higher increase of GBHA-Ca^{2+} is observed.

320

Table 1. GBHA-Ca^{2+} changes induced by lowering of extracellular sodium ion concentration. The values are expressed as percentage of the GBHA-Ca^{2+} level at 144 mM [Na$^+$]$_o$.

Na$^+$	144 – 120 mM		100 – 25 mM		144 – 0 mM	144 – 0 mM
substitute	%	%/mM	%	%/mM	%	%/min
1 Li$^+$	28	1.17	11	0.14	50	5.7
2 choline	37	1.55	38	0.51	114	9.7
3 sucrose	30	1.27	28	0.38	(108)	(9.3)
4 K$^+$	27	1.14	28	0.37	87	7.35

On lowering [Na$^+$]$_o$ from 120 to 25 mM clear differences are seen between the GBHA-Ca^{2+} values attained at different [Na$_o^+$] with the various substitutes. When Na$^+$ is replaced by choline, higher GBHA-Ca^{2+} values are achieved than with sucrose or K$^+$, however, replacement by Li$^+$ induced only a small further increase of GBHA-Ca^{2+} and the levels achieved are lower than seen with sucrose or K$^+$. Gradual replacement of Na$^+$ by sucrose causes a similar concentration dependent increase of GBHA-Ca^{2+} as seen when substituting with K$^+$. However, when sucrose is used as a substitute (and NaHCO$_3$ is replaced by choline bicarbonate) the GBHA-Ca^{2+} level increased disproportionately, reaching a comparable level as seen on complete replacement of Na$^+$ by choline (closed circles in Figure 1 and data in brackets in Table 1). This effect of choline bicarbonate and the higher GBHA-Ca^{2+} level observed on graduate replacement of Na$^+$ with choline strongly suggests that choline has its own specific effect. The nature of this effect we have not investigated.

The increase of GBHA-Ca^{2+} per mM Na$^+$ decrease, occuring when choline, sucrose or K$^+$ is used as a substitute is between 120 and 144 mM Na$^+$, about 3 times higher than the GBHA-Ca^{2+} increase seen between 25 and 100 mM Na$^+$.

For Li^+ this value is 8. The reason for the relatively higher sensitivity to the lowering of Na^+ between 144 and 120 mM is unknown. Furthermore, the reason for the only very small increase of GBHA-Ca^{2+} induced on replacement of Na_o^+ by Li^+ at Na^+ lower than 120 mM is also unknown. It has been shown that the accumulation of ^{45}Ca by islets in a Na^+ deficient medium containing Li^+ as a substitute for Na^+ is about half of that seen in a medium containing choline or sucrose[1]. Whether Li^+ inhibits the influx of Ca^{2+} or induces Ca^{2+} efflux by Li-Ca exchange is not known. The increase of GBHA-Ca^{2+} on complete replacement of extracellular Na^+ by Li^+ is also about half of that seen with choline or sucrose (Table 1). Another remarkable observation is the almost identical increase of GBHA-Ca^{2+} when Na^+ is replaced by sucrose or by K^+. When Na^+ is replaced by K^+, on the one hand the Na-Ca exchange is decreased, which will result in inhibition of Ca^{2+} extrusion, and on the other hand K^+ causes membrane depolarization which also inhibits Na-Ca exchange, and moreover, increases Ca^{2+} influx through the voltage dependent calcium channels. However, the combined effects of opening of the calcium channels and inhibition of the Na-Ca exchange does not cause higher GBHA-Ca^{2+} than seen with sucrose as substitute.

The intracellular Ca^{2+} content of β-cells is the resultant of Ca^{2+} influx and Ca^{2+} extrusion by the Ca-pump and the Na-Ca exchange process. Intracellular Ca^{2+} is compartmentalized and the intracellular organelles take up Ca^{2+} from, or release Ca^{2+} into the cytosol. Whether the increase of GBHA-Ca^{2+} is mainly due to inhibition of Na-Ca exchange cannot be deduced from these data. Inhibition of Na-Ca exchange across the cell membrane would increase the cytosolic Ca^{2+} concentration. However, removal of Na^+ which causes a large increase of the GBHA-Ca^{2+} content of β-cells, most likely does not appreciably increase the cytosolic Ca^{2+} concentration as insulin secretion is not enhanced (not shown). It is generally supposed that enhanced cytosolic $[Ca^{2+}]$ induces insulin secretion. Therefore, it seems more likely that Ca^{2+} is accumulated in one or more subcellular compartment(s). The supposed increase of Ca^{2+} in the subcellular compartment(s) is probably due to the decrease of intracellular $[Na^+]$ resulting from the removal of extracellular Na^+. The reduced $[Na^+]_i$ decreases the Na^+ dependent release of Ca^{2+} from intracellular stores[2,3], which results in an increased accumulation of ionized Ca^{2+} in these stores. Previously we have shown that GBHA-Ca^{2+} strongly decreases on degranulation, which suggests that GBHA-Ca^{2+} is mainly localized in the

secretory vesicles[8]. A very speculative explanation for the different
sensitivity to changes in $[Na^+]_o$ above and below 120 mM may be that the
β-cell contains two Ca^{2+} stores with different sensitivities to Na^+.

REFERENCES

1. B. Hellman, T. Andersson, P.-O. Breggren, and P. Rorsman, Calcium and
    pancreatic β-cell function. XI. Modification of ^{45}Ca fluxes by Na^+
    removal, Biochem. Med. 24:143 (1980).
2. B. Hellman and E. Gylfe, Glucose inhibits ^{45}Ca efflux from pancreatic
    β-cells also in the absence of Na^+-Ca^{2+} countertransport, Biochim.
    Biophys. Acta 770:136 (1984).
3. B. Hellman, T. Honkanen, and E. Gylfe, Glucose inhibits insulin release
    induced by Na^+ mobilization of intracellular calcium, FEBS Lett.
    148:289 (1982).
4. A. Herchuelz and W.J. Malaisse, Regulation of calcium fluxes in rat
    pancreatic islets: Dissimilar effects of glucose and of sodium ion
    accumulation, J. Physiol. 302:263 (1980).
5. A. Herchuelz, A. Sener and W.J. Malaisse, Regulation of calcium fluxes
    in rat pancreatic islets. Calcium extrusion by sodium-calcium
    countertransport, J. Membr. Biol. 57:1 (1980).
6. J.P. Reeves and C.C. Hale, The stochiometry of the cardiac
    sodium-calcium exchange system, J. Biol. Chem. 259:7733 (1984).
7. G.H.J. Wolters, A. Pasma, J.B. Wiegman, and W. Konijnendijk,
    Glucose-induced changes in histochemically determined Ca^{2+} in β-cell
    granules, ^{45}Ca uptake, and total calcium content of rat pancreatic
    islets, Diabetes 33:409 (1984).
8. G.H.J. Wolters, A. Pasma, J.B. Wiegman, and W. Konijnendijk, Changes in
    histochemically detectable calcium and zinc during
    tolbutamide-induced degranulation and subsequent regranulation of
    rat pancreatic islets, Histochem. 78:325 (1983).

MOBILIZATION OF DIFFERENT POOLS OF GLUCOSE-INCORPORATED CALCIUM IN

PANCREATIC β-CELLS AFTER MUSCARINIC RECEPTOR ACTIVATION

B. Hellman, E. Gylfe and P. Bergsten

Department of Medical Cell Biology
Biomedicum
University of Uppsala
S-751 23, Uppsala
Sweden

The introduction of new techniques and access to clonal lines of insulin-secreting cells have resulted in a re-evaluation of how glucose affects Ca^{2+} movements in the pancreatic β-cells. Contrary to previous views that calcium mobilization from intracellular stores is an important factor in glucose stimulated insulin release[30,35] and that the sugar inhibits the extrusion of Ca^{2+} from the β-cells[20,35], it was demonstrated that glucose has the opposite effects[14-16]. The action of glucose on the cytoplasmic Ca^{2+} regulating insulin release can consequently be regarded as reflecting the balance between increased entry of Ca^{2+} into the β-cells and the enhanced removal of the ion from the cytoplasm following intracellular trapping and stimulated outward transport. Although each of these three important movements of Ca^{2+} are stimulated, individual differences exist with regard to the latency of the effects and their sensitivity to the glucose stimulus. Moreover, with the prolongation of the exposure to glucose the promotion of the intracellular buffering of Ca^{2+} becomes less pronounced due to a limited capacity for sequestration.

To elucidate how the opposing effects of glucose on the cytoplasmic Ca^{2+} determine the function of the β-cells it is important to identify the pools involved in the intracellular sequestration of the ion. So far, labelling of the organelles in the intact β-cell with ^{45}Ca has suggested a role for the mitochondria in the sequestration process[1,18], an idea compatible with the observation that ^{45}Ca incorporated in response to glucose is mobilized after raising intracellular Na$^+$ [19]. This communication deals with the calcium mobilization obtained after muscarinic receptor activation with particular reference to the question whether glucose promotes high affinity uptake of Ca^{2+} by the endoplasmic reticulum.

MATERIALS AND METHODS

Studies of Isolated Islets

Pancreatic islets were collagenase isolated from ob/ob-mice taken from a local colony. These islets contain more than 90% β-cells, responding adequately to glucose and other stimulators of insulin release. After loading of the islets with ^{45}Ca, the radioactive efflux was determined as previously described[10]. The basal medium used for the efflux studies with ^{45}Ca was a Hepes buffer physiologically balanced in cations with Cl$^-$ as the sole anion[13].

Studies of a Clonal Line of Tumour-transformed β-cells

Insulin-releasing cells of the rat line RINm5F were used for measuring the cytoplasmic Ca^{2+} activities with quin-2[34] and the net fluxes of Ca^{2+} with arsenazo III[12]. In the latter type of studies pH was clamped at 7.4 by additions of 50 μl samples of 2 M NaOH.

RESULTS AND DISCUSSION

Effects of Carbamylcholine on the Efflux of ^{45}Ca

Measurements of the unidirectional efflux of ^{45}Ca from islets perifused with a Ca^{2+}-deficient medium provide a sensitive means for studying mobilization of intracellular calcium. Using this approach muscarinic receptor activation was found to induce a biphasic stimulation of the efflux of ^{45}Ca from β-cell-rich pancreatic islets isolated from ob/ob-mice[14,17]. The efflux pattern was strikingly different in the presence and absence of glucose. As shown in Fig. 1, the addition of carbamylcholine to a Ca^{2+}-deficient perifusion medium devoid of glucose

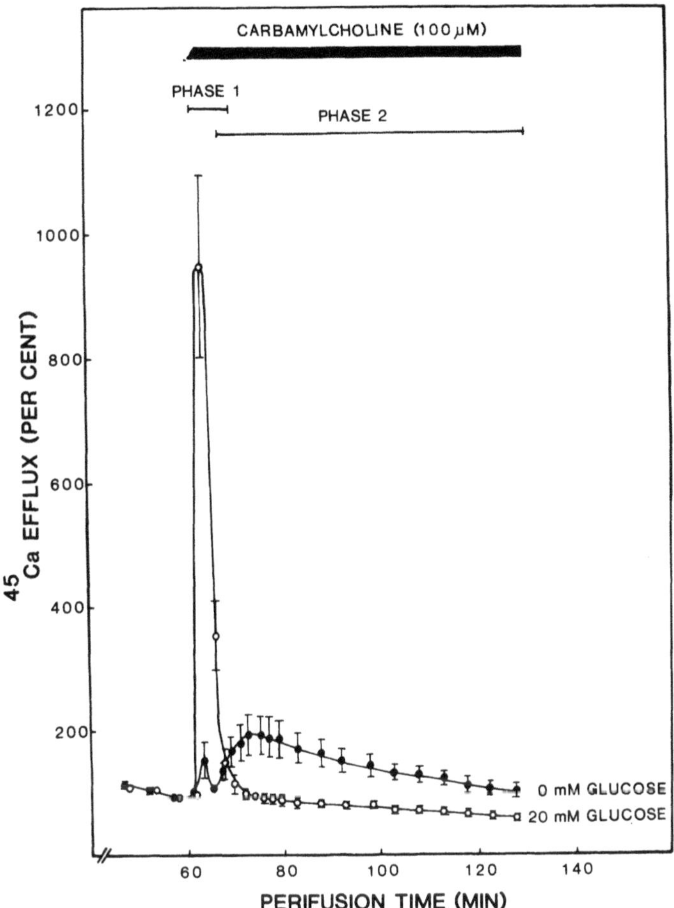

Fig. 1. Effect of carbamylcholine on ^{45}Ca efflux in the presence and
absence of glucose. Islets from ob/ob-mice were loaded for 90 min
with ^{45}Ca in a glucose-containing medium and perifused in the
presence (o) or absence (●) of the same concentration of the sugar
(20 mM) with a Ca^{2+}-deficient medium supplemented with 0.5 mM
EGTA. Carbamylcholine was added to a concentration of 100 μM
during the period indicated by the horizontal black bar. Mean
values ± SE for 5 experiments.

resulted in a small initial peak (phase 1) followed by a secondary rise
approaching a maximum after 10-15 min (phase 2). However, in the presence
of 20 mM glucose, the addition of carbamylcholine induced prominent initial
stimulation without subsequent appearance of phase 2.

Further analyses of the ^{45}Ca efflux provided evidence supporting the
idea that the biphasic nature of the carbamylcholine stimulation reflected
mobilization of different pools of calcium incorporated in response to
glucose. By varying the time sequence for the addition of the sugar it was

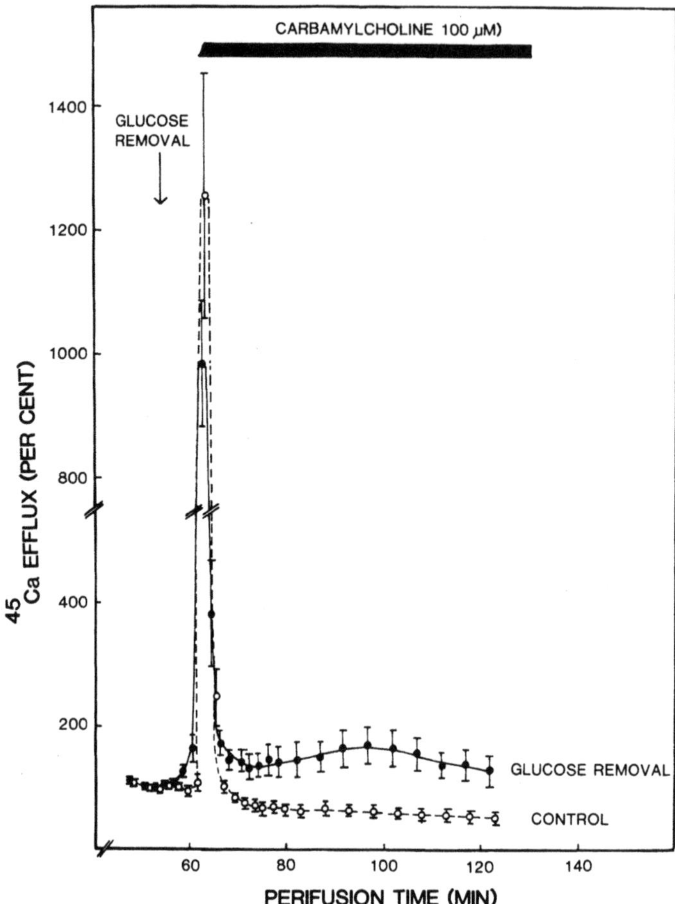

Fig. 2.   Effect of glucose removal on carbamylcholine stimulation of ^{45}Ca efflux.   Islets from ob/ob-mice were loaded for 90 min with ^{45}Ca in a glucose-containing medium and perifused in the presence of the same concentration of the sugar (20 mM) with a Ca^{2+}-deficient medium supplemented with 0.5 mM EGTA.   Carbamylcholine was added to a concentration of 100 μM during the period indicated by the horizontal black bar.   Glucose was either removed (●) or not (o) from the medium 6 min before the addition of carbamylcholine. Mean values ± SE for 5 experiments.

apparent that phase 1 stimulation required glucose during the period preceding the addition of carbamylcholine but not concurrently with it. The exposure to carbamylcholine resulted, for example, in substantial initial stimulation also six min after omission of glucose (Fig. 2).   The dependence of phase 2 stimulation on glucose was different.   Whereas the sugar had a suppressive action when included in the perifusion medium, it had to be present during the loading of the islets with ^{45}Ca.

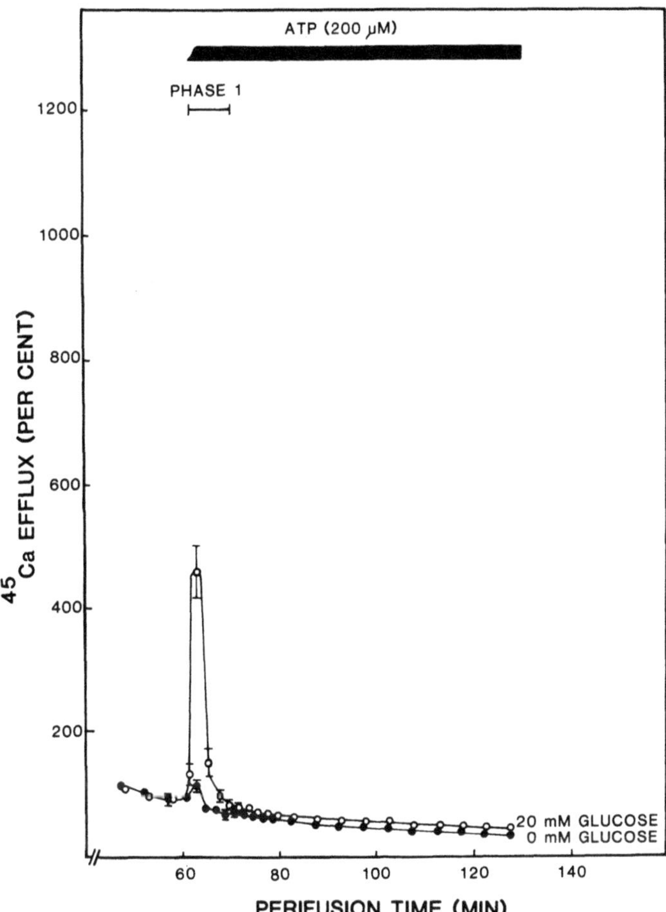

Fig. 3.   Effect of ATP on ^{45}Ca efflux in the presence and absence of ^{45}Ca glucose. Islets from <u>ob/ob</u>-mice were loaded for 90 min with ^{45}Ca in a glucose-containing medium and perifused in the presence (o) or absence (•) of the same concentration of the sugar (20 mM) with a Ca^{2+}-deficient medium supplemented with 0.5 mM EGTA. ATP was added to a concentration of 200 µM during the period indicated by the horizontal black bar. Mean values ± SE for 4 experiments.

A prominent initial stimulation of ^{45}Ca efflux was obtained not only after activation of muscarinic receptors but also in the presence of other external stimuli supposedly promoting phosphoinositide breakdown. Agonist binding to the purinergic P_2-receptor has been reported to induce hydrolysis of phosphatidylinositol 4,5-bisphosphate in both liver[6] and blood platelets[24]. Fig. 3 shows how ATP stimulation of the P_2-purinoceptor affects the efflux of ^{45}Ca from islets perifused with a Ca^{2+}-deficient medium. It is evident that also this nucleotide initiates rapid stimulation in the presence of glucose. However, contrary to what is seen after muscarinic receptor activation a phase 2 effect did not appear in the

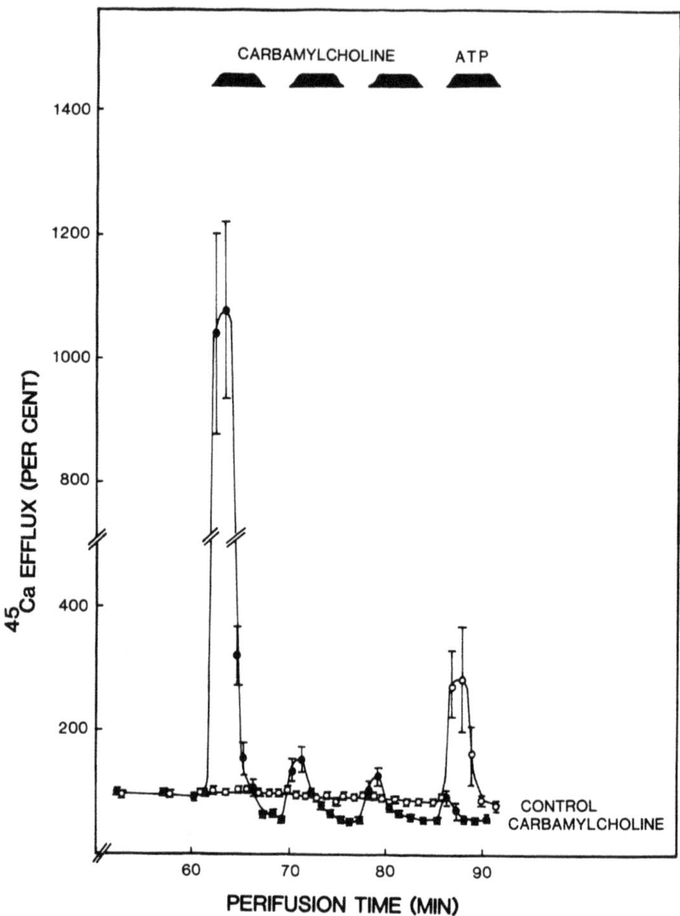

Fig. 4. Effect of repeated exposure to carbamylcholine on subsequent ATP
stimulation of ^{45}Ca efflux. Islets from ob/ob-mice were loaded
for 90 min with ^{45}Ca in a glucose-containing medium and perifused
in the presence of the same concentration of the sugar (20 mM)
with a Ca^{2+}-deficient medium supplemented with 0.5 mM EGTA. As
indicated by the horizontal black bars the islets were exposed (●)
or not (o) to three 4 min pulses of 100 μM carbamylcholine before
a final pulse of 200 μM ATP. Mean values ± SE for 5 experiments.

absence of glucose. When exposure to the agonists was interrupted with
intervals of perifusion with a glucose-containing medium, it was possible
to induce repeated peaks of initial ^{45}Ca efflux. Fig. 4 presents results
of experiments with three periods of perifusion with 100 μM carbamylcholine
accompanied by the presence of 200 μM ATP. It can be seen that the
previous exposure to carbamylcholine results in a disappearance of most of
the calcium otherwise mobilized by ATP. From these and other studies of
how carbamylcholine and ATP affect the ^{45}Ca efflux it can be concluded that
different receptors related to polyphosphoinositide breakdown can mediate a

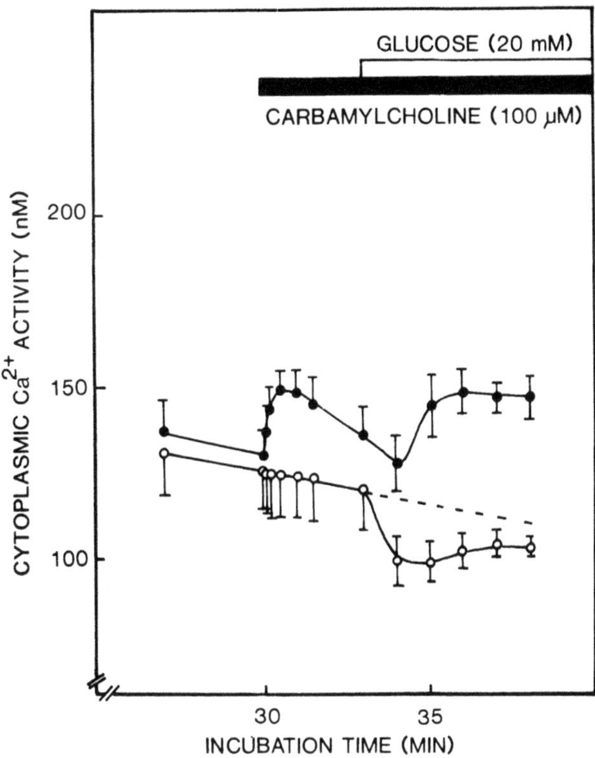

Fig. 5. Effects of carbamylcholine and glucose on the cytoplasmic Ca^{2+} activities of RINm5F cells at a physiological concentration of extracellular Ca^{2+}. Cells were detached with trypsin, incubated for 60 min at 37°C and then loaded for 60 min with 15 µM of the tetraacetoxymethyl-ester of quin-2 dissolved in RPMI 1640 medium. After repeated washing in the absence of glucose in a salt-balanced medium containing 1.28 mM Ca^{2+}, the quin-2 loaded cells were suspended in the same type of medium and later used for measurements of fluorescence (339/492 nm) during constant stirring at 37°C. After 30 min of exposure of the cells to the glucose-free medium carbamylcholine was added (●) or not (o) at a final concentration of 100 µM. In both cases further addition of glucose was made 3 min later. The dotted line indicates the expected activities of cytoplasmic Ca^{2+} in the absence of glucose. The validity of calculating these values from linear extrapolations of the fluorescence signal in each experiment was ascertained separately. Mean values ± SE for 6 experiments.

similar event of initial calcium mobilization dependent on previous exposure to glucose.

Procedures for raising the intracellular Na$^+$ activity have been found to be effective in mobilizing a slowly exchangeable pool of glucose-incorporated calcium probably located in the mitochondria[19]. It is therefore of interest to note, that Gagerman et al.[9] observed carbamylcholine-

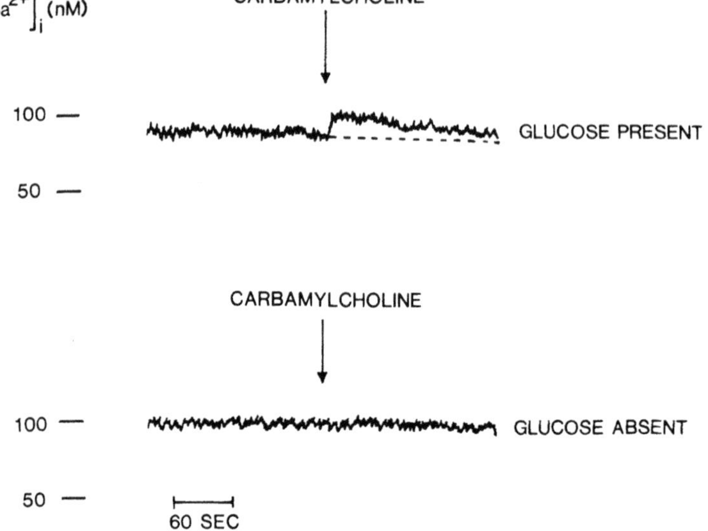

Fig. 6. Effect of carbamylcholine on the cytoplasmic Ca^{2+} activities of
RINm5F cells suspended in medium nominally devoid of Ca^{2+}. The
experiments were performed essentially as described in Fig. 5 with
cells loaded with 7.5 µM of the tetraacetoxymethylester of quin-2.
Results are presented as fluorescent traces obtained in
representative experiments with cells incubated in a
Ca^{2+}-deficient medium in the presence (upper trace) or absence
(lower trace) of 20 mM glucose. After 27 min exposure to the
Ca^{2+}-deficient medium EGTA was added at a final concentration of
0.5 mM followed by 100 µM carbamylcholine 5 min later.

induced uptake of ^{22}Na in islets isolated from ob-ob-mice. Since it was
apparent from our studies that removal of extracellular Na^+ resulted in a
disappearance of carbamylcholine-induced phase 2 stimulation, it seems
likely that this phase can be accounted for by a rise of intracellular Na^+.
The observation of glucose suppression of phase 2 stimulation can be taken
as a further argument for this view, since evidence has been provided that
the sugar also counteracts the mobilization of calcium obtained when
intracellular Na^+ is raised by removal of K^+ or addition of ouabain and
veratridine[19].

Effects of Carbamylcholine on the Cytoplasmic Ca^{2+} Activity

Measurements of the cytoplasmic Ca^{2+} activity in the insulin-releasing
RINm5F cells with the quin-2 indicator provided further evidence for a role
of glucose in preparing the β-cells to respond to muscarinic receptor
activation. When the cells were suspended in medium containing 1.28 mM

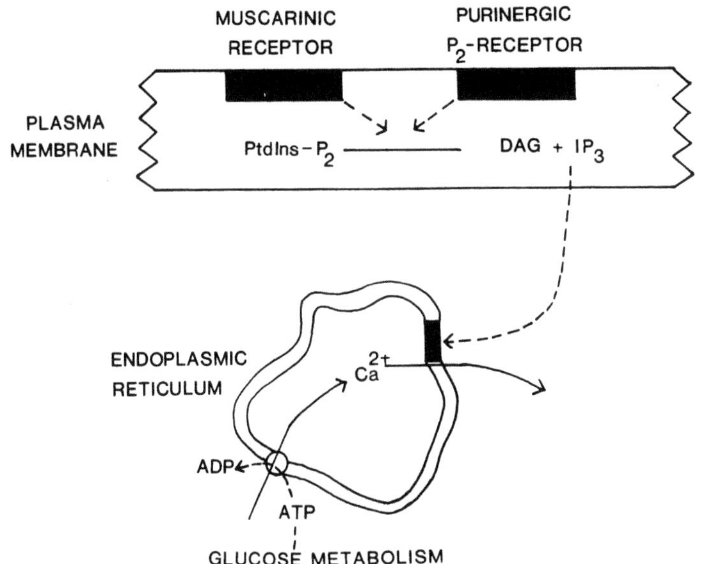

Fig. 7. Diagram illustrating hydrolysis of phosphatidylinositol 4,5-bis-phosphate (PtdIns-P$_2$) accounted for by phospholipase C activation mediated by muscarinic and P$_2$-purinoceptors. Whereas the resulting diacylglycerol (DAG) sensitizes the secretory machinery to the Ca^{2+} signal by activating protein kinase C, the 1,4,5 isomer of inositol trisphosphate (IP$_3$) is supposed to mobilize Ca^{2+} from the endoplasmic reticulum. If phospholipase C activation results in formation of inositol phosphates other than 1,4,5-trisphosphate, the net result may even be sequestration of Ca^{2+} in the intracellular organelles due to the diacylglycerol activation of protein kinase C.

Ca^{2+}, the addition of 100 µM carbamylcholine initiated a prompt increase of the cytoplasmic Ca^{2+} activity. This effect was especially pronounced in the presence of 20 mM glucose, but carbamylcholine had a distinct action also when the cells had been deprived of glucose for 30 min. Unlike the case in normal β-cells, the glucose effect on the voltage-dependent Ca^{2+}-channels in the RINm5F cells is not sufficient to overcome the intracellular sequestration of Ca^{2+}. Consequently, glucose lowers the cytoplasmic Ca^{2+} activity in these cells even at relatively high concentrations of extracellular Ca^{2+} [34]. As shown in Fig. 5 activation of the muscarinic receptors made the cells respond to glucose with increase of the cytoplasmic Ca^{2+} activity. The observation that exposure to carbamylcholine makes glucose able to raise cytoplasmic Ca^{2+} in the RINm5F cells has its counterpart in the stimulation of insulin release from normal β-cells. Whereas activation of muscarinic receptors in itself has little effect on insulin release, it results in considerable potentiation of the secretagogic action of glucose [9].

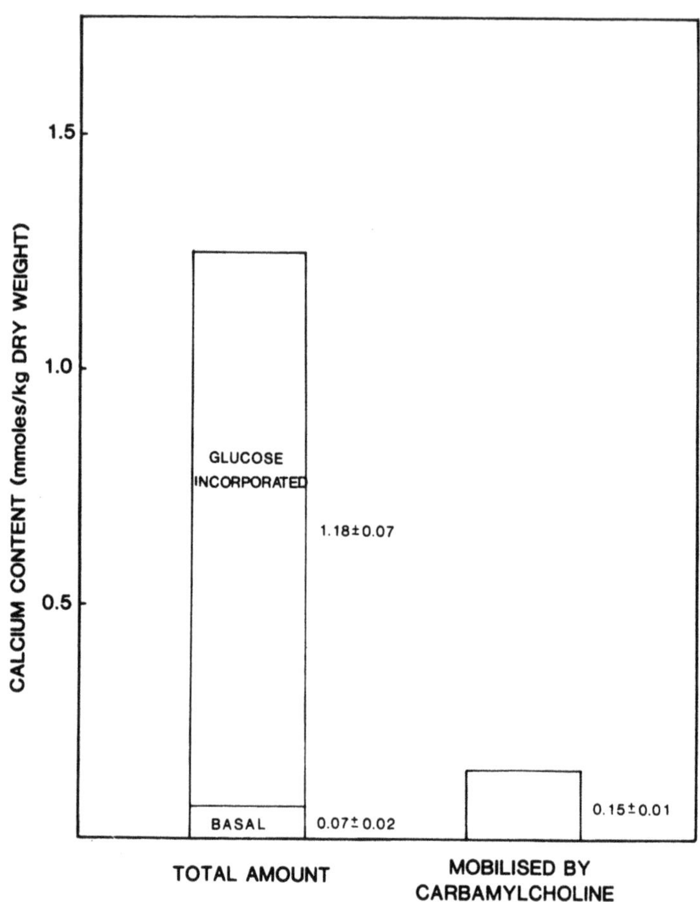

Fig. 8. Calcium contents of RINm5F cells suspended in Ca^{2+}-deficient media before and after exposure to glucose and the amounts mobilized after subsequent addition of carbamylcholine. Cells initially depleted of calcium by equilibration in medium containing 10-20 μM Ca^{2+} were allowed to incorporate calcium to a steady state value in the presence of 20 mM glucose before being exposed to 100 μM carbamylcholine. The cellular contents of calcium were estimated from spectrophotometric dual wavelength recordings of the extracellular Ca^{2+} concentration employing the indicator arsenazo III. The basal content of calcium is defined as the amounts released by subsequent additions of antimycin A and the ionophore A-23187 to cells not exposed to glucose and carbamylcholine. Results are presented as mmoles calcium per kg protein. Mean values ± SE for 7-24 experiments.

In support for the view that muscarinic receptor activation results in mobilization of calcium from intracellular stores, it was possible to confirm the reports of Prentki and Wollheim[30] that carbamylcholine increases cytoplasmic Ca^{2+} also in a medium nominally devoid of this cation. Our studies indicated that this effect disappeared when the cells had been deprived of glucose for 30 min (Fig. 6). The increase of cyto-

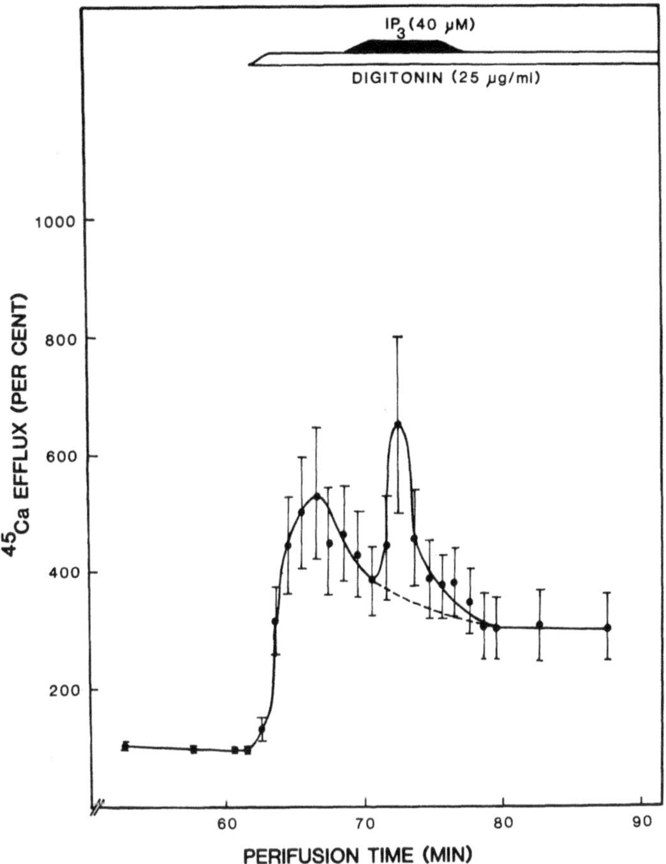

Fig. 9.   Effect of inositol 1,4,5-trisphosphate on ^{45}Ca efflux from islets
with permeabilized cells. Islets from <u>ob/ob-mice</u> were loaded for
90 min with ^{45}Ca in a glucose-containing medium and perifused at
18°C in the presence of the same concentration of glucose (20 mM)
with medium deficient both in Na^{+} (replaced with K^{+}) and Ca^{2+} and
containing 0.1 mM EGTA. Digitonin and inositol 1,4,5-trisphosphate
(IP$_3$) were added during the periods indicated by the horizontal
bars. Mean values ± SE for 5 experiments.

plasmic Ca^{2+} obtained with carbamylcholine consequently exhibited a similar

dependence on glucose as the rapid efflux of ^{45}Ca from islets perifused

with a Ca^{2+}-deficient medium.

Effects of Carbamylcholine on Net Fluxes of Ca^{2+}

It can be concluded from the studies described so far, that external

stimuli mediating hydrolysis of inositol phospholipids induce rapid

mobilization of intracellular calcium when the β-cells have been exposed to
glucose. This effect is probably accounted for by the formation of
inositol 1,4,5-trisphosphate, a water-soluble compound found to mobilize
intracellular calcium from permeabilized RINm5F cells[5,22]. Fig. 7 illustrates schematically how receptor-induced activation of phospholipase C
results in hydrolysis of phosphatidylinositol 4,5-bisphosphate with the
formation of diacylglycerol and inositol trisphosphate. In accordance with
observations made in various types of cells[2], including those from a rat
insulinoma[31,32], the endoplasmic reticulum can be regarded as the source of
calcium sensitive to inositol 1,4,5-trisphosphate. Whereas activation of
phospholipase C invariably results in formation of the protein kinase C
stimulating diacylglycerol, the additional product may well be substances
without Ca^{2+} mobilizing actions such as the 1,3,4-trisphosphate isomer of
inositol or its mono- or bisphosphates. As discussed below, activation of
phospholipase C without significant formation of the active inositol
1,4,5-trisphosphate may even result in effects opposite to mobilization of
intracellular calcium.

Glucose was found to be important in preparing the β-cells to respond
rapidly to carbamylcholine and other stimulators of inositol phospholipid
breakdown with mobilization of intracellular calcium. Although this
requirement may reflect the need of energy for providing the precursor of
the active inositol 1,4,5-trisphosphate, the sugar may also be important in
promoting accumulation of calcium in the intracellular pool affected. The
latter alternative deserves particular attention, since it implicates
glucose promotion of high affinity uptake of calcium into an intracellular
pool also during the perifusion with the Ca^{2+}-deficient medium. Indeed, we
have recently been able to demonstrate by measurements of net fluxes of
Ca^{2+} with dual wavelength spectrophotometry that activation of the
muscarinic receptors in the insulin-releasing RINm5F cells results in
mobilization of calcium incorporated in response to glucose[11]. The studies
were performed with the aid of the Ca^{2+} indicator arsenazo III using cells
initially depleted of calcium by incubation in a phosphate-free medium
containing only 10-20 μM Ca^{2+}. The addition of glucose resulted in a
stimulated net uptake of Ca^{2+}, reaching saturation after 20 min. At this
time the cellular content of calcium had increased from a basal value of
0.07 ± 0.02 mmol per kg protein by 1.18 ± 0.07. Carbamylcholine had no
effect on the net fluxes of Ca^{2+} when added prior to the glucose exposure.
However, after the cells had incorporated calcium to a steady state level

in the presence of glucose, subsequent exposure to 100 μM carbamylcholine
resulted in the mobilization of twice as much calcium as originally present
(Fig. 8). In addition to demonstrating that carbamylcholine initiates
efflux of calcium incorporated in response to glucose, it was evident that
most of this calcium originated from an intracellular pool insensitive to
mitochondrial poisons. The opposing actions of glucose and carbamylcholine
on this pool remained after removal of extracellular Na^+.

Effects of Inositol 1,4,5-trisphosphate on ^{45}Ca Efflux from Permeabilized Islets

Endoplasmic reticulum is characterized by high affinity uptake of Ca^{2+}
sensitive to mobilization by inositol 1,4,5-trisphosphate. We have now
tested whether pre-exposure to glucose is a requirement for inositol
1,4,5-trisphosphate mobilisation of ^{45}Ca from permeabilized cells. In view
of the lability of the calcium pool under consideration, these experiments
had to be performed at room temperature. Fig. 9 illustrates the pulse of
initial mobilization of ^{45}Ca obtained, when the intracellular phosphate
messenger was added to a perifusion medium containing the permeabilizing
agent digitonin. The absence of effect in similarly designed experiments
performed in the absence of glucose during the loading with ^{45}Ca and
subsequent perifusion supports the idea, that inositol 1,4,5-trisphosphate
selectively mobilizes ^{45}Ca incorporated in response to glucose.

Comparisons of the Effects of Carbamylcholine and Glucose

Whereas radioisotope studies of the unidirectional efflux of Ca^{2+} as
well as measurements of the cytoplasmic Ca^{2+} activity and the net fluxes of
Ca^{2+} indicate that muscarinic receptor activation results in mobilization
of intracellular calcium, glucose appears to have the opposite effect.
There is evidence that glucose promotes intracellular sequestration of Ca^{2+}
in the presence of both low and high extracellular concentrations of the
ion[14-16]. It can be argued that a hypothetical action of glucose in
mobilizing intracellular calcium requires previous exposure to glucose in
analogy to phase 1 stimulation of ^{45}Ca efflux obtained when adding
carbamylcholine to a Ca^{2+}-deficient medium. However, even under conditions
of pre-exposure to glucose a further rise of its concentration has been
found to result only in inhibition of the efflux of ^{45}Ca. Differences
exist between the effects of glucose and carbamylcholine not only on the
β-cell handing of Ca^{2+} but also with regard to that of K^+ and PO_4^{3-}.[26,29]

As with exposure to cholinergic agents the presence of glucose has been reported to induce hydrolysis of inositol phospholipids in the islets by activating phospholipase C[3,4,23,27,28]. Although this effect was originally proposed to be dependent on extracellular Ca^{2+}[23,33], other studies have indicated that the ability of glucose to initiate polyphosphoinositide breakdown[8] with formation of inositol trisphosphate[27,28] remains unaffected in a Ca^{2+}-deficient medium supplemented with EGTA. At first glance it seems difficult to reconcile glucose stimulation of phosphoinositide turnover with a promoting effect of the sugar on intracellular sequestration of calcium. Glucose stimulation of phosphoinositide breakdown may involve pools other than those affected by carbamylcholine. Moreover, there is no conclusive evidence that the exposure to glucose actually results in increased amounts of the Ca^{2+}-mobilizing inositol 1,4,5-trisphosphate. The measurements of this compound reported so far, apparently include also the inactive 1,3,4-trisphosphate isomer[2,21]. If glucose stimulation of the phospholipase C activity results essentially in formation of water-soluble inositol phosphates other than 1,4,5-trisphosphate, it is even possible that this process contributes to the sequestration of calcium in the organelles via the diacylglycerol stimulation of protein kinase C. The latter enzyme has been proposed to have a negative feedback role in agonist-induced Ca^{2+} mobilization[7], implying that it might faciliate ATP-dependent Ca^{2+} uptake by intracellular organelles[25].

SUMMARY

Muscarinic receptor activation resulted in a biphasic mobilization of Ca^{2+} from isolated pancreatic islets. Glucose was essential for preparing the β-cells to respond with the initial stimulatory phase. This effect seems to depend on the ability of the sugar to promote active sequestration of Ca^{2+} in the endoplasmic reticulum.

ACKNOWLEDGEMENTS

The authors are indebted to Professor Michael Berridge, Department of Zoology, University of Cambridge, England, for helpful discussions regarding inositol 1,4,5-trisphosphate. The work has been supported by the Swedish Medical Research Council (12x-562), the Swedish Diabetes Association and the Nordic Insulin Foundation.

REFERENCES

1. T. Andersson, P.-O. Berggren, E. Gylfe, and B. Hellman, Amounts and
    distribution of intracellular magnesium and calcium in pancreatic
    β-cells, Acta Physiol. Scand. 114:235 (1982).
2. M.J. Berridge and R.F. Irvine, Inositol trisphosphate, a novel second
    messenger in cellular signal transduction, Nature 312:315 (1984).
3. L. Best and W.J. Malaisse, Stimulation of phosphoinositide breakdown
    in rat pancreatic islets by glucose and carbamylcholine, Biochem.
    Biophys Res. Commun. 116:9 (1983).
4. L. Best and W.J. Malaisse, Nutrient and hormone neurotransmitter
    stimuli induce hydrolysis of polyphosphoinositides in rat
    pancreatic islets, Endocrinology 115:1814 (1984).
5. T.J. Biden, M. Prentki, R.F. Irvine, M.J. Berridge, and C.B. Wollheim,
    Inositol 1,4,5-trisphosphate mobilizes intracellular Ca^{2+} from
    permeabilized insulin-secreting cells, Biochem. J. 223:467 (1984).
6. J.A. Creba, C.P. Downes, P.T. Hawkins, G. Brewster, R.H. Michell and
    C.J. Kirk, Rapid breakdown of phosphatidylinositol 4-phosphate and
    phosphatidylinositol 4,5-bisphosphate in rat hepatocytes stimulated
    by vasopressin and other Ca^{2+}-mobilizing hormones, Biochem. J.
    212:733 (1983).
7. A.H. Drummond, Bidirectional control of cytosolic free calcium by
    thyrotropin-releasing hormone in pituitary cells, Nature 315:752
    (1985).
8. M.E. Dunlop and R.G. Larkins, The role of calcium in phospholipid
    turnover following glucose stimulation in neonatal rat cultured
    islets, J. Biol. Chem. 259:8407 (1984).
9.. E. Gagerman, J. Sehlin and I.B. Täljedal, Effects of acetylcholine on
    ion fluxes and chlortetracycline fluorescence in pancreatic islets,
    J. Physiol. 300:505 (1980).
10. E. Gylfe and B. Hellman, Calcium and pancreatic β-cell function.
    2. Mobilization of glucose-sensitive ^{45}Ca from perifused islets
    rich in β-cells, Biochim. Biophys Acta 538:249 (1978).
11. E. Gylfe and B. Hellman, Glucose-stimulated sequestration of calcium
    in clonal insulin-releasing cells. Evidence for an opposing effect
    of muscarinic receptor activation, Biochem. J. 233:865 (1986).

12. E. Gylfe, T. Andersson, P. Rorsman, H. Abrahamsson, P. Arkhammar, P. Hellman, B. Hellman, H.K. Oie, and A.F. Gazdar, Depolarization-independent net uptake of calcium into clonal insulin-releasing cells exposed to glucose, Biosci, Rep 3:927 (1983).

13. B. Hellman, The significance of calcium for glucose stimulation of insulin release, Endocrinology 97:392 (1975).

14. B. Hellman, β-cell cytoplasmic Ca^{2+} balance as a determinant for glucose-stimulated insulin release, Diabetologia 28:494 (1985).

15. B. Hellman and E. Gylfe, Glucose regulation of insulin release involves intracellular sequestration of calcium, in: "Calcium in Biological Systems", R.P. Rubin, G.B. Weiss, and J.W. Putney, Jr., eds., pp 93-99, Plenum Publishing Corp, New York (1985).

16. B. Hellman and E. Gylfe, Calcium and the control of insulin secretion, in: "Calcium and Cell Function", W.Y. Cheung, ed., vol. 6:pp 253-326, Academic Press, New York (1985).

17. B. Hellman and E. Gylfe, Mobilization of different intracellular calcium pools after activation of muscarinic receptors in pancreatic β-cells, Pharmacology, in press(1986).

18. B. Hellman, T. Andersson, P.-O. Berggren, P. Flatt, E. Gylfe, and K.D. Kohnert, The role of calcium in insulin secretion, in: "Hormone and Cell Regulation", J. Dumont and J. Nunez, eds, vol. 3: pp 69-96, Elsevier/North Holland Biomedical Press, Amsterdam (1979).

19. B. Hellman, T. Honkanen, and E. Gylfe, Glucose inhibits insulin release induced by Na^+ mobilization of intracellular calcium, FEBS Lett. 148:289 (1982).

20. A. Herchuelz and W.J. Malaisse, Calcium movements and insulin release in pancreatic islets, Diabete. Metab. 7:283 (1981).

21. R.F. Irvine, E.E. Anggard, A.J. Letcher, and C.P. Downes, Metabolism of inositol 1,4,5-trisphosphate and inositol 1,3,4-trisphosphate in rat parotid glands, Biochem. J. 229:505 (1985).

22. S.K. Joseph, R.J. Williams, B.E. Corkey, F.M. Matschinsky, and J.R. Williamson, The effect of inositol trisphosphate on Ca^{2+} fluxes in insulin-secreting tumor cells, J. Biol. Chem. 259:12952 (1984).

23. S.G. Laychock, Identification and metabolism of polyphosphoinositides in isolated islets of Langerhans, Biochem. J. 216:101 (1983).

24. N.L. Leung, J.D. Vickers, R.L. Kinlough-Rathbone, H.-J. Reimers, and J.F. Mustard, ADP-induced changes in $[^{32}P]$phosphate labeling of phosphatidylinositol-4,5-bisphosphate in washed rabbit platelets made refractory by prior ADP stimulation, Biochem. Biophys. Res. Commun. 113:483 (1983).

25. C.J. Limas, Phosphorylation of cardiac sarcoplasmic reticulum by a calcium-activated phospholipid-dependent protein kinase, _Biochem. Biophys. Res. Commun._ 96:1378 (1980).

26. P.C.F. Mathias, A.R. Carpinelli, and W.J. Malaisse, Ionic response to cholinergic agents in pancreatic islets (abstract), _Diabetologia_ 27:308A (1984).

27. W. Montague, N.G. Morgan, G.M. Rumford, and C.A. Prince, Effect of glucose on polyphosphoinositide metabolism in isolated rat islets of Langerhans, _Biochem. J._ 227:483 (1985).

28. N.G. Morgan, G.M. Rumford, and W. Montague, Studies on the role of inositol trisphosphate in the regulation of insulin secretion from isolated rat islets of Langerhans, _Biochem. J._ 228:713 (1985).

29. M. Nenquin, P. Awouters, F. Mathot, and J.C. Henquin, Distinct effects of acetylcholine and glucose on ^{45}calcium and ^{86}rubidium efflux from mouse pancreatic islets, _FEBS Lett._ 176:457 (1984).

30. M. Prentki and C.B. Wollheim, Cytosolic free Ca^{2+} in insulin secreting cells and its regulation by isolated organelles, _Experientia_ 40:1052 (1984).

31. M. Prentki, T.J. Biden, D. Janjic, R.F. Irvine, M.J. Berridge, and C.B. Wollheim, Rapid mobilization of Ca^{2+} from rat insulinoma microsomes by inositol-1,4,5-trisphosphate, _Nature_ 309:562 (1984).

32. M. Prentki, B.E. Corkey and F.M. Matschinsky, Inositol 1,4,5-trisphosphate and the endoplasmic reticulum Ca^{2+} cycle of a rat insulinoma cell line, _J. Biol. Chem._ 260:9185 (1985).

33. R.S. Rana, R.J. Mertz, A. Kowluru, J.F. Dixon, L.E. Hokin, and M.J. MacDonald, Evidence for glucose-responsive and unresponsive pools of phospholipids in pancreatic islets, _J. Biol. Chem._ 260:7861 (1985).

34. P. Rorsman, P.-O. Berggren, E. Gylfe, and B. Hellman, Reduction of the cytosolic calcium activity in clonal insulin-releasing cells exposed to glucose, _Biosci, Rep._ 3:939 (1983).

35. C.B. Wollheim and G.W.G. Sharp, Regulation of insulin release by calcium, _Physiol. Rev._ 61:914 (1981).

EFFECT OF THE ORDER OF APPLICATION OF NEURAL INPUTS ON INSULIN SECRETION

L.A. Campfield, F.J. Smith, J.E. Settle
and R. Sohaey

Departments of Physiology and Biomedical Engineering
The Medical School and
The Technological Institute
Northwestern University
Chicago and Evanston
Illinois 60611 and 60201

Sympathetic and parasympathetic neural inputs to the pancreatic β-cell have been shown to modify insulin secretion by many investigators (for reviews see[3,5,9-12]). Electrical stimulation of sympathetic nerves to the pancreas or exposure of the pancreas to norepinephrine inhibited insulin secretion at intermediate and high glucose concentrations, while stimulation of parasympathetic nerves or exposure to acetylcholine increased insulin secretion at low and intermediate glucose concentrations[1-4,5,8,9,12]. We have previously determined the interaction of autonomic neural inputs during simultaneous application of selected concentrations of exogenous acetylcholine and norepinephrine on insulin secretion from isolated pancreatic islets in the presence of an intermediate glucose concentration[3]. We found three response regimes: net stimulation (parasympathetic dominance), net inhibition (sympathetic dominance) and no net effect (cancellation). We concluded that insulin secretion was a complex function of both the absolute and relative magnitudes of neural inputs in addition to the local glucose concentration.

We have investigated the effects of sequential application of autonomic neural inputs on insulin secretion. We report here the effects of the order of application of acetylcholine or/and norepinephrine on

insulin secretion from perifused, isolated pancreatic islets in the
presence of an intermediate glucose concentration.

METHODS

Pancreatic islets were isolated from ad lib fed, female Wistar rats
(200-250 gm) by a modification of the collagenase technique[7]. Thirty
freshly isolated islets of approximately equal size were loaded into 25 mm
diameter perifusion chambers fitted with cellulose acetate membranes with 5
micron diameter pores (Gelman Instrument Company, Ann Arbor, Michigan,
USA). Chambers were perifused with Krebs-Ringer bicarbonate buffer, pH =
7.4, containing 0.3% bovine serum albumin which was continuously
equilibrated with 95% O_2/5% CO_2. Chambers were placed in a water bath to
maintain islet temperature at 37°C.

The volumetric flowrate of the buffer ranged from 0.6 to 1.3 ml/min
and was measured frequently by four minute timed collections. One minute
aliquots of the effluent were collected and selected samples were frozen
and stored -20°C for insulin assay.

Islets were perifused with 5 mM glucose for the initial 60 minutes,
then with 10 mM glucose for the next 150 or 180 minutes, followed by 5 mM
for the final 60 minutes. Following 60 minutes of 10 mM glucose, islets
were exposed to either 30 µM acetylcholine or 0.3 µM norepinephrine for 30
minutes followed by combined application of the same concentrations of
acetylcholine and norepinephrine for 30 minutes. In some experiments, a
third 30 minute period of neurotransmitter exposure was added. During this
period the islets were exposed to the second neuro-transmitter alone. The
same neurotransmitter concentrations were used.

Insulin secretion rate was calculated as follows:

$$\text{Insulin secretion rate (pg/min/islet)} = \frac{\text{Insulin conc (pg/ml) x Flowrate (ml/min)}}{\text{Number of islets/chamber}}$$

Incremental insulin secretion rate, over the 5 mM glucose baseline,
was calculated during each experimental period: the two or three periods of
neurotransmitter exposure and the pre- and post-neurotransmitter 10 mM

glucose periods. Incremental insulin secretion rate was then integrated
over the last 20 minutes of each experimental period and the percent change
from the average insulin secretion rate during the pre- and post-neurotransmitter 10 mM glucose periods was calculated. The resulting data was
averaged across all perifusions using the same order of neurotransmitter
application. Insulin concentrations were measured using a competitive
protein binding assay with rat insulin as standard[3]. Statistical
comparisons were made using Student's t-test for unpaired sample
populations.

RESULTS AND DISCUSSION

The mean percent changes in integrated insulin secretion during single
and combined neurotransmitter exposure are shown in Figure 1. When 30 μM
acetylcholine was applied first, insulin secretion rate rapidly increased
by 176 ±25% and then fell to 10 mM glucose control levels (17 ±14%) during
combined exposure to 30 μM acetylcholine and 0.3 μM norepinephrine (Figure
1a). When 0.3 μM norepinephrine was applied first, insulin secretion rate
was promptly inhibited by 62 ±6% and then increased to 45 ±7% below 10 mM
glucose control levels during combined exposure to 30 μM acetylcholine and
0.3 μM norepinephrine (Figure 1, bottom).

Comparison of insulin secretion rates during combined exposure to
identical acetylcholine and norepinephrine inputs revealed a significant
difference (17 ±14%) (acetylcholine first) vs. -45 ±7% (norepinephrine
first); p <0.05) depending on the order of presentation of neurotransmitter
inputs.

Although each neurotransmitter had its expected effect on insulin
secretion when applied alone[1-4,6,8,9,12] and when added in the presence of
the other neurotransmitter[3], the different rates of insulin secretion in
the presence of both neurotransmitters suggests that the potency of the
second neurotransmitter was decreased following exposure of the β-cell to
the first neurotransmitter alone. These results led us to postulate that
the potency of each neural input was a function of the magnitude of both
sympathetic and parasympathetic neural inputs. The experiment illustrated
in Fig 2 was designed to test this hypothesis.

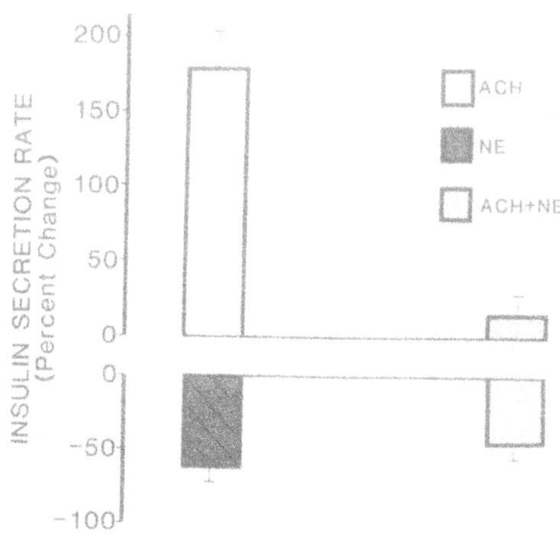

Fig. 1. Integrated insulin secretion during neurotransmitter exposure -
Experiment 1. Mean percent change in integrated insulin
secretion rate during single (left) or combined (right)
neurotransmitter exposure. Incremental insulin secretion rate
over 5 mM glucose baseline was integrated during last 20 minutes
of both single and combined neurotransmitter exposure and the
percent change from the mean insulin secretion rate during pre-
and post-neurotransmitter 10 mM glucose periods was calculated and
averaged across experiments. The acetylcholine concentration used
was 30 µM, while norepinephrine concentration was 0.3 µM. The
number of individual experiments for the acetylcholine first
design was 12, while for the norepinephrine first design was 8.
Note significantly different insulin secretion rates in response
to identical input during combined neurotransmitter exposure.

The mean percent changes in integrated insulin secretion during single
(first and third periods) and combined neurotransmitter exposure are shown
in Figure 2. During the first (single) and combined neurotransmitter
exposure, changes in integrated insulin rates similar to those seen in
experiment 1 were observed. When 30 µM acetylcholine was applied first,
insulin secretion rate rapidly increased by 123 ±15% and then fell to 10 mM
glucose control levels (3 ±9%) during combined exposure (Figure 2; top).
When 0.3 µM norepinephrine was applied first, insulin secretion rate was
promptly inhibited by 65 ±4% and then increased to 55 ±12% below 10 mM
glucose control levels during combined exposure (Figure 2, bottom).

When 0.3 µM norepinephrine was applied following exposure to acetylcholine alone and combined acetylcholine and norepinephrine, insulin
secretion rate decreased from 10 mM glucose control levels to 34 ±9% below
control. When 30 µM acetylcholine was applied following exposure to
norepinephrine alone and combined acetylcholine and norepinephrine, insulin

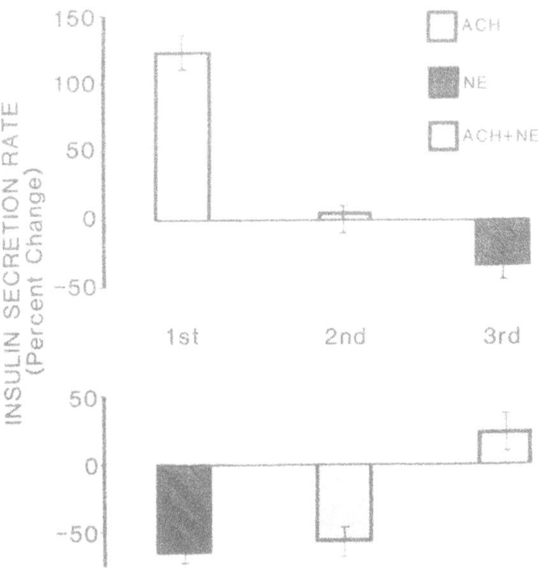

Fig. 2. Integrated insulin secretion during neurotransmitter exposure –
Experiment 2. Mean percent change in integrated insulin
secretion rate during single (first and third) and combined
(second) neurotransmitter exposure. The acetylcholine
concentration used was 30 μM, while the norepinephrine
concentration was 0.3 μM. The number of individual experiments
for the acetylcholine first design was 11, while for the
norepinephrine first design was 9. Note significant decrease in
potency of both acetylcholine and norepinephrine during their
exposure period compared to the first period.

secretion rate increased from 55 ±12% below to 24 ±10% above the 10 mM

glucose control levels (Figure 2, right side). In each case, following

combined exposure, the effect or potency of the second neurotransmitter

alone was significantly less than the effect of the same neurotransmitter

on insulin secretion when applied in the absence of previous neurotrans-

mitter exposure (acetylcholine: 24 ±10% vs 123 ±15%; norepinephrine: -34

±9% vs. -65 ±4%; p <0.05). Therefore, these results indicated that the

potencies of acetylcholine and norepinephrine were not constant but rather

were functions of previous neurotransmitter exposure. These results with
one pair of neurotransmitter concentrations were consistent with the
hypothesis that the potency of each neural input was a function of both
sympathetic and parasympathetic neural inputs.

DISCUSSION

These data suggest that both the order and temporal sequence of
sympathetic and parasympathetic neural inputs may be important determinants
of glucose-induced insulin secretion rate. Thus, insulin secretion in
response to identical inputs may be different due to the temporal history
of application of previous neural inputs. These studies imply the
existence of a β-cell memory for autonomic neural inputs as well as for
glucose.

Since pancreatic islets are innervated by both branches of the
autonomic nervous system and nerve endings have been observed in close
association with β-cells[9,12], it may be assumed that β-cells in vivo
receive temporal sequences of sympathetic and parasympathetic inputs in
which the absolute and relative magnitudes are functions of time. This
study implies that the β-cell performs temporal as well as spatial
integration of neurotransmitter inputs and that instantaneous insulin
secretion rate is conditioned by both its neurotransmitter as well as
glucose input time history.

Possible mechanisms for the observed decrease in neurotransmitter
potency by previous neurotransmitter exposure include: 1) interaction at
the receptor level; 2) interaction at the post-receptor level; 3) interaction at the level of stimulus-secretion coupling; or 4) a combination of
these possibilities[3]. Identification of the mechanism or mechanisms must
await further experimentation.

These studies have elucidated additional complexities of the β-cell:
1) autonomic neural inputs to the β-cell interact in a complex way; 2) both
the order and temporal sequence of neural inputs may be important
determinants of insulin secretion rate; and 3) a β-cell memory for neural
inputs may exist. We conclude that the effects of autonomic neural inputs
on insulin secretion are functions of the magnitudes (absolute and
relative), timing and order of application of sympathetic and
parasympathetic neural inputs in addition to the local glucose
concentration.

ACKNOWLEDGEMENTS

We would like to thank Kathy Harsch for preparation of the manuscript.
These studies were supported, in part, by National Science Foundation Grant
DCB-82-07625.

REFERENCES

1. R.N. Bergman and R.E. Miller, Direct enhancement of insulin secretion
   by vagal stimulation of the isolated pancreas, Am. J. Physiol.
   225:481 (1973).
2. I.M. Burr, A.E. Slonim, and R. Sharp, Interactions of acetylcholine
   and epinephrine on the dynamics of insulin release in vitro, J.
   Clin. Invest. 58:230 (1976).
3. L.A. Campfield and F.J. Smith, Neural control of insulin secretion:
   interaction of norepinephrine and acetylcholine, Am. J. Physiol.
   (Regulatory Integrative Comp. Physiol.) 244:R629 (1983).
4. L.A. Frohman, J. Ezdinli, and R. Javid, Effect of vagotomy and vagal
   stimulation on insulin secretion, Diabetes 16:443 (1967).
5. J.E. Gerich, M.A. Charles, and G.M. Grodsky, Regulation of pancreatic
   insulin and glucagon secretion, Annu. Rev. Physiol. 38:353 (1976).
6. L. Girardier, J. Seydoux, and L.A. Campfield, Control of A and B cells
   in vivo by sympathetic nervous input and selective hyper- or
   hypoglycemia in dog pancreas, J. Physiol. Paris 72:801 (1976).
7. P.E. Lacy and M. Kostianovsky, Method for the isolation of intact
   islets of Langerhans from the rat pancreas, Diabetes 16:35 (1967).
8. W.J. Malaisse, F. Malaisse-Lagae, P.H. Wright, and J. Ashmore, Effects
   of adrenergic and cholinergic agents upon insulin secretion in
   vitro, Endocrinology 80:975 (1976).
9. P.H. Smith, S.C. Woods, and D. Porte, Jr., Control of the endocrine
   pancreas by the autonomic nervous system and related neural
   factors, in: "Integrative Functions of the Autonomic Nervous
   System", C.M. Brooks, K. Koizumi, and A. Sato, eds., pp 84-97,
   Elsevier/North Holland, Amsterdam, Netherlands.
10. R.H. Unger, R.E. Dobbs, and L. Orci, Insulin glucagon and somatostatin
    in the regulation of metabolism, Annu. Rev. Physiol. 40:307 (1978).
11. C.B. Wollheim and G.W.G. Sharp, Regulation of insulin release by
    calcium, Physiol. Rev. 61:914 (1981).
12. S.C. Woods, and D. Porte, Jr., Neural control of the endocrine
    pancreas, Physiol. Rev. 54:596 (1974).

MUSCARINIC RECEPTORS AND THE CONTROL OF GLUCOSE-INDUCED ELECTRICAL ACTIVITY
IN THE PANCREATIC β-CELL

I. Palafox, J.V. Sanchez-Andres, S. Sala,
R. Ferrer and B. Soria

Departmento de Fisiologia
Facultad de Medicina
Alicante, Spain

There is ample physiological evidence for the presence of muscarinic
acetylcholine receptors in pancreatic β-cells[7,8,9,12]. Stimulation of
parasympathetic nerves can cause hypoglycemia and insulin release, an
effect which may participate in the so-called cephalic phase of insulin
secretion. Since the role of parasympathetic innervation in health and
disease is not obvious, a more accurate characterization of this receptor
may be of interest. The aims of the present work are to test whether
muscarinic receptors may be directly demonstrated in pancreatic islets of
mice as judged by binding of the muscarinic antagonist ^{3}H-quinuclidinyl-
benzilate (^{3}H-QNB) and to study the effects of cholinergic agonists on
glucose-induced electrical activity, in order to search for the ionic
mechanisms involved in muscarinic action.

METHODS

Ten to 14 week old albino mice were used in this study. The mice were
fasted during a period of 12 to 24 h in order to increase the number of
visible islets per pancreas. Membranes were obtained from microdissected
islets free of exocrine tissue. During microdissection, over a period of
about 10 hours, the islets were bathed with a modified Krebs-Henseleit
solution containing in mM: 120 NaCl, 25 NaHCO, 2.5 CaCl, 1.1 MgCl. The
solution was kept at 37°C and was continuously equilibrated with 95% O_2 and
5% CO_2 to give a pH of 7.5. After careful microdissection the islets were
gently washed with a solution of grade V collagenase (Sigma). Groups of 300
to 400 islets were stored overnight at -20°C.

Islet membranes were obtained according to the following procedure:
islets were homogenated in 10 volumes of modified Krebs (1000 rpm, 10
times), sonicated (1 min), centrifuged (5 min, 1000 g), the pellet was
discarded and the supernatant centrifuged (20 min, 40,000 g), the
supernatant discarded and the pellet sonicated (1 min). This method yields
approximately 1 µg of membrane protein per islet. Protein content of the
membranes was determined using a modification of the Lowry procedure.

Binding to islet membranes was carried out following a filtration
technique: 0.3 ml of membranes (0.1 µg/ml) were incubated with 0.6 ml of
Krebs solution, pH 7.5 containing ^{3}H-QNB. After incubation at 37°C for
1 h, the samples were filtrated in a Millipore scintered glass filtration
apparatus through Whatman GF/B filters. The filters were washed twice with
5 ml ice cold buffer and placed in scintillation vials, 8 ml of
scintillation mixture (PPO 5 g, bis-MSB 0.5 g in a mixture of toluene and
triton X-100, 2:1) were added to each vial and left overnight at room
temperature before counting (efficiency 34.5%).

Membrane potential of the mouse β-cell was recorded as described
previously[1,5]. To analyze the membrane potential fluctuations, membrane
potentials were recorded on magnetic tape (RACAL Store 4), ditigized using
a NORLAND 3001A processing system at 5 ms/point and stored in data blocks
of 2048 points. The digitized data were analyzed and the variance was
calculated as described previously[2,3,6]. The volume of the perifusion
chamber was 0.04 ml and the time needed to reach the chamber after the
switch to a new solution was approximately 2 s. All the experiments were
conducted at 37°C. Acetylcholine (ACh), carbamylcholine (Cbch), and
atropine (Atr) were from Sigma.

RESULTS AND DISCUSSION

Fig. 1 shows specific binding (i.e. displaceable by atropine) of
^{3}H-QNB to islet membranes. The high value of B_{max} indicates that islet
membranes are rich in muscarinic receptors. B_{max} is higher than that
observed in some brain territories, such as rat brain septum and
hippocampus, and are similar to that found in rat brain cortex[16]. However,
the variability in our results is large because of the small amount of
proteins available to be used in the binding procedure. Thus, the value

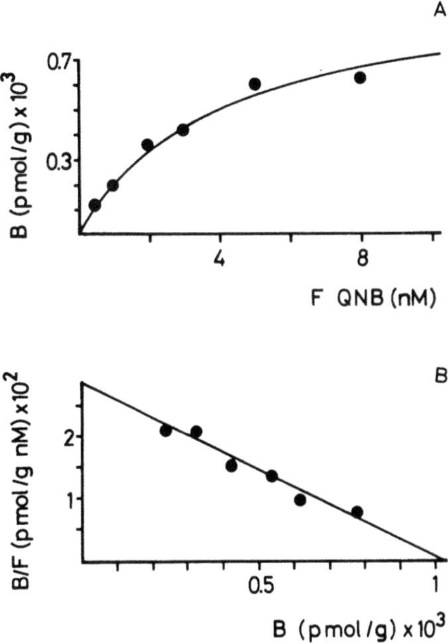

Fig. 1. A) Specific binding (displaceable by 5 μM atropine) of ^{3}H-QNB to islet membranes. Each point is the average of 4 experiments performed in triplicate;variability from the mean was 15%. Curve represents best fit using a non-linear regression program. B_{max}: 1000 ± 146 pmol/g of membrane protein. K_d: 4.12 ± 1.42 nM. B) Scatchard plot of the data in A.

reported has to be taken as a rough estimate of the concentration of receptors in mouse β-cell (β-cells make up approximately 80% of the whole islet). On the other hand, the K_d reported here, about 4 nM, is high compared to those reported in other tissues[16], indicating a relatively low affinity site for muscarine on the β-cell.Nevertheless, K_ds of approximately 8 nM and 2 nM have been reported for human neutrophils[4] and rat lymphocytes[11], respectively. The value reported here may be high due to the method of islet isolation (in vitro incubation for up to 10 h during the microdissection procedure) and membrane preparation and/or it may reflect a physiological situation (mice were fasted). Recently Grill and Ostenson[8] have reported specific binding of ^{3}H-methylscopolamine to pancreatic islets of the rat. Scatchard analysis of their data indicate a high affinity binding site and another possible low affinity binding site for scopolamine.

<u>Fasting Changes the Sensitivity of the Electrical Response of the β-Cell to</u>
<u>Muscarinic Agonists</u>

Fig. 2 shows the effect of carbamylcholine on glucose-induced electrical activity in β-cells from fasted (Fig. 2, A-C) and fed (Fig. 2, D-E) mice. As can be seen, in islets from fasted mice, addition of 10 μM Cbch enhances the duration of the active phases, or "bursts" of action potentials, and reduces the duration of the silent phases. Increasing Cbch to 20 μM induced continuous spike activity. Atropine, 5 μM, inhibited these effects of Cbch (Fig. 2C). The concentrations of Cbch are higher than those used to produce continuous firing in the β-cell as reported previously[7]. In mice having free access to the standard diet, 1 μM carbamylcholine induces continuous firing (Fig. 2 E). Fasting <u>per se</u> for 24-36 h does not change the pattern of glucose induced electrical activity in the mouse pancreatic β-cell (the burst pattern at the beginning of Trace

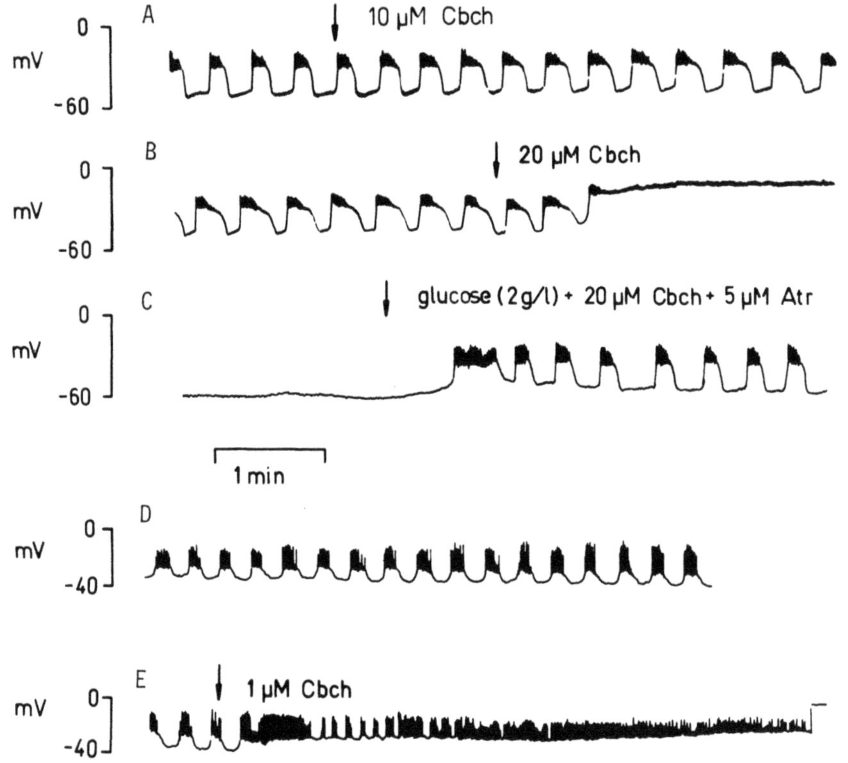

Fig. 2.   Effects of carbamycholine on 11 mM glucose-induced electrical activity. Traces A-C: mouse was fasted during 24 h. Glucose was absent at the beginning of Trace C. Traces D-E: mouse had free access to food.

A and in Trace D are within the normal range). The relatively high
concentrations of agonists needed in fasted animals to produce an effect
are consistent with the high K_d measured in binding experiments. Whether
or not the high K_d for ^{3}H-QNB represents a physiological situation and/or
the muscarinic receptor undergoes modulation in different experimental
conditions are points needing further investigation. These preliminary
experiments suggest that the nutritional state of the mouse may change the
response of the β-cell to muscarinic agonists. Ostenson and Grill[12] have
also recently observed a 32% decrease in islet muscarinic receptor binding
in fasted animals.

Ionic Mechanisms Underlying Muscarinic Effects

Figure 3 shows the changes in the input resistance in the presence of
different glucose concentrations. In the absence of the agonist, input
resistance increased, concomitantly with a depolarization of about 10 mV
when passing from 0 to 5.6 mM glucose, corresponding to a decrease in the
K-conductance[1]. However, when carbamylcholine was present, the input
resistance of the cell decreased during the depolarization induced by a
change from 0 to 5.6 mM glucose. In the presence of higher concentrations
of glucose (11 to 22 mM) the input resistance is also lowered by

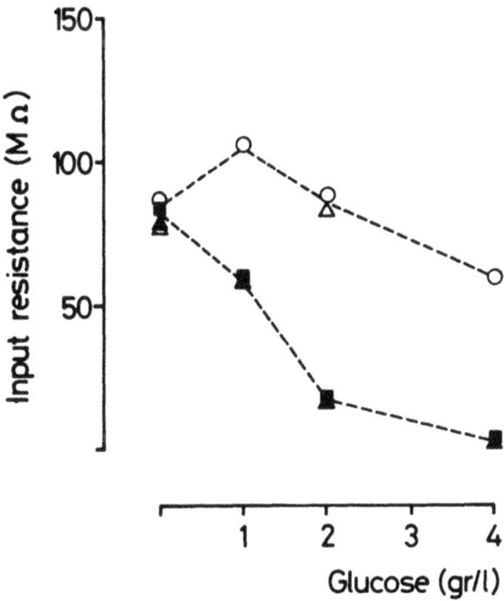

Fig. 3. Effect of carbamylcholine on the input resistance of the mouse
β-cell as a function of glucose concentration. Glucose
concentration is given in g/l and corresponds to mM as follows:
1=5.6; 2=11; 3=16.7; 4=22. Input resistance was measured before
and one minute after addition of Cbch. Control (open circles),
10 µM Cbch (filled squares), 100 µM Cbch (filled triangles) and
Cbch 100 µM + atropine 50 µM (open triangles).

carbamylcholine. This effect is reversed by atropine. Similar results
were obtained with acetylcholine (0 to 20 µM). Simultaneous decrease of
the input resistance together with cell depolarization might be explained
by an increase in specific membrane conductances produced by muscarinic
agonists[16].

The effect of acetylcholine on membrane potential and input resistance
has been assayed in other secretory, albeit exocrine cells[13]. Membrane
potential changes induced by acetylcholine and carbamylcholine on exocrine
cells are usually poliphasic. Two types of responses have been observed:
a) hyperpolarization due to an increase in K conductance and b) depolarization due to a less specific conductance increase. Thus, in pancreatic
acinar cells[14], ACh causes simultaneous depolarization and decrease of the
input resistance as well as an increase in $^{45}Ca^{2+}$ efflux. In rat lacrimal
glands muscarinic activation increases intracellular calcium[10]. Internal
Ca^{2+} activates three types of channels: one selective for K, one selective
for Cl^- and one selective for cations. A similar mechanism operating in
the pancreatic β-cell will activate preferentially conductances promoting
depolarization: either g_{Cl} or g_{Na} (through the conductance selective for
cations).

Membrane potential variance may be also used to test the state of the
membrane conductance. There is strong evidence suggesting that excess
voltage noise is well correlated with calcium channel activity[2,3]. The
variance during the first minute after agonist application was measured in
the presence of 22 mM glucose. The results obtained were: 9.03×10^{-5}
$volts^2$ in control conditions, 9.53×10^{-5} $volts^2$ when 10 µM Cbch was added
and 11.90×10^{-5} $volts^2$ in the presence of 100 µM Cbch. These results
clearly show that in the presence of stimulatory concentrations of agonist
the variance increases, and, thus, indicate that Ca-channel activity may be
enhanced by carbamycholine. This is consistent with the observation that
in islets maintained for 46 h in tissue culture, acetylcholine stimulates
^{45}Ca uptake and insulin release at subthreshold concentrations of
glucose[15]. Since both effects were blocked by atropine, it was concluded
that ACh stimulates the early phase of insulin release by enhancing Ca^{2+}
influx from extracellular fluid.

In conclusion, this study confirms the presence of muscarinic binding
sites in islet cell membranes. The data reported here indicate a relatively
high receptor density with a relatively low binding affinity. The electrophysiological measurements suggest that muscarinic receptor affinity may be

modulated by starvation. Furthermore, these measurements indicate that
cholinergic stimulation of the β-cell may increase membrane conductance to
Na^+, Cl^- and/or Ca^{2+}, a mechanism fundamentally distinct from glucose
action, which is primarily to reduce membrane conductance to K^+.

ACKNOWLEDGEMENTS

This study was supported in part by the CAICYT (Grant n 3167-84) and
the U.S.-Spain Joint Committee for Scientific and Technological
Cooperation. J.V.S.-A. is a recipient of a FISS fellowship.

REFERENCES

1. I. Atwater, B. Ribalet, and E. Rojas, Cyclic changes and resistance of
      the β-cell membrane induced by glucose in islets of Langerhans from
      mouse, J. Physiol. 278:117 (1978).
2. I. Atwater, C.M. Dawson, G.T. Eddleston, and E. Rojas, Voltage noise
      measurements across the pancreatic β-cell membrane: calcium-channel
      characteristics, J. Physiol. 314:195 (1981).
3. C.M. Dawson, I. Atwater, and E. Rojas, Potassium-induced insulin
      release and voltage noise measurements in single mouse islets of
      Langerhans, J. Membr. Biol. 64:33 (1982).
4. B.H. Dulis, M.A. Gordon and I.B. Wilson, Identification of muscarinic
      binding sites in human neutrophils by direct binding, Molec.
      Pharmacol. 15:28 (1979).
5. R. Ferrer, B. Soria, C.M. Dawson, I. Atwater and E. Rojas, Effects of
      Zn^{++} on glucose-induced electrical activity and insulin release
      from mouse pancreatic islets, Amer. J. Physiol. 246:C520 (1984).
6. R. Ferrer, S. Sala, J.V. Sanchez-Andres, and B. Soria, Further
      evidence that Zn^{++} blocks voltage-dependent Ca^{++} channels in the
      mouse pancreatic β-cell, Biochem. Soc. Trans. 13:680 (1985).
7. E. Gagerman, L.A. Idahl, H.P. Meissner, and I.B. Taljedal, Insulin
      release, cGMP, cAMP and membrane potential in acetyl-
      choline-stimulated islets, Amer. J. Physiol. 235:E493 (1978).
8. V. Grill and C.G. Ostenson, Muscarinic receptors in pancreatic islets
      of the rat. Demonstration and dependence of long-term glucose
      environment, Biochem. Biophys. Acta 756:159 (1983).

9. W. Malaisse, F. Malaisse-Lagae, P.H. Wright and J. Ashmore, Effects of adrenergic and cholinergic agents upon insulin secretion in vitro, Endocrinology 80:975 (1967).

10. A. Marty, Y.P. Tan, and A. Trautman, Three types of calcium dependent channel in rat lacrimal gland, J. Physiol. 357:293 (1984).

11. W. Maslinski, E. Grabczewska, and J. Ryzewski, Acetylcholine receptors of rat lymphocytes, Biochem. Biophys. Acta 633:269 (1980.

12. C.G. Ostenson and V. Grill, Cholinergic binding to pancreatic islets in the fasted, fed and diabetic state is regulated by prevailing blood glucose, Diabetes Res. Clin. Pract. Suppl. 1:S427 (1985).

13. M.L. Roberts and O.H. Petersen, Membrane potential and resistance changes induced in salivary gland acinar cells by micro-iontophoretic application of acetylcholine and adrenergic agonists, J. Membr. Biol. 39:197 (1978).

14. J. Singh and O.H. Petersen, The effects of L-alanine and acetylcholine on membrane potential, ^{45}Ca, and ^{86}Rb efflux and amylase secretion in isolated mouse pancreas, Quart. J. Exp. Physiol. 69:531 (1984).

15. C.B. Wollheim, E.G. Siegel, and G.W.G. Sharp, Dependency of acetylcholine induced insulin release on Ca^{++} uptake by rat pancreatic islets, Endocrinology 107(4):924 (1980).

16. Subtypes of Muscarinic Receptors II: Proceedings of the Second International Symposium. Eds: R. Levine, N. Birdsell, A. Giachetti, R. Hammer, L. Iversen, D. Jenden and R. North. TIBS, February Supplement, Elseveir Science Publishers, U.K. ISSN:01656147 (1986).

ELECTROPHYSIOLOGICAL EVIDENCE FOR HISTAMINERGIC MODULATION OF PANCREATIC
β-CELL FUNCTION

A. Marques, R. Ferrer, C. Ripoll and B. Soria

Departamento de Fisiologia
Facultad de Medicina
Alicante, Spain

Although it is known that amines affect insulin secretion, their
effects on glucose-induced insulin secretion are difficult to interpret for
several reasons: 1) differential effects of alpha- and beta-receptor
stimulation by catecholamines, 2) induction of catecholamine secretion by
serotonin, 3) changes of capillary blood flow by histamine or 4) degranulation of β-cells with a marked dilatation of the rough endoplasmic
reticulum, induced by the histamine and serotonin antagonist
cyproheptadine. Membrane potential recording provides a direct assay for
the effects of these substances on pancreatic β-cell membrane properties.
This technique has been successfully used to further understand the effects
of catecholamines on the β-cell membrane[7]. The aim of the present study
was to determine the role of the putative neurotransmitter, histamine, on
the control of membrane potential in the mouse pancreatic β-cell.

METHODS

The electrophysiological methods used here have been previously
described[1,3]. To analyze membrane potential variance, the records were
digitized with a Norland 3001A processing digital oscilloscope in data
blocks of 1024 points, each point representing 5 ms. The digitized data
were analyzed and the variance calculated as described previously[4].

RESULTS AND DISCUSSION

Fig. 1 shows the effects of histamine, Hist, and histaminergic
agonists on 11 mM glucose-induced electrical activity. The presence of
histamine in the superfusion fluid results in an inhibition of the
electrical activity with fast onset and fast recovery (Fig. 1A). There is
some variability from cell to cell. The inhibitory effect consists in a
reduction of time in the active phase and a transient hyperpolarization.
This transient inhibition is accompanied by a decrease in the input
resistance (results not shown) suggesting that activation of potassium
conductance may be involved in this effect. Such an effect has been
previously reported in molluscan neurones, where histamine released by
presynaptic stimulation produces an increase in K conductance[5]. The fast
onset and fast recovery response of the β-cell has also been observed in
the presence of catecholamines[7]. It has been postulated that rapid
degradation by the tissue enzymes could explain such an effect. To test
this possibility and also to assess the predominant type of receptor

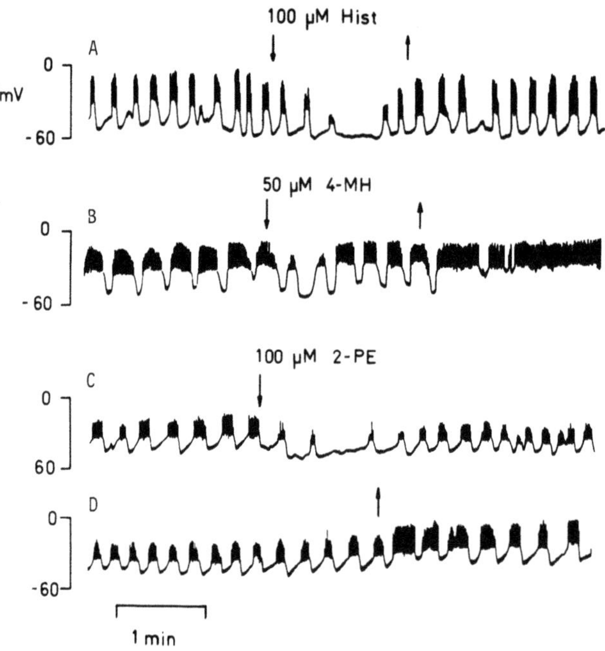

Fig. 1. Effects of histamine, Hist, 4-methylhistamine, 4-MH, and
2-pyridylethylamine, 2-PE, on glucose-induced electrical activity.
Traces A, B and C are from different islets. Trace D is a
continuation of Trace C. Glucose was 11 mM in all experiments.
Hist, 4-MH and 2-PE were added and removed as indicated by the
arrows over each trace.

involved, the H1 agonist, 2-pyridylethylamine (2-PE) and the H2 agonist,
4-methylhistamine (4-MH) were tested. Both 4-MH (Fig. 1B) and 2-PE (Fig.
1C) produced an effect similar to that produced by histamine.

Histamine effects on pancreatic β-cell are concentration dependent
(Fig. 2) and are partially reversed by cimetidine, an H2 antagonist, but
not by chemizol, an H1 antagonist. The response to both agonists, 4-MH and
2-PE, is partially reversed by 1 x 10^{-4} cimetidine. Chemizol (5 x 10^{-5} M)
fails to reverse the effects of 2-pyridylethylamine. Simultaneous addition
of both antagonists do not reverse the effect either (results not shown).
The fast inhibitory response could be mediated by H2 receptors, as
suggested by its partial blockade with cimetidine. However, depressant
effects of histamine are mimicked not only by H2 agonists like
4-methylhistamine, but also by compounds predominantly active at H1
receptors like 2-pyridylethylamine.

At high glucose concentrations, when the β-cell is stimulated to give
continuous spike activity, addition of histamine induces a burst pattern
(Fig. 3), similar to the burst pattern induced by slightly lowering the
glucose concentration. Unlike the effects of noradrenaline, where

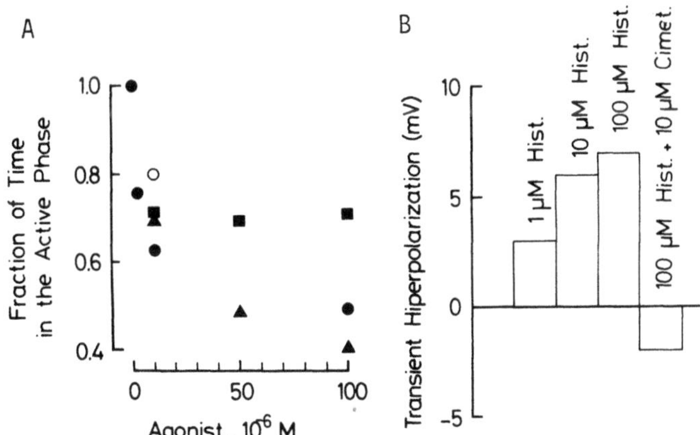

Fig. 2. Concentration dependent effects of histamine and histaminergic
agonists on electrical activity (A) and β-cell membrane potential
(B) during transient inhibition. Histamine (filled circles),
2-pyridylethylamine (filled triangles), 4-methylhistamine (filled
squares) and histamine + 50 μM cimetidine (open circles).
A: measurements made during first two minutes after addition of
histaminergic agonist during apparent inhibition.
B: maximum change measured during first minute after addition of
histamine.

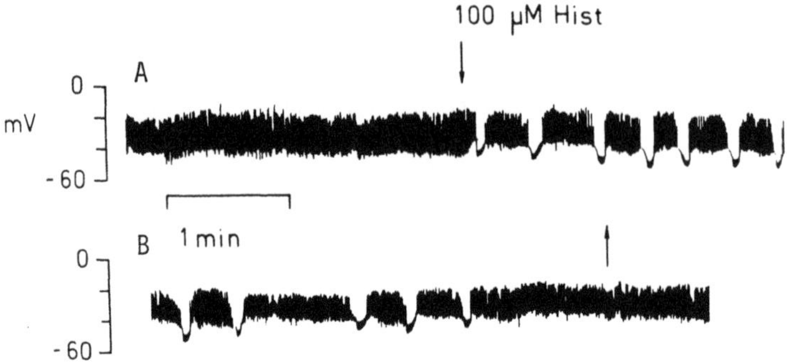

Fig. 3. Effects of 100 µM histamine on 22 mM glucose-induced electrical
activity. Histamine was added and removed as indicated by the
arrows over the traces. Trace B is a continuation of Trace A.

increasing glucose concentration over-rides the inhibition[7], histamine
appears to be as effective at 22 as at 11 mM glucose.

A frequent observation (8 of 10 cells tested) is the appearance of a
potentiation of activity after withdrawal of the histaminergic agonist Fig.
1B and D). This potentiation persists during 10-20 min after withdrawal of
the agonists and may reflect long-term effects.

Together with these short-term effects, after a brief exposure to
histamine or to the H1 or H2 agonists tested, changes appeared in the
electrical activity that were not removed by cimetidine or agonist
withdrawal. Fig. 4 illustrates the long-term effects of 100 µM histamine
on action potentials induced by stimulatory concentrations of glucose. The
main effect observed is a decrease in the rate of repolarization of the
action potential without significant changes in the rate of depolarization.
An inhibition of the voltage dependent K-channels responsible for action
potential repolarization could produce such an effect. Nevertheless, block
of voltage dependent K-channels by TEA produces an increase in both size
and duration of β-cell calcium spikes (Fig. 4c) and after exposure to
histamine the average size of the action potentials does not change Fig. 4,
a and b). Alternative explanations for the effects of histamine on glucose
induced action potentials are that Ca-channel inactivation is reduced or
that coupling between cells in the islet is increased. The repolarizing
component of the action potentials is modified such that small spikes
appear on the repolarizing phase, similar to those observed in other
tissues where excitation depends on Ca-channels[10]. Graded electrogenesis
in the pancreatic β-cell has been described during glucose-induced
electrical activity[3,9].

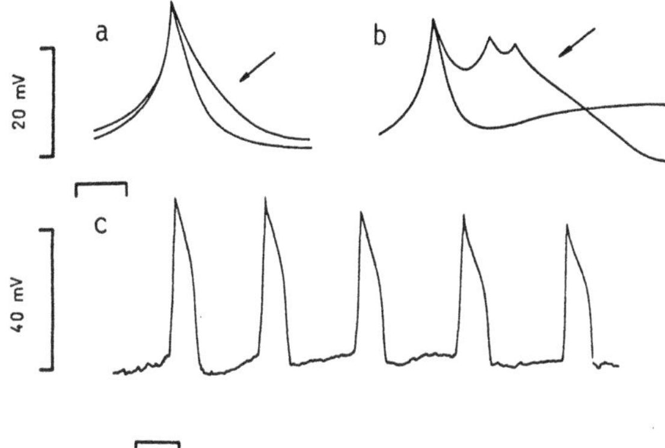

Fig. 4. Long-term effect of 100 µM histamine on action potentials induced
by 22 mM glucose (a) and 11 mM glucose (b). For comparison,
effects of 20 mM TEA (c) on 11 mM glucose-induced electrical
activity are shown. Spikes were stored, digitized at 5 ms per
point and averaged. Arrows indicate the average spike after
exposure to histamine. Calibration: upper, 100 ms, lower 1 s.

Voltage dependent channels may be modulated by neurotransmitters. In
cultured dorsal root ganglion cells a group of substances (GABA, serotonin,
noradrenaline, enkephalin and somatostatin) have been reported to decrease
voltage-activated calcium current and in _Aplysia californica_ neurons
serotonin produces an increase in the voltage dependent calcium
conductance. In pancreatic β-cell it may be postulated that histamine
modulates calcium currents.

Ionic mechanisms may be studied analyzing the variance of the membrane
potential. Excess voltage-noise induced by either stimulatory
concentrations of glucose or high concentrations of K^+ in the pancreatic
β-cell is well correlated with Ca-channel activity[1,2]. Ca-channel
blockers, such as Zn^{2+} and Co^{2+}, produce a significant reduction in the
excess-voltage noise of the 22 mM glucose-induced electrical activity[4].
Histamine, 100 µM, produces a transient inhibition in the variance of the
membrane potential (Fig. 5B). This effect is consistent with the transient
inhibition of glucose-induced electrical activity. Nevertheless, at lower
concentrations, histamine induces the unexpected effect of increasing the
variance of membrane potential fluctuations (Fig. 5A).

The results described clearly show that histamine affects the ionic
permeabilities underlying glucose induced electrical activity in the

β-cell. The more dominant effect, a transient inhibition, may be due to an
activation of potassium permeability in the pancreatic β-cell. Together
with the inhibitory effects, histamine induces an excitation of membrane
channels more apparent at lower concentrations of histamine and after its
removal. Histamine responses of the β-cell are similar to those described
in central neurons[8]. The simple view that excitatory responses are
mediated by H1 receptors and inhibitory responses by H2 receptors is not
fully substantiated by experimental evidence[6].

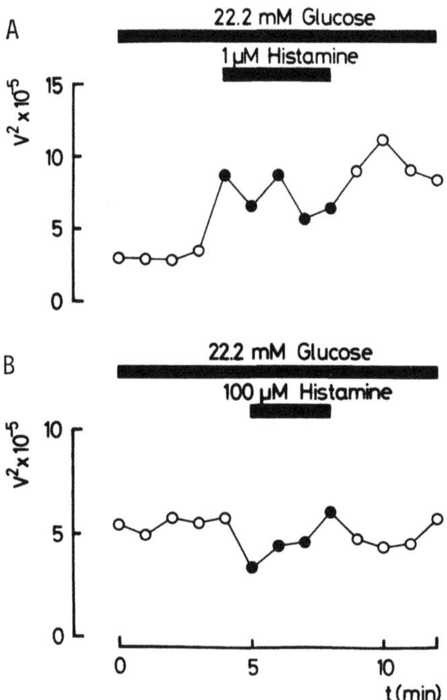

Fig. 5. Effect of histamine on the membrane potential fluctuations
recorded in the presence of 22 mM glucose. A: effect
of 1 µM histamine; B: effect of 100 µM histamine.

In conclusion, histamine effects on pancreatic β-cell are complex, are
not easily explained as H1 or H2 mode of response and probably involve
several mechanisms that have to be explored.

ACKNOWLEDGEMENTS

Supported by the Spanish CAICYT (grant n 3167-84) and the U.S.-Spain
Joint Committee for Scientific and Technological Cooperation.

REFERENCES

1. I. Atwater, C.M. Dawson, G.T. Eddleston, and E. Rojas, Voltage noise
    measurements across the pancreatic β-cell membrane: calcium-channel
    characteristics, J. Physiol. 314:195 (1981).
2. C.M. Dawson, I. Atwater, and E. Rojas, Potassium-induced insulin
    release and voltage noise measurements in single mouse islets of
    Langerhans, J. Membr. Biol. 64:33 (1982).
3. R. Ferrer, B. Soria, C.M. Dawson, I. Atwater and E. Rojas, Effects of
    Zn^{++} on glucose-induced electrical activity and insulin release
    from mouse pancreatic islets, Am. J. Physiol. 246:C520 (1984).
4. R. Ferrer, S. Sala, J.V. Sanchez-Andres, and B. Soria, Further
    evidence that Zn^{++} blocks voltage-dependent Ca^{++} channels in the
    mouse pancreatic β-cell, Biochem. Soc. Trans. 13:680 (1985).
5. T. Gotow, Modulation of the histaminergic inhibitory synaptic
    potential in the Onchidium neuron by cyclic nucleotides, Japan. J.
    Physiol. 34:1135 (1984).
6. H.L. Haas, P. Wolf, and J.C. Nussbaumer, Histamine: action on
    supraoptic and other hypothalamic neurons of the cat, Brain Res.
    88:166 (1975).
7. S. de Sa Santana, R. Ferrer, E. Rojas, and I. Atwater, Effects of
    adrenaline and noradrenaline on glucose-induced electrical activity
    of mouse pancreatic β-cell, Quart. J. exp. Physiol. 68:247 (1983).
8. B.S.R. Sastry and J.W. Phyllis, Evidence for an ascending inhibitory
    histaminergic pathway to the cerebral cortex, J. Physiol. 54:782
    (1976).
9. B. Soria, R. Ferrer, I. Atwater, and E. Rojas, Graded electrogenesis
    in pancreatic β-cell, in: "Biophysics of the Pancreatic β-cell",
    Eds: I. Atwater, E. Rojas and B. Soria, Plenum Press, N.Y. this
    volume. (1986).
10. R. Llinas and M. Sukimori, Electrophysiological properties of in vitro
    Purkinge cell somata in mammalian cerebellar slices, J. Physiol.
    305:171 (1980).

EFFECT OF MELATONIN ON INSULIN SECRETION FROM

ISOLATED RAT ISLETS OF LANGERHANS

J.M. Pou, T. Cervera, M. Codina
and A. de Leiva

Facultad de Medicina
Universidad Autonoma de Barcelona
Barcelona, Spain

The aim of this paper was to evaluate the possible interaction
between β-cells and some indolamines. The effects of melatonin (MEL)
on insulin secretion from rat islets were evaluated. Rat islets were
obtained by enzymatic degestion of the pancreas from Sprague-Dawley
rats with collagenase (Boehringer).

We incubated islets with 2 or 20 mM glucose plus different MEL
concentrations for 2 hours. MEL concentrations used ranged from 40 to
320 pg/ml (physiological levels in rats). When MEL was added to the
incubation medium in the second hour with 20 mM glucose, after an
initial exposure for one hour to 2 mM glucose, MEL slightly stimulated
insulin secretion. Insulin secretion stimulated for the first hour
with 20 mM glucose was only increased by MEL at the concentration of
80 pg/ml; however, in the second hour, MEL increased insulin secretion
in an inverse dose dependent fashion (see Table I).

It is concluded that: a) MEL produces a two-fold increase
in insulin secretion; b) MEL added to the incubation medium has a
potentiating effect of insulin secretion induced by glucose; c) the
effect of MEL on insulin secretion is very slow. Thus, a relationship
between the endocrine secretion of the pineal gland and islets of
Langerhans has been demonstrated.

Table I: Effects of Melatonin on Glucose-evoked Insulin Secretion

	IRI (pmol/ islet/60 min in 2nd hour	P
20 mM glucose + MEL 0 pg/ml	794 ± 124	
20 mM glucose + MEL 320 pg/ml	841 ± 195	N.S.
20 mM glucose + MEL 160 pg/ml	1162 ± 201	0.05
20 mM glucose + MEL 80 pg/ml	1471.5 ± 226	0.001
20 mM glucose + MEL 40 pg/ml	1666 ± 263	0.001

N.S.: non-significant

P: probability from t-test

CALCIUM REGULATION OF MEMBRANE FUSION DURING HORMONE SECRETION

H.B. Pollard, K.W. Brocklehurst, E.J. Forsberg,
A. Stutzin, G. Lee, and A.L. Burns

Laboratory of Cell Biology
NIADDK, National Institutes of Health
Bethesda, Maryland 20892

INTRODUCTION

Calcium is only one of a number of second messengers that are
becoming increasingly implicated in processes regulating hormone
secretion by exocytosis. The most completely studied calcium-dependent secretory system is the chromaffin cell from the adrenal medulla,
a system with many operational similarities to the insulin secreting
β-cell from islets of Langerhans. Inasmuch as we have had the opportunity to compare these two systems in detail we will attempt to make
a number of hopefully interesting comparisons between data obtained
from the chromaffin cell and certain data of relevance from the β-cell
system.

The crucial problems in analysis of calcium-dependent hormone
secretion include understanding the mechanisms by which a specific
receptor can induce elevation of cytosolic calcium concentration. Is
the calcium brought in from the medium or mobilized from intracellular
calcium stores? Is the calcium uniformly distributed in the cytosol
or localized to specific areas? Does receptor binding directly induce
elevation of free calcium concentration? Or are there intervening
messenger systems involving cyclic nucleotides or inositol phosphates,
generation of which themselves may depend on receptor dependent activation of a GTP binding protein? And, finally, once calcium levels
are elevated how does this get translated into granule movement and
membrane fusion leading to exocytosis? This of course is not the end
of the series of questions, for the process must be turned off in some

manner. The importance of this last question is manifest by the fact
that in the chromaffin cell, at least, physiologic secretion terminates long before the cytosolic calcium concentration returns to its
resting baseline level.

Some of the answers to these problems may lie in recent work
from our laboratory involving phospholipase C and protein kinase C
activities in secreting chromaffin cells. These enzymes may be involved in generating signals for secretion including elevation of
cytosolic calcium, or making the secretion system more sensitive to
calcium. The ultimate act, membrane fusion, may also be related to
recent studies by ourselves and others on a class of calcium binding
proteins called "synexins." These proteins bind to secretory vesicle
membranes and plasma membranes, and under certain circumstances fuse
secretory vesicle membranes to one another.

PROTEIN KINASE C AND PHOSPHOLIPASE C

Protein kinase C is an enzyme whose activity depends on calcium,
acidic phospholipids such as phosphatidylserine, and diglyceride. The
diglyceride is presumably generated by a specific phospholipase C
which attacks various phosphatidylinositols to generate the diglyceride and inositol phosphates. As an experimental convenience the
relatively insoluble diglycerides have been replaced by phorbol esters
such as TPA. Indeed, in chromaffin cells, TPA is able to potentiate
secretion of catecholamines induced by ionophores such as A23187,[15]
thus supporting the concept that protein kinase C might indeed be
involved in regulating some aspects of secretion. Furthermore, in
digitonin permeabilized chromaffin cells TPA can potentiate calcium
evoked secretion of catecholamines.[2] This preparation presumably
models the exocytosis process.[9,28] Recent work from our laboratory
has also demonstrated that the chromaffin cell can generate inositol
phosphate products when stimulated to secrete under certain conditions. These results thus further support the concept that diglycerides can be generated during physiologic secretion,[10] presumably due
to phospholipase C action.

Now convinced of the possible value of protein kinase C for
regulating at least some types of secretion we have proceeded to study
the properties of this enzyme from adrenal medulla.[3] The purified
enzyme proved to be sensitive to TPA and to be activated by both
calcium and phosphatidylserine. Not only did TPA increase the maximal

activity of the enzyme but it also made the enzyme more sensitive to
calcium (Table I).

Table I. Comparison of Protein Kinase-C in β-Cell Tumor and Adrenal
          Medulla

Parameter	HIT-T15 β-cell[a]	Bovine Adrenal Medulla[b]
Molecular weight:		
Gel Filtration	120,500	110,000
SDS-Page	81,300	78,000
$K_{1/2}$, PS	18 μg/ml	18 μg/ml
$K_{1/2}$, Ca^{2+}	3.9 μM	13 μM
$K_{1/2}$, diolein	2.5 μg/ml	1.5 μg/ml
Specific Activity,	34.7 nmol P/min/mg	6,500 nmol P/min/mg
Purification, fold	250	1,400

a. Data from reference 16.

b. Unpublished data (K.W. Brocklehurst, G. Lee and H.B. Pollard)

We were also able to make a series of detailed comparisons of
the adrenal medullary enzyme with protein kinase C from an insulinoma
cell line.[16] As shown in Table I, the enzyme from both cell lines had
similar molecular weights by either gel filtration or SDS gel electrophoresis. The fact that both the enzymes appeared uniformly smaller
by the latter technique indicated that the proteins were somewhat
assymetric. The half-maximal activation concentrations for phosphatidylserine, 18 μg/ml, and diolein, ca. 2 μg/ml, were also consistent
with a close relationship between the two enzymes. However, the
half-maximal calcium activation values were slightly different.

The exact values of the calcium ion concentration needed to
activate the protein kinase C, between 3 and 18 μM, might seem at
variance with the values of $[Ca^{2+}]_i$ reported for these cells when
activated with secretogogue. These values, based primarily on quin-2
studies[8] have been in the range of 0.1-0.2 μM. However, there is an
increasing appreciation that the calcium ion concentration in such
cells is not homogeneous. Indeed, local concentrations have been

estimated to be as high as "tens to hundreds of micromoles (per liter)" over quite brief time periods in studies on the squid stellate ganglion.[22]

SYNEXIN AND MEMBRANE FUSION

Following the elevation of $[Ca^{2+}]_i$, contact and fusion must occur between the secretory vesicle membrane and the plasma membrane. In addition "piggyback" or compound exocytosis also occurs, particularly in chromaffin cells, pancreatic β-cells and other endocrine

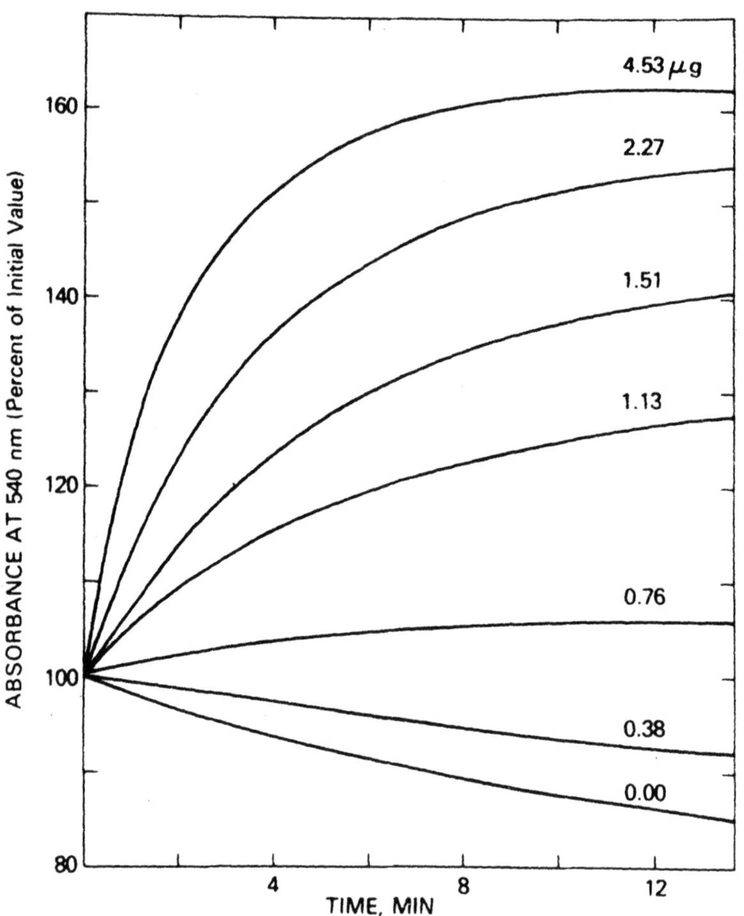

Fig. 1. Time course of turbidity changes induced in a chromaffin granule suspension by addition of various amounts of synexin. Initial absorbance was 0.3 at 540 nm. Synexin was obtained from the Ultragel AcA34 column (Fraction III). pCa=3.4, pH 6.0. The figure is from[10].

372

cells. In the latter case, a secretory vesicle within the cell fuses
with a vesicle membrane that is itself still attached to the plasma
membrane. In β-cells and chromaffin cells whole chains of such
vesicles are often found, ultimately attached to the plasma membrane.
Knowing that such processes could occur we searched for factors in the
cytosol that might cause secretory vesicles to attach and fuse to each
other in a calcium-dependent manner. One was found,[4,7] and termed
"synexin," from the Greek word "synexos," meaning "a coming together"
or a "meeting."

Synexin is a 47,000 dalton calcium binding protein, widely
distributed throughout the body, that aggregates intact chromaffin
granules only if calcium is also present. An example of a typical
assay of synexin is shown in Fig. 1. Upon the addition of
arachidonic acid, or other cis-unsaturated fatty acids, the granule
aggregates fuse into much larger vesicles, 1-5 microns in diameter.
Identical structures of this sort are found in secreting chromaffin
cells,[19] leading us to the conclusion that synexin might be at least
one mediator of exocytotic membrane fusion. Fusion can also be followed turbidometrically, as shown in Fig. 2.

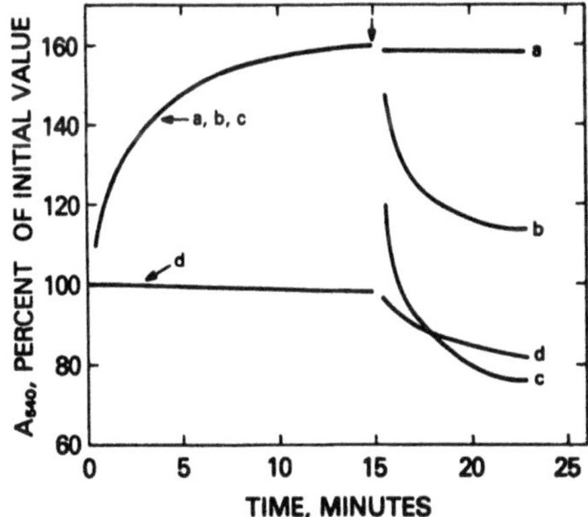

Fig. 2. Turbidity (A540) of chromaffin granule suspension undergoing
aggregation and fusion. From 0-15 min a suspension (A540=0.3, 80
µg/ml protein) is incubated with (traces a, b, c) or without (trace d)
15 µg/ml of synexin. At 15 min 4 µg/ml of a fatty acid is introduced:
trace a, elaidic acid (trans 18:1); trace b, oleic acid (cis 18:1);
traces c and d, arachidonic acid (cis 20:4). The figure is from
Biophys. J. 37:119-120, 1982.

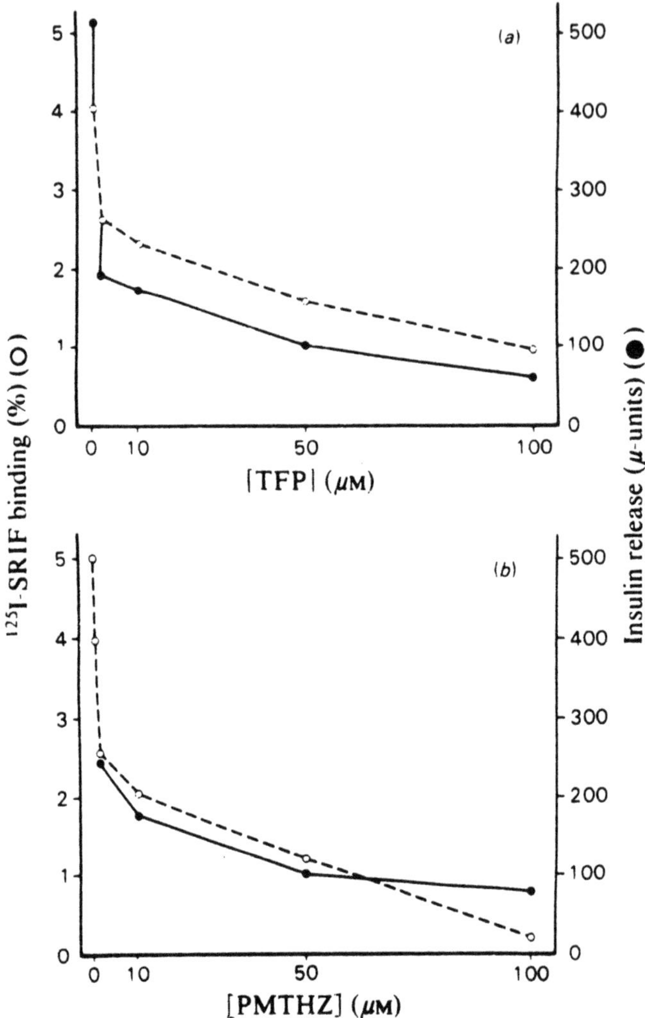

Fig. 3. Inhibition of insulin secretion and somatostatin-receptor
recruitment by TFP (a) and PMTHZ (b). 10 mM stock solutions of pheno-
thiazines were freshly prepared in water and then diluted in Krebs-
Ringer bicarbonate buffer before use. Phenothiazine at these concentrations had no influence on either the insulin radioimmunoassay or
the somatostatin receptor assay. (Abbreviation used: SRIF, somatostatin). Results for both insulin release and somatostatin-receptor
recruitment represent the change in the difference between results
obtained for buffer containing 300 mg and 30 mg of glucose/dl. The
mean values of 10 experiments are presented. The figure is from.[26]

Pharmacologic evidence for this conclusion also exists. Phenothiazine drugs such as trifluoperazine and promethazine block synexin
activity _in vitro_ with equivalent potency and similarly block secretion from chromaffin cells.[18] In cases of calmodulin activation or
local anesthetic action, trifluoperazine is two orders of magnitude

more potent than promethazine. Thus, as summarized in Table III, the
pattern of inhibition of secretion parallels that of inhibition of
synexin activity.

We have also performed some studies of relevance to the question
of synexin action in islets by testing the potency of these drugs as
inhibitors of glucose-induced insulin release. We found that promethazine was always slightly more potent than trifluoperazine as an
inhibitor, using either normal Sprague-Dawley rats or _Psommomys_
obesus, the sand rat from the Dead Sea, as a source of islets.[26] Such
data are shown in Fig. 3 for standard rats. Both drugs were three-
fold more potent as inhibitors in the psommomys islet preparation than
in the preparation from the Sprague-Dawley rat. We concluded that a
synexin-like activity could also be the molecular basis of secretion
in these islet preparations.

Table II. Synexin-Like or Related Proteins

Protein	K_d	Action	Reference
Synexin	47,000	Aggregates granules, $K_{1/2}$, Ca^{2+} = 200 μM	4
ν-Synexin	38,000	Aggregates granules at $[Ca^{2+}]$ ≤1 μM	17[+]
Synexin II	56,000s	Similar to synexin	23
ν-Synexin	51,000	Similar to synexin	17[++]
Calelectrin	34,000	Like synexin, found in Torpedo	8
Calelectrin	32,500	Like synexin, found in mammalian tissues	24,25
Synhibin	67,000	Blocks synexin	27

[+]Lee and Pollard, personal communication as described in review.[17]
[++]C. Creutz, personal communication as described in review.[17]

MECHANISM OF SYNEXIN ACTION

On the basis of such encouraging physiologic and pharmacologic
data, it has become important to learn how synexin acts at the membrane
level. Of great relevance to this question is the fact that synexin
does bind to intact chromaffin granules in a strictly calcium dependent manner.[4,5] Furthermore, in the exclusive presence of calcium
synexin has been shown to aggregate and fuse liposomes prepared from
acidic phosopholipids,[13] and to bind to columns prepared from
chromaffin granule lipids attached to cyanogen bromide activated
Sepharose beads.[6] In our own laboratory we have developed a solid
phase assay for interaction of synexin with defined phospholipids.[21]
In this case, we also found that synexin could bind to acidic
phospholipids only if calcium were also present. The inescapable
conclusion is that the receptor(s) to which synexin binds in intact
chromaffin granules must include acidic phospholipids. Indeed, these
types of lipids coat the cytosolic face of both granules and plasma
membranes for chromaffin and other cell types.

On the other hand there are some important differences between
the action of synexin on pure phospholipid dispersions and intact
chromaffin granules. Prominently, synexin spontaneously fuses
liposomes but only aggregates granules. Indeed, in order to induce
fusion of aggregated granules arachidonic acid or some other cis-
unsaturated fatty acid must also be added, albeit at low concentration
(ca. 5 μM,[7]). Chromaffin cells, like many other secreting cell
types, also produce significant quantities of arachidonic acid during
secretion.[11,12] This means that the fatty acid dependent fusion
process is likely to be of physiologic significance. Thus the main
conclusion must be that more is involved in synexin induced fusion of
intact chromaffin granules than simply the binding of synexin to
acidic phospholipids accessible at the granule membrane surface.

We have derived one clue to these additional events from recent
studies of synexin's ability to aggregate and fuse chromaffin granule
ghosts rather than intact granules. In this system, fusion, measured
by the mixing of the contents of specifically loaded ghosts prepared
by freeze and thaw in liquid nitrogen, occurs in a strictly synexin
dependent manner.[23] However, it is independent of calcium concentration. The fusion process is also blocked by pretreatment of ghosts by
trypsin, an observation not true of synexin dependent aggregation and
fusion of intact granules.

376

Thus the modification of the granule into a ghost by our procedure appears to allow the synexin molecule not only to bind the membrane in a calcium independent manner but also to fuse the membranes of neighboring ghosts without added fatty acid. We have indeed verified that synexin binds to freeze-thawed ghosts, but not intact chromaffin granules, in a calcium-independent manner using an ^{125}I-synexin binding assay. This can only mean that a profound change in the structure of the membrane has occurred. The physiologic importance of this observation can be appreciated immediately if one considers that the state of the granule immediately after fusion with the plasma membrane is essentially ghost-like. Indeed during compound exocytosis fusion of intact granules with these ghost-like granule membrane residues in the plasma membrane must occur. The caveat here is that the technique may change the granule properties in some important ways.

A POSSIBLE MODEL FOR SYNEXIN ACTION

However, the difference between synexin action on granules and these ghosts may tell us something very important about the specific mechanism of synexin action on granules. Imagine a native granule membrane surface, coated with acidic phospholipid head groups, to which synexin can gain access by it's calcium regulated phospholipid binding site. Having gained such a foothold on the granule surface the synexin molecule must further insinuate itself into the substance of the membrane, presumably in order to rearrange the bilayer. With intact granules it uses the good offices of arachidonic acid to achieve this end. But when presented with the ghost membrane rather than a granule membrane, synexin is able to dispense with the phospholipid binding intermediate and the arachidonic acid helper. This may mean that there is a specific site, cryptic in the granule membrane but immediately accessible in our ghost preparation, to which synexin must bind in order to effect fusion. Binding to the fusion site apparently does not involve calcium activation of synexin, if we trust the ghost fusion data. Indeed, calcium is not even required for arachidonic acid induced fusion of synexin aggregated intact granules, a fact long appreciated. This putative fusion site, possibly protein in nature, we might call "Φ," the Greek letter for the first letter in the word "fusion." We choose to use this Greek letter following the etymology of synexin itself. Perhaps the chief mission of synexin, insofar as fusion is concerned, is ultimate union with Φ, and Ca^{2+}-mediated binding to phospholipids may be only a means to this end.

It is also possible that a recent study by Bental, et al.,[1] in
which spontaneous fusion between granule ghosts and liposomes was
observed, might be related to accessible Φ factors in ghosts and their
possible interaction with unencumbered phospholipids. Complete fusion
must, after all, involve not only synexin and calcium, the charged
head groups on the phospholipids, and putative Φ (fusion) factors but,
also the hydrophobic substance of the membrane itself. (See Footnote 1).

OTHER SYNEXINS AND RELATED MOLECULES

A variety of synexin-like proteins have now been described
subsequent to our original report, and are listed in Table II. Two of
these, synexin II[17] and ν-synexin[8] are slightly bigger than synexin
and aggregate intact granules in a calcium-dependent manner. ν-Synexin
is the only synexin found in muscle (hence the name, "ν" for
muscle) but is found in smaller amounts in other tissues studied.

Table III. Influence of Phenothiazine Drugs on Calcium-Related Systems

System	I.D. 50 Values, uM		
	TFP	PMTHZ	PMTHZ/TFP
Synexin, 400 μM Ca^{++}	4	16	4
Chromaffin cell secretion			
i) 62 μM Nicotine	1	1.5	1.5
ii) 10 μM Veratridine	1.5	3.6	2.3
Calmodulin			
i) PDE Activation[a]	10	340	34
ii) Myosin Super-			
precipitation	18	310	17
Membrane Stabilization			
(local anesthetic) Effects			
i) RBC Lysis[b]	1	20	20
ii) Five Cases of Virus			
Induced Cell Fusion	2-5	73-540	37-108

a. PDE is phosphodiesterase

b. RBC is red blood cell

This concept has recently been further expanded in molecular terms in
an article entitled "Synexin and chromaffin granule membrane fusion:
A novel 'hydrophobic bridge' hypothesis for driving and directing the
fusion process" by H.B. Pollard, E. Rojas and A.L. Burns, to be
published in the Annals of the New York Academy of Science, 1986. The
model is based on new biophysical and genetic data on the synexin
molecule.

Anti-synexin antibodies also react with v-synexin. By comparison, little is known about synexin II.

The calelectrins are an immunologically distinct set of proteins, first found in Torpedo electric organ, which also aggregate chromaffin granules in the presence of calcium. In this sense they are "synexins" as well. Like synexin they also bind to anionic phospholipids.[24,25] These proteins were first discovered in Whittaker's laboratory[27] and their co-existence with synexin in the same cells constitutes a mystery. Why indeed have so many different synexins in the same cells? Might some of the regulatory enzymes discussed in the first section of this article differentiate among these synexins?

In addition to these synexin-like proteins, now numbering four or five, we have recently discovered a separate protein which inhibits synexin activity. We called this protein "synhibin."[20] It appears to block synexin by a competitive mechanism. Creutz[8] has also reported that synhibin interferes with calelectrin activity. The role of synhibin remains obscure, but the cell clearly cares enough about various synexin activities to have at least one specific inhibitor available for their control. It is possible that the eventual reduction of secretion, in the face of continued high intracellular calcium concentrations (vide supra), may, at least in part, be related to activity of such inhibitors.

The last protein we will discuss in this section is a new synexin-like protein, termed v-synexin. It aggregates intact granules and supports arachidonic acid fusion of granule aggregates only if calcium is absent. Indeed v-synexin activity is blocked by elevated calcium concentrations if they are much greater than 1 μM. The role of this protein is uncertain, but its very existence points up the fact that granule membrane contact and fusion mechanisms may in fundamental essence have calcium independent or indeed calcium suppressed steps. v-synexin may also prove quite valuable in investigation of the calcium independent Φ (fusion) sites, presumably cryptic in intact chromaffin granules.

CONCLUSION AND PROLOGUE

Hormone secretion by exocytosis appears to be controlled by changes in the cytosolic free calcium concentration. However, the

gap between changes in calcium and final secretion events is separated
by a set of as yet poorly understood intermediate events. Some of
these events may include phospholipase C activated processes in which
products of the enzyme, IP, IP_2 and IP_3, promote mobilization of
calcium, which together with the other product, diglyceride, can
activate protein kinase C. The data discussed in this paper are
indeed consistent with this concept and with the possibility that
products of phospholipase C and protein kinase C action are involved
in hormone secretion. However, it is also clear that the nature of
this action remains obscure. The observations so far are necessary
but not sufficient to make the case because we have no knowledge of
whether any phosphorylated proteins are directly involved in the
membrane fusion process. However, we remain optimistic.

By contrast, there do exist a set of proteins, the synexins and
synexin-related proteins, which actually mediate in vitro the very
membrane contact and fusion events occurring in secreting cells.
Perversely, none of them appear to be controlled by kinase reactions,
inositol phosphates or diglycerides, although calcium itself exerts
critical control on the properties of these proteins. For that reason
we have directed an enormous amount of our attention to these proteins, anticipating that some overlap between popular control elements
and actual membrane contact and fusion events may eventually be detected. It is of course possible that the popular control elements
alluded to and the synexins may prove to have little in common besides
calcium. By careful study we will hopefully determine which situation
is correct.

The future of such studies, at least with the synexins, lies in
cloning and sequencing synexin and its relatives. From this data, the
mechanism of calcium action and of membrane fusion may be deduced.
Furthermore, site-directed mutagenesis will allow these functional
deductions to be tested and insertion of anti-sense messages into
cells will allow production of phenotypic deletions. At present, we
have made some progress toward these goals by cloning synexin from a
λgt11 library of cDNA's prepared from bovine adrenal medulla, using
polyclonal and specific monoclonal antibodies to synexin to identify
the correct clones.

380

REFERENCES

1. M. Bental, P. I. Lelkes, J. Scholama, D. Hockstra, and J.
    Wilschut, Calcium independent, protein-mediated fusion of
    chromaffin granule ghosts with liposomes, Biochim. Biophys.
    Acta 774:296 (1984).

2. K. W. Brocklehurst and H. B. Pollard, Enhancement of Ca^{2+}-in-
    duced catecholamine release by the phorbol ester TPA in
    digitonin-permeabilized cultured bovine adrenal chromaffin
    cells, FEBS Lett. 183:107 (1985).

3. K. W. Brocklehurst, K. Morita, and H. B. Pollard, Characteriza-
    tion of protein kinase C and its role in catecholamine
    secretion from bovine adrenal medullary cells, Biochem. J.
    228:35 (1985).

4. C. E. Creutz, C. J. Paroles, and H. B. Pollard, Identification
    and purification of an adrenal medullary protein (synexin)
    that causes calcium-dependent aggregation of isolated
    chromaffin granules. J. Biol. Chem. 253:2858 (1978).

5. C. E. Creutz and D. C. Sterner, Calcium dependence of the bind-
    ing of synexin to isolated chromaffin granules, Biochem.
    Biophys. Res. Commun. 114:355 (1983).

6. C. E. Creutz, L. G. Dowling, J. J. Sando, C. Villar-Palasi, J.
    H. Whipple, and W. J. Zaks, Characterization of the chromo-
    bindins-soluble proteins that bind to the chromaffin granule
    membrane in the presence of Ca^{2+}, J. Biol. Chem. 258:14664
    (1983).

7. C. E. Creutz, Cis-unsaturated fatty acids induce the fusion of
    chromaffin granules aggregated by synexin, J. Cell Biol.
    91:247 (1981).

8. L. G. Dowling and C. E. Creutz, Comparison of synexin isotypes
    in secretory and non-secretory tissues, Biochem. Biophys.
    Res. Commun. 132:382 (1985).

9. L. A. Dunn and R. W. Holz, Catecholamine secretion from
    digitonin-treated adrenal medullary chromaffin cells, J.
    Biol. Chem. 258:4989 (1983).

10. E. J. Forsberg, E. Rojas, and H. B. Pollard, Muscarinic receptor
    enhancement of nicotine-induced catecholamine secretion may
    be mediated by phosphoinositide metabolism in bovine adrenal
    chromaffin cells, J. Biol. Chem., 261;4915 (1986).

11. R. A. Frye and R. W. Holz, The relationship between arachidonic
     acid release and catecholamine secretion from cultured
     bovine adrenal chromaffin cells, *J. Neurochem.* 43:146
     (1984).

12. A. Hotchkiss, H. B. Pollard, J. Scott and J. Axelrod, Release of
     arachidonic acid from adrenal chromaffin cell cultures
     during secretion of epinephrine, *Fed. Proc.* 40:256 (1981).

13. K. Hong, N. Duzgunes, and D. Papahadjopoulos, Modulation of
     membrane fusion by calcium binding proteins, *Biophys. J.*
     37:297 (1982).

14. D. E. Knight and N. T. Kestenen, Evoked transient intracellular
     free Ca^{++} changes and secretion in isolated bovine adrenal
     medullary cells, *Proc. R. Soc. Lond. B.* 219:177 (1983).

15. K. Morita, K. W. Brocklehurst, S. M. Tomares, and H. B. Pollard,
     The phorbol ester TPA enhances A23187-but not carbachol or
     high K^+-induced catecholamine secretion from cultured bovine
     adrenal chromaffin cells, *Biochem. Biophys. Res. Commun.*
     129:511 (1985).

16. J. M. Lord and S. J. H. Ashcroft, Identification and characteri-
     zation of Ca^{2+}-phospholipid-dependent protein kinase in rat
     islets and hamster B-cells, *Biochem. J.* 219:547 (1984).

17. W. F. Odenwald and S. J. Morris, Identification of a second
     synexin-like adrenal medullary and liver protein that en-
     hances calcium-induced membrane aggregation, *Biochem.
     Biophys. Res. Commun.* 112:147 (1983).

18. H. B. Pollard, J. H. Scott, and C. E. Creutz, Inhibition of
     synexin activity and exocytosis from chromaffin cells by
     phenothiazine drugs, *Biochem. Biophys. Res. Commun.*
     113:908 (1983).

19. H. B. Pollard, C. E. Creutz, V. M. Fowler, J. H. Scott, and C.
     J. Pazoles, Calcium-dependent regulation of chromaffin
     granule movement, membrane contact, and fusion during
     exocytosis, *Cold Spring Harbor Symp. Quant. Biol.* 46:819
     (1982).

20. H. B. Pollard and J. H. Scott, Synhibin: a new calcium depend-
     ent membrane binding protein that inhibits synexin-induced
     chromaffin granule aggregation and fusion, *FEBS Lett.*
     150:201 (1982).

21. H. B. Pollard, R. Ornberg, M. Levine, E. Heldman, K. Morita, P. Lelkes, and J. Heldman, Ultrastructural and biochemical aspects of biosynthesis and secretion of hormones, in: "Endocrinology," F. Labrie and L. Pronix, ed., Elsevier, Amsterdam, pp. 383-386 (1984).

22. S. M. Simon and R. R. Llinas, Compartmentalization of the submembrane calcium activity during calcium influx and its significance in transmitter release, Biophys. J. 48:485 (1985).

23. A. Stutzin, A fluorescence assay for monitoring and analyzing fusion of biological membrane vesicles in vitro, FEBS Lett. 197:274 (1986).

24. T. C. Sudhof, J. H. Walker, and J. Obrocki, Calelectrin self-aggregates and promotes membrane aggregation in the presence of calcium, EMBO J. 1:1167 (1982).

25. T. C. Sudhof, M. Ebbecke, J. H. Walker, U. Fritsche, and C. Bonstead, Isolation of mammalian calelectrins: a new class of ubiguitous Ca^{2+}-regulation proteins, Biochemistry 23:1103 (1984).

26. K. E. Sussman, H. B. Pollard, J. W. Leitner, R. Nesher, J. Adler, and E. Cerasi, Differential control of insulin secretion and somatostatin receptor recruitment in isolated islets, Biochem. J. 214:225 (1983).

27. J. H. Walker, Isolation from cholinergic synapses of a protein that binds to membranes in a calcium dependent manner, J. Neurochem. 39:815 (1982).

28. S. P. Wilson and N. Kirshner, Calcium-evoked secretion from digitonin-permeabilized adrenal medullary chromaffin cells. J. Biol. Chem. 258:4994 (1983).

THE INSULIN SECRETORY GRANULE: FEATURES AND FUNCTIONS IN COMMON

WITH OTHER ENDOCRINE GRANULES

J.C. Hutton, M. Peshavaria, H.W. Davidson, K. Grimaldi
R. Pogge Von Strandmann and K. Siddle

Department of Clinical Biochemistry
University of Cambridge
Addenbrooke's Hospital
Hills Road
Cambridge CB2 2QR, U.K.

The insulin secretory granule of the pancreatic β-cell although
simple in morphology is a complex biochemical entity containing
upwards of a hundred different proteins (for review see (1)). Besides
the primary secreted products which constitute about 80% of the total
granule protein, there are minor cosecreted peptides, proteases
involved in processing of proinsulin, ion pumps in the granule membrane and granule surface proteins involved in the movement of the
organelle in the cell and its fusion with the plasma membrane during
exocytosis.

Probably most of these proteins, like insulin, are synthesized as
pre-proteins with hydrophobic leader sequences on membrane bound
ribosomes, and enter the secretory pathway at the level of the endoplasmic reticulum. As with most proteins following this route it is
anticipated that the pre-sequence would be removed co-translationally
and that further modifications such as glycosylation and sulphation
will occur during the transit to and through the Golgi apparatus.
Granule assembly probably occurs at the trans-face of the Golgi as is
evidenced at the ultrastructural level by the formation of condensing
vacuoles containing secretory material. During this process the
granule proteins synthesized _de novo_ are sorted out from proteins
destined for other parts of the cell and possibly combined with
material recycled from previously-used granules. Subsequent maturation

of the granule involves changes in the membrane typified by the
transient insertion of clathrin molecules[21] and the insertion or
activation of ion transporting proteins which result in the acidification of the granule interior and the intragranular accumulation of
divalent cations.[15,20] At this stage the proteolytic conversion of
proinsulin and possibly other granule matrix proteins occurs, this
being the product of the congregation of the substrates and enzymes
involved and the establishment of a suitable ionic environment within
the organelle.

In terms of its pathway of biogenesis and the function of individual protein constituents much of what can be said of the insulin
secretory granule also applies to granules isolated from other endocrine tissues which release polypeptide hormones by a regulated
exocytotic mechanism. Whether individual protein constituents within
different granule types are identical is however generally not known.
This stems principally from the difficulty of isolating sufficient
quantities of granules from tissues secreting polypeptide hormones for
biochemical studies and the fact that most endocrine cells occur in
organs with multiple endocrine functions. Much of our current understanding ironically stems from studies on the chromaffin granule of
the adrenal medulla which, although actively secreting polypeptides,
has as its principal function the storage and release of biogenic
amines.

The imbalance of information has begun to be redressed by the
introduction of a number of tumour models derived from endocrine
tissues which provide gram quantities of homogeneous cell populations. Paramount among these has been an X-ray induced insulinoma
which has proved a useful source for the purification and characterization of proteins of the insulin granule.[3]

In our laboratory we have studied a proton translocating ATPase,[16]
membrane substrates for Ca^{2+}-dependent phosphorylation reactions,[2] a
membrane Ca^{2+}-inhibitable phosphatidylinositol kinase,[25] a proinsulin
endopeptidase,[12] cathepsin B-like proteins,[6] a carboxypeptidase[5] and
betagranin, a protien which shares a common N-terminus with bovine
adrenal medullary chromogranin A.[13] A number of these proteins have
been purified to homogoeneity and antisera raised to them.[4,17] In
addition, monoclonal antibodies recognising three different granule
membrane antigens have been prepared.[10]

With such antisera it is now possible to examine whether the
normal endocrine tissue from which the tumours were derived contain
these proteins in the same molecular form and quantity and to examine
whether the common functions of different granules reside in identical
or perhaps subtly different forms of the same proteins.

To examine these questions we have examined the distribution of
four insulin secretory granule antigens in a range of other tissues
with particular emphasis on their subcellular localization in other
endocrine cells.

MATERIALS AND METHODS

Insulin secretory granules,[14] adrenal chromaffin granules[23] and a
mixed secretory granule fraction from whole pituitary[19] were prepared
from NEDH strain rats by published methods. Betagranin[13] and carboxypeptidase E[7] were purified from insulin granules by published procedures and immunization of guinea pigs and rabbits achieved by
subcutaneous injection (10-100 µg/animal) of the antigen in Freunds
adjuvant at 4-6 week intervals. Insulin secretory granule membranes
were prepared by hypotonic lysis of purified granules and used to
immunize mice of the BALB/c strain. Monoclonal antibodies recognising
insulin secretory granule membrane peptides were identified and
prepared by conventional procedures.[8,9]

NEDH rat tissue samples and secretory granule preparations were
subjected to single dimensional SDS polyacrylamide gel electrophoresis,[18] and the separated proteins transferred to nitrocellulose
membranes.[26] The transferred proteins were incubated for 16 h at 4°C
with antisera (1:500 -1:1000 dilution) or ascites fluid (1:5000)
followed by a second anti IgG antibody, either conjugated to horse
radish peroxidase (polyclonal sera) or radiolabelled with [125]I.
Antibody binding was detected by chromogenic substrate reaction
(diaminobenzidine) or autoradiography respectively.

RESULTS

Carboxypeptidase E

Carboxypeptidase E (CPE) is an acidic metalloenzyme (pH opt. 5.5)
which is involved in the removal of C-terminal basic amino acids

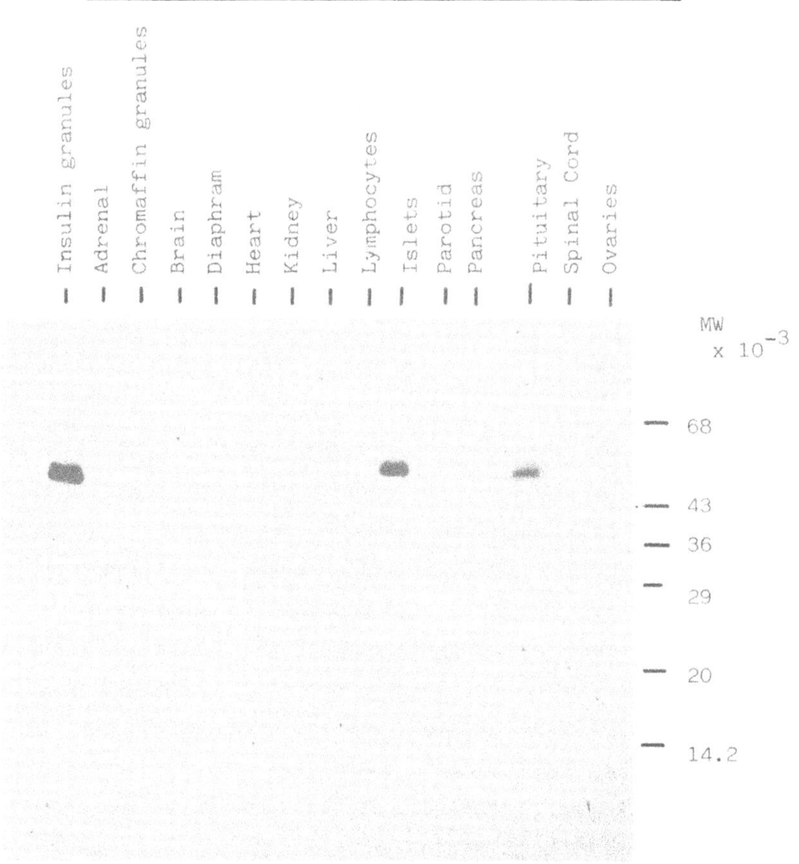

Fig. 1. Tissue distribution of insulin secretory granule carboxypeptidase E activity. Samples (c.a. 50 µg protein) were electrophoresed on SDS polyacrylamide gels (15.5% acrylamide 0.08% bisacrylamide) and the protein transferred electrophoretically to nitrocellulose membranes. Immunoreactive peptides were detected using an
immunoperoxidase-based procedure.

generated by cleavage of proinsulin with an endopeptidase with a
trypsin-like specificity. The enzyme was isolated as a monomer of
50,000 daltons from insulinoma and antisera raised to it reacted on
immunoblotting with a single peptide in insulin secretory granules of
identical molecular weight (Fig. 1). A protein of identical molecular
size was identified in islet tissue and in pituitary but not in a wide
range of other tissues of non-endocrine origin. Adrenal chromaffin
granules showed a weakly immunoreactive protein of the same size.
Pituitary granules showed in addition to the 50,000 dalton protein an
immunoreactive form of 55,000 (data not shown). This may coincide

with a membrane bound form of the enzyme which has been reported.[24]
Immunofluorescence studies showed that all the major islet cell types
contained CPE, that the protein was widely distributed among pituitary
cells of both lobes but by no means in all cells. In the adrenal it
was localized in the medulla. The distribution of CPE thus was
consistent with its role in post translational proteolytic processing
of protein precursors where biologically-active peptides are deline-
ated by pairs of basic amino acid residues. The apparently lower
content of adrenal tissue cannot be explained simply on the basis of
its localization to the medullary region since isolated granules were
likewise poorly immunoreactive.

Betagranin

Betagranin has been isolated from insulinoma tissue as three
closely-related 21,000 dalton peptides each having the same N-terminus
as adrenal medullary chromogranin A (approx 80,000 daltons). Of the
tissues studied (see Fig. 1) immunoreactivity was found only in
adrenal, islets and anterior pituitary.

The immunoreactive form in islets was identical to that in
insulinoma granules (Fig. 2) being comprised of a major doublet of
21-22,000 dalton together with a lesser quantity (approx 5%) of a
protein of 85,000 daltons. In adrenal tissue the major immunoreactive
forms were of 72,000 and 85,000 daltons with very small amounts of a
21,000 dalton protein. Differences were also observed in the major
molecular form present .in a low density versus a higher density
chromaffin granule fraction. Anterior pituitary contained both high
and lower molecular weight forms im approximately equal proportion.
Immunohistochemical studies showed the distribution within these
tissues to be similar to that of CPE. Antibodies directed toward
bovine adrenal medullary chromogranin A gave essentially the same
pattern as anti-betagranin antisera, however the latter appeared to
recognise the lower molecular weight forms more readily.

SGM 80, SGM 110

SGM 80 and 110 are antigens recognised in immunoblots by monoclonal antibodies GM4 and GM10 respectively which have been raised to
insulinoma secretory granule membranes. These antigens are relatively

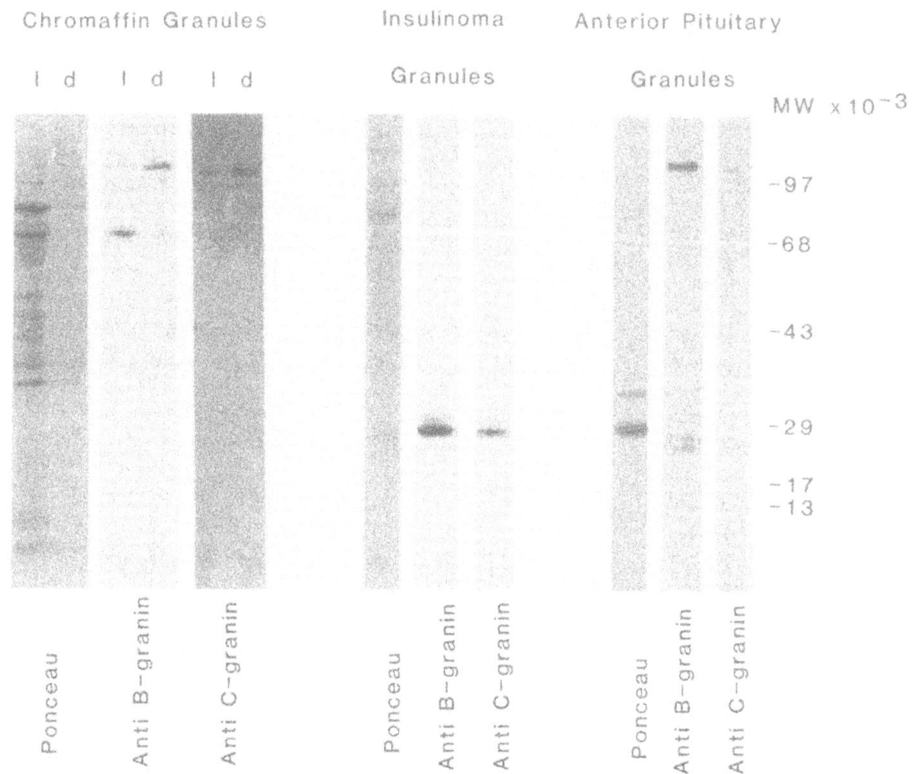

Fig. 2. Betagranin and chromogranin A immunoreactivities in different granule types. Samples (c.a. 50 µg protein) were electrophoresed on SDS polyacrylamide gels (12.5% acrylamide 0.908% bisacrylamide) and the proteins transferred electrophoretically to nitrocellulose membranes. In the case of isolated rat chromaffin granules the granule population identified on the basis of catecholamine content was subfractionated into low (l) and high (d) density components. The spectrum of proteins transferred were visualized by staining with Ponceau S and the immunoreactive peptides detected by an immuno-peroxidase-based procedure. Guinea pig anti rat betagranin (B-granin) and rabbit anti bovine adrenal chromogranin A (C-granin) antisera were used.

rare proteins as judged by our inability to correlate immunoreactivity with Coomassie blue staining proteins on electrophoretograms. They are consequently only detectable in tissue subcellular fractions and not in the tissue as a whole.

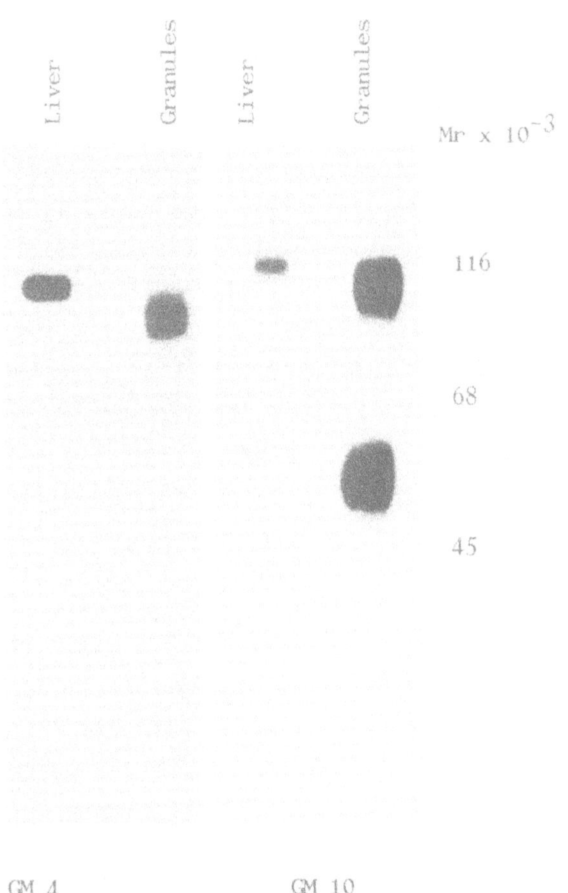

Fig. 3. Monoclonal antibody binding to insulin secretory granules and liver lysosomal membranes. Samples (c.a. 50 µg protein) were electrophoresed on SDS polyacrylamide gels (10% acrylamide 0.08% bisacrylamide) and the proteins transferred electrophoretically to nitrocellulose membranes. The binding of the two classes of monoclonal antibodies (GM4 and GM10) were detected by an autoradiographic procedure.

Antibody GM4 which recognises an 80,000 dalton insulinoma granule protein (Fig. 3) recognised proteins of identical molecular size and with similar intensity in anterior pituitary and adrenal chromaffin granules (data not shown). It also reacted with a protein of slightly higher molecular size in a liver lysosomal membrane preparation. The

antigen in this case migrated as a much less diffuse band on electrophoresis which might suggest a difference in the degree of glycosylation rather than a difference in size of the peptide backbone of the
molecule.

Antibody GM10 which recognises two proteins of 50 and 110,000
daltons in insulinoma granules recognised antigens of similar size in
anterior pituitary and adrenal chromaffin granules. Biosynthesis
experiments have suggested that the 50,000 dalton protein is derived
post-translationally from the larger molecule. It was interesting to
note in this regard that whereas the immunoreactivity of these forms
in anterior pituitary and insulinoma granules was approximately
equivalent, the higher molecular weight form predominated in chromaffin granules.

GM10 like GM4 recognised an antigen in liver lysosomal membrane.
The molecular form was slightly larger than the 110,000 dalton
insulinoma antigen and migrated as a tighter band on electrophoresis.
A form equivalent to the 50,000 dalton antigen was not seen.

DISCUSSION

Obviously secretory granules from different endocrine cell types
show marked compositional differences best exemplified by the nature
of the secreted product itself. Further differences are clearly
apparent when electrophoretic profiles of different granule types are
considered. Indeed most simple electrophoretic analyses fail to
demonstrate any common components.[14,22] It must be remembered
however that in many studies of this kind different species have been
used. Also any tissue-specific differences in post-translational
modification of proteins would result in molecular weight discrepanci-
es between tissues leading to an erroneous conclusion. In this regard
the use of immunological techniques appear to have considerable
advantages in the search for common structural features among different granule antigens.

Of the four antigens considered two are predominantly matrix
proteins (CPE and betagranin) and two are associated with the granule
membrane (SGM 80 and SGM 110). All were first identified in
insulinoma tissue and their granule-specific localization ascertained.
All were present in normal islet tissue although significant differ-

ences were observed especially in the case of betagranin immunoreactivity (data not shown). The distribution of the insulin granule
matrix proteins appeared to be restricted to other endocrine-tissue.
The granule membrane antigens however also appeared in liver suggesting that they may be involved in other intracellular membrane traffic.
Their apparent granular localization in insulinoma tissue might occur
by virtue of the relative high proportion of granule membrane to other
membrane structures in the cell.

A further interesting feature of the data was the presence of
different molecular forms of the antigens in different endocrine
tissues. The chromogranin A related peptides for example ranged
from higher molecular forms in adrenal medullary tissue to predominantly low molecular weight species in insulin secreting tissues. It
is tempting to speculate that these differences relate to differences
in the proteolytic processing of a common higher molecular weight
precursor. Different biological activities might reside in different
portions of the chromogranin molecule thus permitting a different
biological response depending on the cell in question and its physiological status. The biological functions of the chromogranin A-
related peptides however remain elusive and much more needs to be
learned about the structural and genetic relationship of the different
tissue forms.

Given that in these studies no attempt was made to select common
antigens at the outset one must conclude there are probably many more
similarities between endocrine secretory granules than hitherto
suspected. Some of these similarities would seem to reside in shared
functions such as a common role in exocytosis and pro-protein prote-
olysis. However in the case of the proton-translocating ATPase it
would seem that the same activity serves different roles in different
granule types.[16] It might be argued in this respect that what we
observe using an immunological approach are the vestiges of an arche-
typal organelle whose basic functions have evolved to fulfill quite
diverse physiological functions.

ACKNOWLEDGEMENTS

These studies were supported by The British Diabetic Association,
The Medical Research Council of Great Britain and The Wellcome Trust.
Mrs. J. Eastwell is thanked for help in preparing the manuscript.

REFERENCES

1. H. Blaschko, R. S. Comline, F. H. Schneider, M. Silver, and A. D.
   Smith, Secretion of a chromaffin granule protein, chromo-
   granin, from the adrenal gland after splanchnic stimulation,
   Nature 215:58 (1967).

2. K. W. Brocklehurst and J. C. Hutton, Ca^{2+}-dependent binding of
   cytosolic components to insulin secretory granules results
   in Ca^{2+}-dependent protein phosphorylation, _Biochem. J._
   210:533 (1983).

3. W. L. Chick, S. Warren, R. N. Chute, A. A. Like, V. Lauris, and
   K. C. Kitchen, A transplantable insulinoma in the rat, _Proc._
   Natl. Acad. Sci. USA 74:628 (1977).

4. H. W. Davidson and J. C. Hutton, Purification of an insulin
   secretory granule carboxypeptidase, _Diabetes Res. Clin._
   Practice Suppl. 1, S122 (1985).

5. K. Docherty and J. C. Hutton, Carboxypeptidase activity in the
   insulin secretory granule, _FEBS Lett._ 162:137 (1983).

6. K. Docherty, J. C. Hutton, and D. F. Steiner, Cathepsin B-related
   proteases in the insulin secretory granule, _J. Biol. Chem._
   259:6041 (1984).

7. L. D. Fricker and S. H. Snyder, Purification and characterization
   of enkephalin convertase, an enkephalin-synthesizing carboxypeptidase, _J. Biol. Chem._ 258:10950 (1983).

8. G. Galfre and C. Milstein, Preparation of monoclonal antibodies:
   Strategies and procedures, _Methods in Enzymology_ 73:3
   (1981).

9. J. W. Goding, Antibody production by hybridomas, _J. Imm. Methods_
   39:285 (1980).

10. K. A. Grimaldi, J. C. Hutton, and K. Siddle, Characterization of
    monoclonal antibodies to the insulin secretory granule
    membrane, _Diabetes Res. Clin. Practice_ Suppl. 1 S207
    (1985).

11. J. C. Hutton, Secretory granules, _Experientia_ 40:1091 (1984).

12. J. C. Hutton, H. W. Davidson, K. A. Grimaldi, and K. Siddle,
    Proteins of the insulin secretory granule, _Excerpta Medica._
    Int. Congress Series (in press, 1986).

13. J. C. Hutton, F. Hansen, and M. Peshavaria, Beta-Granins: 21 KDa
    co-secreted peptides of the insulin granule closely related

to adrenal medullary chromogranin A, <u>FEBS Lett.</u> 188:336 (1985).

14. J. C. Hutton, E. J. Penn, and M. Peshavaria, Isolation and characterization of insulin secretory granules from a rat islet cell tumor, <u>Diabetologia</u> 23:365 (1982).

15. J. C. Hutton, E. J. Penn, and M. Peshavaria, Low-molecular-weight constituents of isolated insulin secretory granules. Bivalent cations, adenine nucleotides and inorganic phosphate, <u>Biochem. J.</u> 210:297 (1983).

16. J. C. Hutton and M. Peshavaria, Proton-translocating Mg^{2+}-dependent ATPase activity in insulin-secretory granules, <u>Biochem. J.</u> 204:161 (1982).

17. J. C. Hutton and M. Peshavaria, Beta-granins; 21000 NW peptides co-secreted with insulin which are related to adrenal medullary chromogranin A, <u>Diabetes Res. Clin. Practice</u> Suppl. 1 S254 (1985).

18. U. K. Laemmli, Cleavage of structural proteins during the assembly of the head of bacteriophage T4, <u>Nature</u> 227: 680 (1970).

19. Y.- P. Loh, W. W. Tam, and J. T. Russell, Measurement of delta pH and membrane potential in secretory vesicles isolated from bovine pituitary intermediate lobe, <u>J. Biol. Chem.</u> 259:8238 (1984).

20. L. Orci, The insulin factory: a tour of the plant surroundings and a visit to the assembly line. The Minkowski Lecture 1973 revisited, <u>Diabetologia</u> 28:528 (1985).

21. L. Orci, P. Halban, M. Amherdt, M. Ravazzola, J. D. Vassalli, and A. Perelet, A clathrin-coated, Golgi-related compartment of the insulin secreting cell accumulates proinsulin in the presence of monensin, <u>Cell</u> 39:39 (1984).

22. R. W. Rubin and B. C. Pressman,

<u>J. Cell Biol.</u> 95:393a (1982).

23. A. D. Smith and H. Winkler, A simple method for the isolation of adrenal chromaffin granules on a large scale, <u>Biochem. J.</u> 103:480 (1967).

24. S. Suplattapone, L. D. Fricker, and S. H. Snyder, Purification and characterization of a membrane-bound enkephalin-forming carboxypeptidase, "enkephalin convertase", <u>J. Neurochem.</u> 42:1017 (1984).

25. N. E. Tooke, C. N. Hales, and J. C. Hutton, Ca^{2+} sensitive
    phosphatidylinositol 4-phosphate metabolism in a rat beta-
    cell tumour, Biochem. J. 219:471 (1984).
26. H. Towbin, T. Staehelin, and J. Gordon, Electrophoretic transfer
    of proteins from polyacrylamide gels to nitrocellulose
    sheets: procedure and some applications, Proc. Natl. Acad.
    Sci. USA 76L435 (1979).

EFFECTS OF MONENSIN ON GLUCOSE-INDUCED INSULIN RELEASE AND $^{45}Ca^{2+}$ OUTFLOW

A. Kanatsuka, H. Makino, N. Hashimoto
M. Sakurada and S. Yoshida

The Second Department of Internal Medicine
Chiba University School of Medicine
Chiba 280, Japan

INTRODUCTION

The β-cells in the pancreatic islets respond to glucose with a
biphasic insulin release[2]. Protons generated from glucose metabolism have
been proposed to be a coupling factor between metabolic and ionic events in
the cells[1,7]. A rise in cytosol Ca^{2+} concentration is considered to be
essential in stimulus-secretion coupling in the β-cells[4]. Thus, it seems
possible that the biphasic insulin release is a reflection of a biphasic
change in the Ca^{2+} concentration of the cytosol, the change being
controlled by the rate of generation of protons. To examine the role of
protons and calcium in the insulin response to glucose, we examined the
effect of monensin, a carboxylic ionophore[8], on the first and second phase
of glucose-induced insulin release and on the $^{45}Ca^{2+}$ efflux from perifused
rat pancreatic islets.

MATERIALS AND METHODS

Experiments were performed with isolated islets from fed male
Sprague-Dawley rat weighing 300 g[5]. Ten islets, preincubated for 60 min in
Krebs-Ringer bicarbonate buffer containing 5.5 mM glucose, were incubated
for 60 min in 1 ml of the medium, containing 16.7 mM glucose and increasing

concentrations (from 1 to 1000 nM) of monensin at 37°C. The medium was equilibrated with 95% O_2/5% CO_2 to maintain a pH of 7.4. Insulin released into the medium was assayed by radioimmunoassay. To assess the interference of monensin on the glucose metabolism of the islets, $^{14}CO_2$ production rate from $D-(U-^{14}C)$ glucose of the islets was determined. Groups of 30 islets placed in liquid-scintillation counting vials were incubated for 120 min in the medium containing 16.7 mM glucose. 4 μCi/ml $D-(U-^{14}C)$ glucose and 10 or 100 nM monensin. After the incubation, radioactivity absorbed for 30 min by hyamine hydroxide was determined by liquid-scintillation counting. Groups of 100 islets were incubated for 60 min in the medium containing $^{45}Ca^{2+}$ (1.12 mM: 0.2 mCi/ml) and 16.7 mM glucose, at 37°C under 95% O_2/5% CO_2. After the incubation, the islets were perifused with the medium. Total volume of the perifusion system was 0.8 ml. The medium was pumped over the islets at a rate of 0.8 ml/min and, from the 31st to 80th min, the effluent was collected over successive periods of 1 min. Insulin and radioactivity of $^{45}Ca^{2+}$ in the effluent were determined. $^{45}Ca^{2+}$ efflux was expressed in percentage of the mean value found within the same experiment between the 45th and 50th min. Student's t-test was used for statistical analysis.

RESULTS

Monensin, 1-100 nM, dose-dependently inhibited insulin release from the islets incubated for 60 min in the medium containing 16.7 mM glucose (Table 1). Radioactivity of ^{14}C incorporated into $^{14}CO_2$ was not affected by addition of 10 or 100 nM monensin (Table 2).

With a change from 0 to 16.7 mM glucose and simultaneous addition of 100 nM monensin (Fig. 1), the agent had only a slight and nonsignificant inhibitory effect on the first phase of insulin release. In contrast, the second phase of insulin release was markedly blocked. $^{45}Ca^{2+}$ efflux fell during the first 2 min and then rose and reached a peak at 4 min. The agent markedly depressed $^{45}Ca^{2+}$ efflux concomitant with the second phase of insulin release. When monensin was added at the start of the second phase, release was completely inhibited, and the concomitant $^{45}Ca^{2+}$ efflux markedly decreased.

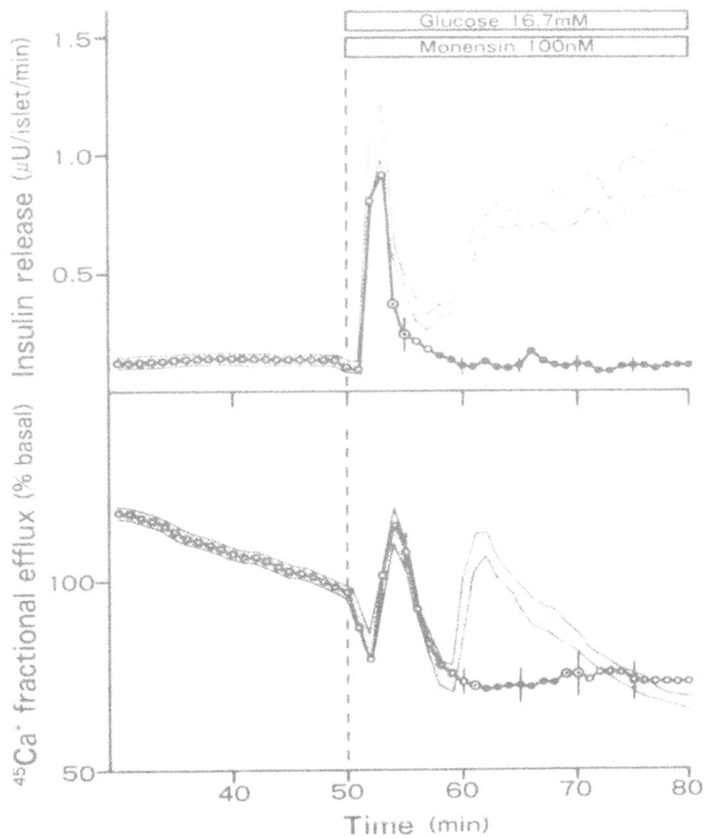

Fig. 1. Effect of monensin, 100 nM, and glucose, 16.7 mM, concentration
from 0 to 16.7 mM on insulin release (upper panel) and $45\,Ca^{2+}$ efflux (lower
panel) from isolated rat pancreatic islets prelabeled with $^{45}Ca2+$. Mean
values ($\pm$ SEM) for insulin release are expressed in μU/islet/min. Mean
values for $45Ca^{2+}$ efflux are expressed as a fractional outflow rate. Solid
lines denote the mean of insulin release and $45Ca2+$ efflux from the islets
perifused with the medium containing monensin (n=5). Shaded area enclosed
by the mean $\pm$ SEM represents insulin release and $^{45}Ca^{2+}$ efflux from the
islets without monensin (n=10). $\bullet$, p <0.01; $\bullet$, p <0.05; o, not
significant; compared with the release and the efflux from the islets
without monensin.

DISCUSSION

Since monensin, 100 nM, did not interfere with CO_2 production from

glucose, the drug appeared to inhibit glucose-induced insulin release at a

site distal to that related to glucose metabolism. Because monensin

depressed $^{45}Ca^{2+}$ efflux during the second phase, it is possible that the

drug inhibited the second phase of insulin release mainly by affecting Ca^{2+}

handling. Monensin is a carboxylic ionophore and acts as a Na^+/H^+

exchanger[8]. Glucose-induced insulin release is accompanied by a slight

Table 1. Effect of monensin on glucose-induced insulin secretion from
incubated islets

| | Glucose (16.7 mM) Monensin (nM) | | | | |
	0	1	10	100	1000
Insulin release	85.9 ± 7.1	$67.1^* \pm 5.0$	$32.1^{**} \pm 2.3$	$19.8^{**} \pm 9.8$	$19.9^{**} \pm 3.8$
(U/islet/hr)	(n=21)	(n=9)	(n=23)	(n=11)	(n=12)

The number of test tubes run in each group is indicated in parentheses.
Mean ± SEM; ** : p <0.01; *: p <0.05 compared with the insulin release
induced with 16.7 mM glucose.

Table 2. Effect of monensin on $^{14}CO_2$ production from D-(U-^{14}C) glucose in
the incubated islets.

| | Glucose (16.7 mM) Monensin (nM) | | |
	0	10	100
^{14}C incorporated into $^{14}CO_2$	825 ± 91	760 ± 197	720 ± 252
(CPM/islet/2 hr)	(n=11)	(n=10)	(n=6)

change in intracellular pH[6]. Therefore, one possible explanation for our
results is that monensin exchanged Na^+ for the protons generated from

glucose metabolism, affecting intracellular pH and thus blocking ionic
events in the islet cells. The inhibitory effect of monensin was produced
in part through perturbation of the proton gradient within the secretory
granule[3].

In contrast, only a slight inhibitory effect of monensin on the first
phase of insulin release and a lack of effect on the $^{45}Ca^{2+}$ efflux was
confirmed. When monensin was added at the start of the second phase,
insulin release was completely inhibited. The lack of effect on first
phase cannot be the result of a slow onset in the action of monensin. The
coupling between glucose metabolism and Ca^{2+} handling during first phase is
possibly independent of the increase in proton concentration due to glucose
metabolism.

REFERENCES

1. A.R. Carpinelli, A. Sener, A. Herchuelz, and W.J. Malaisse, Stimulus-
   secretion coupling of glucose-induced insulin release. Effect of
   intracellular acidification upon calcium efflux from islet cells,
   Metabolism 29:540 (1980).
2. D.L. Curry, L.L. Bennet, and G.M. Grodsky, Dynamics of insulin
   secretion by the perfused rat pancreas, _Endocrinology_ 83:572 (1968).
3. G. Gold, J. Pou, R.M. Nowlain, and G.M. Grodsky, Effect of monensin on
   conversion of proinsulin and secretion of newly synthesized insulin
   in isolated rat islets, _Diabetes_ 33:1019 (1984).
4. G.M. Grodsky and L.L. Bennet, Cation requirements for insulin secretion
   in the isolated perfused pancreas, _Diabetes_ 15:910 (1966).
5. A. Kanatsuka, H. Makino, M. Osegawa, J. Kasanuki, T. Suzuki, and S.
   Yoshida, Is somatostatin a true local inhibitory regulator of
   insulin secretion, _Diabetes_ 33:510 (1984).
6. W.J. Malaisse, A.R. Carpinelli, and A. Sener, Stimulus-secretion
   coupling of glucose-induced insulin release. Timing of early
   metabolic, ionic, and secretory events, _Metabolism_ 30:527 (1981).
7. C.S. Pace, Role of pH as a transduction device in triggering electrical
   and secretory responses in islets β-cells, _Fed. Proc._ 43:2379
   (1984).
8. B.C. Pressman, Biological applications of ionophores, _Annu. Rev.
   Biochem._ 45:501 (1976).

INTERDEPENDENCY OF Ca^{2+} AVAILABILITY AND CYCLIC AMP GENERATION IN THE
PANCREATIC β-CELL

I. Valverde, P. Garcia-Morales, M. Saceda
and W.J. Malaisse

Fundacion Jimenez Diaz
Universidad Autonoma de Madrid
Madrid, Spain
+Laboratory of Experimental Medicine
Brussels Free University
Brussels, Belgium

At least ten distinct coupling factors or second messengers may
participate in the process of insulin release from the pancreatic β-cell.
They include, _inter alia_, reducing equivalents, protons, ATP, diacylglycerol, inositol 1,4,5-triphosphate, polyamine and obviously cyclic AMP and
cytosolic ionized calcium. This report deals briefly with the interdependency of Ca^{2+} availability and cyclic AMP generation in the pancreatic
β-cell.

Calcium-dependency of Nutrient-stimulated Cyclic AMP Formation in Intact Islets

A first modality of interaction between Ca^{2+} and cyclic AMP is
suggested by the finding that calmodulin activates both adenylate cyclase
and phosphodiesterase in islet subcellular fractions[13]. Both the binding
of [^{125}I]calmodulin to a subcellular particulate fraction and the
activation by calmodulin of adenylate cyclase in the same fraction represent Ca^{2+}-dependent processes[15]. Half-maximal activation of the enzyme is
observed at a calmodulin concentration close to 0.1 μM[14]. The latter value
is much lower than the calmodulin content of the islet, which is close to
50 μM. Hence, the immediate regulation of adenylate cyclase by

Ca-calmodulin may depend more on the cytosolic Ca^{2+} activity than on the concentration of the calcium-dependent regulatory protein.

To explore the validity of the latter proposal, we have investigated the calcium-dependency of the stimulant action of various nutrient secretagogues upon cyclic AMP formation in intact islets. In islets exposed to either D-glucose, L-leucine and/or L-glutamine, the nutrient-induced increment in cyclic AMP formation was indeed invariably lower in the absence than presence of extracellular Ca^{2+} [12].

It should be stressed that, even at normal extracellular Ca^{2+} concentration, the glucose-induced increase in cyclic AMP formation by intact islets is quite modest, when compared to the effect of such agents as glucagon and, even more so, forskolin[1]. It is conceivable, therefore, that endogenously formed cyclic AMP plays a greater role in the hormonal modulation of insulin release than in the normal process of glucose-stimulated insulin secretion.

Hormonal Modulation of Adenylate Cyclase Activity

We have recently identified in islet cell membranes the α-subunits of the two guanyl nucleotide regulatory proteins, Ns and Ni, which mediate, respectively, the hormonal stimulation and inhibition of adenylate cyclase activity[6,11].

Cholera toxin (1-20 µg/ml) stimulated, in membranes exposed to $[\alpha-^{32}P]NAD^+$, the ADP-ribosylation of the two forms of the α-subunit of Ns, with prevalence of the heavy over light form. When intact islets were pre-incubated with cholera toxin, the adenylate cyclase activity of a subcellular particulate fraction was increased. The responsiveness of adenylate cyclase to GTP (but not GTPαS) was also augmented. In intact islets, the production of cyclic AMP and the glucose-stimulated release of insulin were also enhanced after pretreatment with cholera toxin[11].

Likewise, exposure of rat pancreatic islet membranes to $[\alpha-^{32}P]NAD^+$ in the presence of _Bordetella pertussis_ toxin (islet-activating protein) revealed the ADP-ribosylation of a peptide with a MW close to 41 kDa, which corresponds to the α-subunit of the guanine nucleotide regulatory protein Ni. Islets removed from rats pretreated with the _Bordetella pertussis_ toxin displayed a specific increase in the adenylate cyclase responsiveness to GTP and were characterized by a resistance to the inhibitory action of α_2-adrenergic agonists upon either adenylate cyclase activity of glucose-induced insulin release[6]. Such agonists (e.g. clonidine) indeed inhibit both adenylate cyclase activity in islet homogenates and cyclic AMP production by intact islets. The latter effect is most evident in islets exposed to forskolin[1].

Effect of Cyclic AMP Upon Calcium Handling by Islet Cells

It is still a matter of debate whether cyclic AMP facilitates Ca^{2+} influx into the β-cell, causes an intracellular redistribution of Ca^{2+} within the β-cell or fails to affect significantly the cytosolic Ca^{2+} activity[4]. We have been unable to detect any major effect of either clonidine or forskolin upon the net uptake of ^{45}Ca by isolated islets. This does not rule out, however, more subtle changes in the handling of ^{45}Ca by the islet cells[2,3]. Nevertheless, there are reasons to believe that cyclic AMP facilitates insulin release independently of any major effect upon the net uptake and intracellular distribution of Ca^{2+}. For instance, in the absence of extracellular Ca^{2+}, and even in the presence of EGTA, forskolin augments insulin release from islets exposed to a high concentration of D-glucose[3]. Thus, cyclic AMP may increase the responsiveness to Ca^{2+} of the effector system controlling insulin release. The latter view is compatible with data derived from the mathematical modelling of cyclic AMP/Ca^{2+} interaction in the β-cell.

Mathematical Modelling of Stimulus-secretion Coupling in the β-cell

A mathematical model was recently proposed to simulate the effect of D-glucose and extracellular Ca^{2+} upon ^{45}Ca net uptake, ^{45}Ca outflow, cyclic AMP formation and insulin release in the β-cell[5]. This model includes such features as the influence of cytosolic Ca^{2+} upon adenylate cyclase activity[7], the stratification of both cytosolic and vacuolar Ca^{2+} pools in a cortical layer and central core[8], the participation of a process of Ca^{2+}-stimulated Ca^{2+} release from the vacuolar system[9], the stimulation by D-glucose of Ca^{2+} uptake by the vacuolar system[10], and the dissociated time course for the glucose-induced changes in distinct Ca^{2+} movements[5]. Although this model may involve a regulatory role for cyclic AMP in either the fractional removal rate of cytosolic Ca^{2+} by the vacuolar system[7] or

fractional outflow rate of Ca^{2+} from the latter system[10], the major
postulated effect of the nucleotide is to increase the slope of the line
relating insulin output to cytosolic Ca^{2+} activity or, in other words, to
enhance the responsiveness to Ca^{2+} of the effector system mediating the
exocytosis of secretory granules. Despite the limitations and risks
inherent to any mathematical modelling of a process as sophisticated as
insulin release, this model may provide useful guiding information for
further studies on the interdependency of Ca^{2+} availability and cyclic AMP
generation in the pancreatic β-cell.

REFERENCES

1. P. Garcia-Morales, S.P. Dufrane, A. Sener, I. Valverde, and W.J.
    Malaisse, Inhibitory effect of clonidine upon adenylate cyclase
    activity, cyclic AMP production, and insulin release in rat
    pancreatic islets, Biosci. Reports 4:511 (1984).

2. V. Leclercq-Meyer, A. Herchuelz, I. Valverde, E. Couturier, J.
    Marchand and W.J. Malaisse, Mode of action of clonidine upon islet
    function. Dissociated effects upon the time course and magnitude
    of insulin release, Diabetes 29:193 (1980).

3. W.J. Malaisse,P. Garcia-Morales, S.P. Dufrane, A. Sener, and I.
    Valverde, Forskolin-induced activation of adenylate cyclase, cyclic
    adenosine monophosphate production and insulin release in rat
    pancreatic islets, Endocrinology 115:2015 (1984).

4. W.J. Malaisse and F. Malaisse-Lagae, The role of cyclic AMP in insulin
    release, Experientia 40:1068 (1984).

5. W.J. Malaisse, Y. Scholler, and V. De Maertelaer, Mathematical
    modelling of stimulus-secretion coupling in the pancreatic β-cell.
    Dissociated time course for the glucose-induced changes in distinct
    Ca^{2+} movements, Diabetes Res. 2:195 (1985).

6. W.J. Malaisse, M. Svoboda, S.P. Dufrane, F. Malaisse-Lagae, and
    J. Christophe, Effect of Bordetella pertussis toxin on
    ADP-ribosylation of membrane proteins, adenylate cyclase activity
    and insulin release in rat pancreatic islets, Biochem. Biophys.
    Res. Commun. 124:190 (1984).

7. W.J. Malaisse, I. Valverde, A. Owen, D. Verhulst and F. Cantraine,
    Mathematical modelling of cyclic $AMP-Ca^{2+}$ interactions in
    pancreatic islets, Diabetes 31:170 (1982).

8. Y. Scholler, V. De Maertelaer, and W.J. Malaisse, Mathematical
    modelling of stimulus-secretion coupling in the pancreatic β-cell.
    I. Dynamics of insulin release, Acta Diabet. Lat. 20:329 (1983).

9. Y. Scholler, V. De Maertelaer, and W.J. Malaisse, Mathematical
    modelling of stimulus-secretion coupling in the pancreatic β-cell.
    II. Calcium-stimulated calcium release, Computers Prog. in Biomed.
    19:119 (1985).

10. Y. Scholler, V. De Maertelaer, and W.J. Malaisse, Mathematical
    modelling of stimulus-secretion coupling in the pancreatic β-cell.
    III. Glucose-induced inhibition of calcium efflux, Biophys. J.
    46:439 (1984).

11. M. Svoboda, P. Garcia-Morales, S.P. Dufrane, A. Sener, I. Valverde, J.
    Christophe, and W.J. Malaisse, Stimulation by cholera-toxin of
    ADP-ribosylation of membrane proteins, adenylate cyclase and
    insulin release in pancreatic islets, Cell. Biochem. Funct. 3:25
    (1985).

12. I. Valverde, P. Garcia-Morales, M. Ghiglione, and W.J. Malaisse, The
    stimulus-secretion coupling of glucose-induced insulin release.
    LIII. Calcium dependency of the cyclic AMP response to nutrient
    secretagogues, Horm. Metab. Res. 15:62 (1983).

13. I. Valverde and W.J. Malaisse, Calmodulin and pancreatic β-cell
    function, Experientia 40:1061 (1984).

14. I. Valverde, A. Sener, P. Lebrun, A. Herchuelz, and W.J. Malaisse, The
    stimulus-secretion coupling of glucose-induced insulin release.
    XLVII. The possible role of calmodulin, Endocrinology 108:1305
    (1981).

15. I. Valverde, A. Vandermeers, R. Anjaneyulu, and W.J. Malaisse,
    Calmodulin activation of adenylate cyclase in pancreatic islets,
    Science 206:225 (1979).

CYCLIC AMP-ANTAGONIST, A SECOND MESSENGER FOR INSULIN ACTION,

INHIBITS GLUCOSE-STIMULATED INSULIN SECRETION IN ISOLATED ISLETS OF

CHINESE HAMSTERS

H.-J. Partke and H.K. Wasner
Diabetes Research Institute
Dusseldorf, Federal Republic of Germany

INTRODUCTION

The intracellular regulator cAMP-antagonist (cPIP) appears well
suited as a mediator of insulin action: cPIP has been isolated, after
stimulation with insulin and adrenalin, from rat livers at a
concentration of 0.6×10^{-7} M.[6] Similarly it can be isolated from
other tissues as heart, fat, kidney, muscle, brain, spleen, suggesting
that cPIP is an ubiquituous regulator. Dose-response curves show
increasing cPIP synthesis with increasing insulin concentrations.
This dose-response relationship parallels the receptor binding of
insulin. Evidence exists that cPIP is composed of prostaglandin E,
myo-inositol and phosphate as primary constituents. Therefore, it has
been named prostaglandyl-inositol cyclic phosphate (cPIP).[7,8].

cPIP inhibits cAMP-dependent protein kinases and activates
phosphoprotein phosphatase.[5]. Thus, by regulating the equilibrium
between the phosphorylated and the dephosphorylated form of
interconvertible enzymes, cPIP, for example, activates mitochondrial
pyruvate dehydrogenase and completely inhibits phosphorylase.

Protein phosphorylation is involved in the regulation of many
cellular processes. Evidence is increasing that phosphorylation-
dephosphorylation mechanisms are also involved in the control of
insulin secretion.[2,3,4] For instance, recently it has been shown in
rat pancreatic islets that the inhibitor phosphodiesterase 3-isobutyl-
1-methyl-xanthine (IBMX), which increases islet-cAMP, leads to increased phosphorylation of islet proteins together with potentiation

of insulin secretion. Conversely, 2-deoxyadenosine, an inhibitor of
adenylate cyclase, blocked IBMX-stimulated phosphorylation and decreased glucose-induced insulin secretion in presence of IBMX.[1]

The aim of this study was to investigate, whether cPIP, a second
messenger of insulin action, exerts any effect on stimulated insulin
secretion.

METHODS

Islets from 5 to 7 month old normal Chinese hamsters of both
sexes were isolated by collagenase digestion and 60 islets per
experiment were perifused in Krebs-Ringer-bicarbonate buffer
supplemented with 0.5% bovine albumine and 2.8 mM glucose (KRB).

Insulin-concentrations in the perifusate were determined by RIA
(Phadeseph-insulin-kit, Pharmacia, Sweden). cPIP was isolated and
purified as described elsewhere.[6]

RESULTS AND DISCUSSION

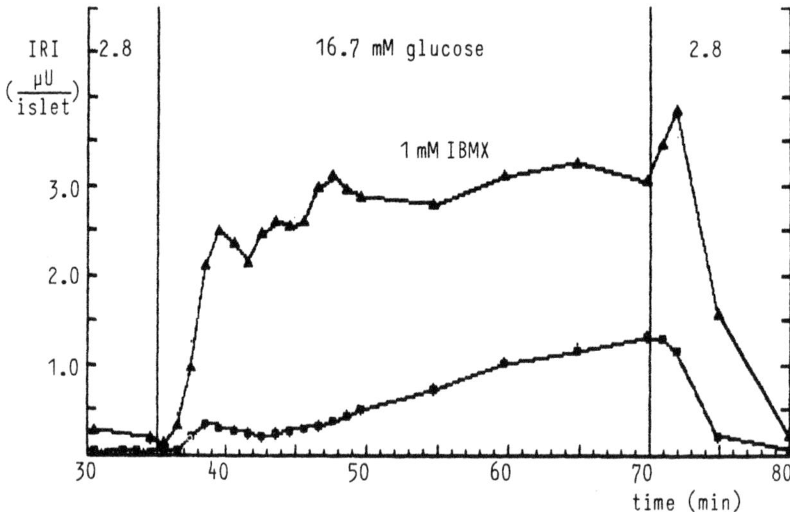

Fig. 1. Insulin secretion in isolated islets from Chinese hamsters
stimulated with 16.7 mM glucose (n=6, filled squares, and 16.7 mM
glucose plus 1 mM IBMX (n=4, filled triangles).

Stimulation by 16.7 mM glucose results in a biphasic insulin
secretion (Fig. 1). 1 mM IBMX added to 16.7 mM glucose results in

410

insulin secretion about five-fold higher than that in its absence.

When islets were stimulated for 25 minutes in the presence or
absence of IBMX, subsequent addition of cPIP strongly inhibited
insulin secretion within 3 min (Fig. 2).

The cPIP-concentration in the medium was approximately 10^{-7} M,
but it is as yet unknown whether the cPIP-concentration inside the
β-cells reaches the cPIP-concentration in the medium. However, it is
known that cPIP, when added to intact hepatocytes easily reaches
enzymes located within the plasma membrane. For instance it inhibits
adenylate cyclase[6] and ATPase.[7] Further aims to investigate are,
whether cPIP, when added to β-cells, not only exerts its action within
the cell membrane, but also reaches the cytosol. The effects of cPIP
on cAMP-levels and protein phosphorylation have yet to be shown in
pancreatic islets.

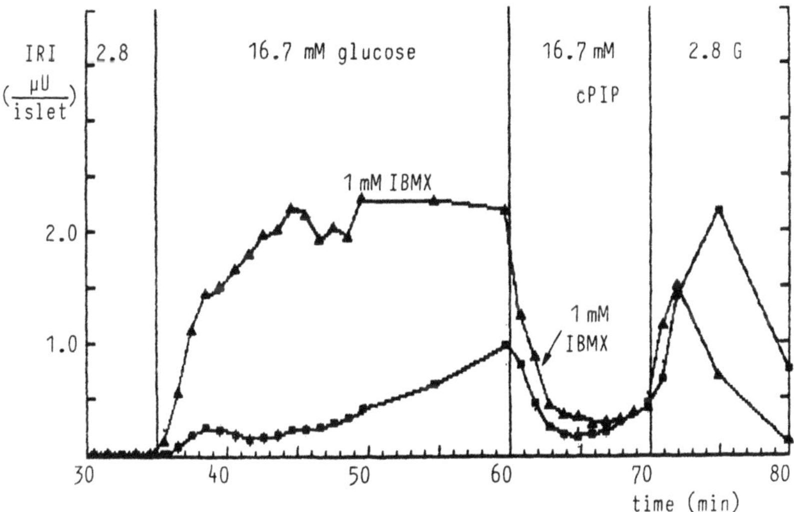

Fig. 2. Inhibition of insulin secretion by addition of cPIP to
isolated islets from Chinese hamsters. 16.7 mM glucose (n=5, filled
squares) and 16.7 mM glucose + 1 mM IBMX (n=3, filled triangles).
cPIP was added in a concentration of 20 to 25 units/ml which is
equivalent to 1-1.25 x 10^{-7} M.

This study shows clearly that cPIP is a very potent inhibitor of
insulin secretion, acting presumably by decreasing phosphorylation of
proteins essential for the release process. This report also confirms
that isolated hamster islets behave in many aspects as isolated rat
islets. The results indicate that cPIP may be a helpful tool in
pancreatic islet research.

REFERENCES

1. M. R. Christie and S. J. H. Ashcroft, Cyclic AMP-dependent protein phosphorylation and insulin secretion in intact islets of Langerhans, Biochem. J. 218:87 (1984).

2. D. E. Harrison and S. J. H. Ashcroft, Effects of Ca^{2+}, calmodulin and cyclic AMP on the phosphorylation of endogenous proteins by homogenates of rat islets of Langerhans, Biochim. Biophys. Acta 714:313 (1982).

3. U. K. Schubart, J. Erlichman, and N. Fleischer, The role of calmodulin in the regulation of protein phosphorylation and insulin release in hamster insulinoma cells, J. Biol. Chem. 255:4120 (1980).

4. S. Suzuki, H. Oka, H. Yasuda, M. Ikeda, p. Y. Cheng, and T. Oda, Effect of glucose on protein phosphorylation in rat pancreatic islets, Biochem. Biophys. Res. Commun. 99:987 (1981).

5. H. K. Wasner, Regulation of protein kinase and phosphoprotein phosphatase by cyclic AMP and cyclic AMP antagonist, FEBS Lett. 57:60 (1975).

6. H. K. Wasner, Biosynthesis of cyclic AMP antagonist in hepatocytes from rats after adrenalin- or insulin-stimulation, FEBS Lett. 133:260 (1981).

7. H. K. Wasner, Prostaglandyl-inositol cyclic phosphate, a second messenger for insulin, Abstr. 2nd Int. Symp. Ins. Recp., Rome, Italy (1983).

8. H. K. Wasner, Prostaglandyl-inositol cyclic phosphate an antagonist to cyclic AMP, in: "Prostaglandines, Leukotrienes and Lipoxines," M. Baily, ed., Plenum Publ. Corp., New York, pp. 251-257 (1985).

PULSATILE INSULIN RELEASE AND ELECTRICAL ACTIVITY

FROM SINGLE ob/ob MOUSE ISLETS OF LANGERHANS

L. M. Rosario, I. Atwater, and A. M. Scott*

Laboratory of Cell Biology and Genetics
NIDDK, National Institutes of Health
Bethesda, MD 20892

*Department of Biophysics
School of Biological Sciences
University of East Anglia
Norwich, NR4 7TJ England

INTRODUCTION

Since Ca-action potentials were first measured in β-cells, it has
been assumed that the resultant Ca^{2+} influx provides a critical Ca^{2+}-
dependent step in the events leading to insulin secretion.[4,6]. The
correlation between electrical activity and insulin release remained
indirect, however, since membrane potentials were recorded from single
β-cells within an isolated islet and insulin release was measured from
batches of collagenase isolated islets (usually more than 100) or from
the perfused pancreas (containing over 1000 islets).

The simultaneous recording of the burst pattern of electrical
activity from two different cells within an islet indicated a close
degree of electrical synchrony between cells of the same islet and
enabled us to view the membrane potential fluctuations of one cell as
representative of the majority of cells in that islet.[5,7] Synchrony
was not always perfect, but for the most part, the β-cells within an
islet were far more similar to each other than to the β-cells from a
neighboring islet exposed to identical conditions. This synchronous
behavior was observed, even in the very large islets studied (all
diameters greater than 500 μm). Further evidence for extensive domains of electrically coupled cells within an islet came from the

recent findings that both K^+ and Ca^{2+} concentrations in the intra-
islet, extracellular, space oscillate in phase with the membrane
potential recorded from one cell in the islet.[9,10] Therefore, since a
single islet appears to behave as a coordinated electrical unit,
"syncitium" it may also behave as a "syncitium" for release.

The first simultaneous measurements of insulin release and electrical activity from single islets with a time resolution close to the
time scale of the β-cell membrane potential oscillations, were reported by us in 1979.[2] These studies indicated that the release of
insulin from a single islet in response to glucose was pulsatile, in
agreement with another observation.[3] However, the high frequency of
the electrical oscillations prevented any conclusions to be drawn on
the possible phase characteristics of the electrical and release
events in the majority of islets studied. Thus, given diffusional
delays of about 20 sec for insulin to leave the islet extracellular
space,[4] it was impossible to correlate oscillations in release with
electrical oscillations showing a period of less than 20 sec. We had
previously determined that the average burst frequency recorded in
islets from the ob/ob mice (Norwich strain) was lower than in islets
from normal mice (1.8 bursts/min instead of 3 bursts/min at 11 mM
glucose).[11] It has also been shown that increasing extracellular Ca
lengthened the burst period by increasing the duration of the silent
phases.[1] Furthermore, increasing extracellular Ca could probably
enhance insulin release and cell-to-cell adhesion (and therefore
coupling). Therefore, in order to improve the time resolution of our
insulin measurements with respect to the electrical measurements, in
this study, we used islets from ob/ob mice and increased Ca concentrations in the perfusion medium.

The aim of this study was to determine if, and how, the pulsatile
insulin release from a single islet was related to the bursts of
electrical activity recorded in the presence of glucose.

METHODS

Homozygous obese (ob/ob) brown mice from a colony maintained in
the animal house of the School of Biological Sciences of the University of East Anglia, Norwich, U.K. were used in the present study.[11]
The electrophysiological methods used here for single and double
microelectrode potential recordings have been described in detail
previously.[5]

The feasibility of the technique for measuring single islet
insulin release simultaneously with the electrical activity of a cell
in the islet depends on several factors. The radioimmunoassay for
insulin was used at its limit of sensitivity by maximizing binding
during long incubations (24 to 72 hours) and by choosing an intermediate antibody concentration. This enabled us to measure the insulin
output from single large micro-dissected mouse islets. The average
volume of these islets is about 20 nl, or 20 times more than the
average rat islet, and they secrete correspondingly more insulin.
This was important in order to obtain sufficient insulin for assay in
the short collection times of 5 or 10 sec. The exchange time for the
perfusion chamber was reduced to a few seconds and the dead space
between stopcock and islet and between islet and collection tubes was
reduced to less than 350 µl. Samples were always assayed in duplicate
and the duplicate variation was less than 10% of the sample-to-sample
variation in 90% of the samples. Details of the technique are described elsewhere.[12]

The basic protocol used for the experiments illustrated in Figs.
3 and 5 and for three other experiments was as follows. The islet was
dissected out of a pancreas within 5 minutes of sacrifice of the
mouse. It was pinned into the perifusion chamber in the presence of
11 mM glucose (2.6 mM Ca) while the fine dissection was completed and
until an electrode could be successfully impaled in a cell (this
procedure was accomplished in less than 20 minutes for all experiments
described here). Glucose was then removed from the medium and 0.5%
albumin added. (Albumin was then present for the remainder of the
experiment.) After 25 minutes, glucose concentration was increased to
16.7 mM. After 20 minutes, $[Ca]_o$ was increased from 2.6 to 5.2 mM,
then to 7.8 and finally to 10.4 mM for 20 minutes at each concentration; or otherwise $[Ca]_o$ was increased directly to 7.8 mM for 30 min.
At the end of these experiments, glucose was removed and $[Ca]_o$ was
reduced to 2.6 mM for 30 minutes to reestablish basal conditions. In
two other experiments (one is illustrated in Fig. 4) an alternative
short protocol was used. The islet was placed in the perifusion
chamber in the presence of 16.7 mM glucose while a cell was impaled.
Then albumin was added and $[Ca]_o$ was increased to either 7.8 or 10.4
mM for 30 min. In both types of protocol, the perfusate was collected
continuously after 15 minutes exposure of the islet to albumin until
the end of the experiment. Samples were taken at one minute intervals
and during the steady state period of the various conditions at 5 or
10 sec intervals. The samples were kept on ice within 20 minutes of

collection and frozen until the time of assay. An event marker recorded the onset of each collection period on one channel of a 4-
channel tape recorder (Racal, Store-4) such that the insulin content
could be accurately matched up to the electrical recording simultaneously made on another channel.

RESULTS

<u>Electrical Synchrony in ob/ob Islets</u>

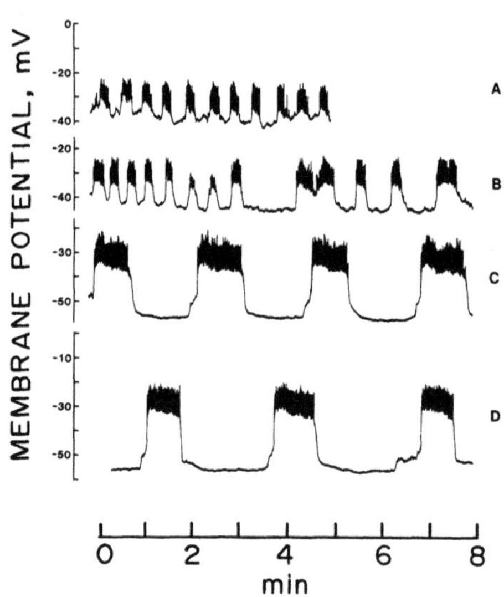

Fig. 1. Effects of extracellular Ca on the burst pattern of glucose-
induced electrical activity in ob/ob β-cells. $CaCl_2$ was added immediately before the experiment from a 1M stock solution to increase the
Ca concentration from 2.6 to 10.2 mM. Ca^{2+} activity was measured at
37 °C using a Ca^{2+}-sensitive electrode after the experiment. A: Calculated Ca concentration, 2.6 mM; Measured Ca^{2+} activity, 1.7 mM. B:
Calculated, 5.2 mM; Measured, 2.5 mM. C: Calculated, 7.8 mM; Measured, 3.07 mM. D: Calculated, 10.2 mM; Measured, 3.47 mM. Glucose
concentration was 11 mM throughout. All records made from the same
cell towards the end of a 12 min exposure to each $[Ca]_o$. Experiment
is representative of 3 similarly performed experiments.

416

Electrical synchrony between β-cells has been well established in normal mouse islets.[5,7] Synchrony of the burst pattern to within 2 sec was observed between cells separated by at least 70 μm in 24 out of 26 islets.[7] Dual electrode studies were performed in ob/ob islets and showed that the ob/ob β-cells were also well synchronized. In 8 cell pairs studied in 5 ob/ob islets, the burst pattern was synchronous to within 2 sec, even with electrode tip separations of over 150 μm (approx. 15 cell diameters) in all pairs studied (results not shown).

Effects of Increasing Extracellular Ca on the Electrical Activity in ob/ob Islet Cells

The burst pattern recorded from β-cells in islets taken from ob/ob mice is similar to that recorded from normal islets. However, the average oscillation period in the presence of 11 mM glucose is longer, about 30 sec instead of 15 sec as in normal islets. This is illustrated in Fig. 1A. Also shown in Fig. 1 (parts B-D) are the effects of increasing extracellular Ca on the burst pattern in an ob/ob islet. As previously observed in normal islets,[1] the amplitude of the underlying oscillation is increased and the burst period is increased while the relative time spent in the active phase is decreased. The effects of increasing extracellular Ca are more marked in ob/ob islets than in normal islets. For example, in three islets studied, the burst amplitude increased from 5 mV to 19 mV when extracellular Ca was increased from 2.6 mM to 10.4 mM. In normal islets, the same increase in Ca increased burst amplitude from 12 mV to 20 mV.[8] Also, in ob/ob islets, 10.4 mM Ca induced a 3-fold increase in the absolute duration of the active phase (not typically seen in normal islets)[1] and a 5-fold increase in the absolute duration of the silent phase.

The electrical response of β-cells in islets from ob/ob mice (Norwich strain) has been characterized in response to glucose and shows some differences from the response of normal cells.[11] The average membrane potential is lower in the ob/ob cells at low glucose concentrations and during the silent phase of the bursts, the burst pattern often persists at high glucose concentrations and the threshold concentration of glucose for induction of electrical activity is lower.[11] One effect of increasing extracellular Ca is to increase the membrane potential during the silent phase in the ob/ob β-cells (Fig. 1). Another effect is to shift the activation threshold towards

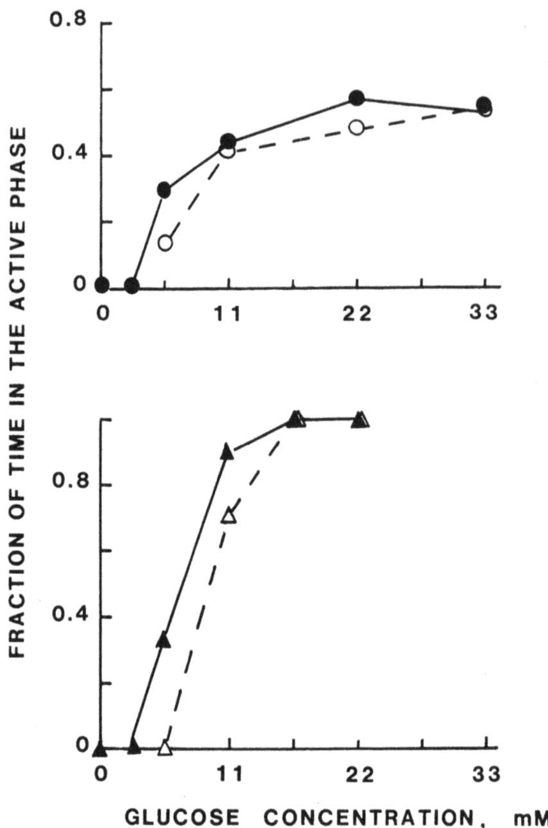

Fig. 2. Effects of increasing external Ca on the fraction of time in the active phase as a function of glucose concentration in ob/ob β-cells. The islets were exposed to increasing concentrations of glucose in the presence of 2.6 mM Ca (continuous lines) and 5.2 mM Ca (dashed lines). Upper and lower parts represent data from two different islets.

higher concentrations of glucose. The results of two experiments illustrating this effect are summarized in Fig. 2. Increasing $[Ca^2]_o$ from 2.6 to 5.2 mM induced a 2.5 mM shift of the mid value of the dose-response curve. Note, also, that the fraction of time in the active phase for the experiment illustrated in the upper part of Fig. 2 never exceeded 0.6. This response, which is almost never observed in normal islets, of some ob/ob islets to high glucose concentrations is not affected by increasing extracellular Ca. In some respects,

extracellular Ca affects electrical activity of the ob/ob β-cell
antagonistically to glucose.

<u>Simultaneous Measurements of Insulin Release and Electrical Activity
in Single ob/ob Islets</u>

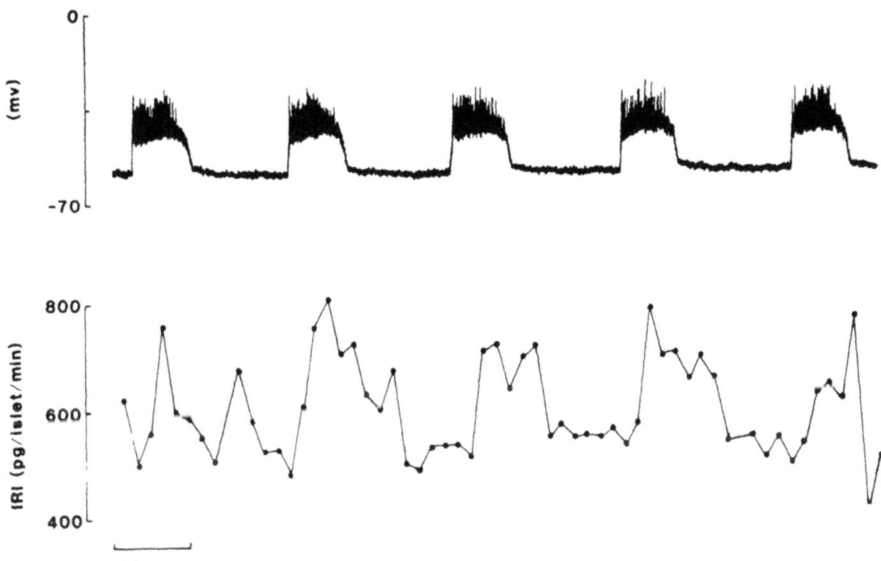

Fig. 3. Oscillations in membrane potential and insulin release from
an ob/ob mouse islet (in phase). Upper record shows a segment of the
membrane potential response to 16.7 mM glucose and 7.8 mM Ca. Lower
trace indicates the level of insulin release measured simultaneously
with membrane potential in 10 sec samples. Dots represent the median
of the duplicates and the lines were drawn to connect the dots. Basal
release from this islet was 334 pg/min (0 glucose) and stimulated
release in the presence of 2.6 mM Ca was 714 pg/min (16.7 mM glucose).
Experiment was performed according to the first protocol described in
Methods.

Three experiments showing the electrical burst pattern recorded
from one cell and the insulin release from the whole islet are illustrated in Figs. 3, 4 and 5. These experiments are representative of
seven experiments performed under similar conditions (as described in
Methods and in the legend accompanying each figure). Glucose concentration was 16.7 mM in all experiments and Ca concentration was 7.8 mM

during the segment illustrated in Fig. 3 and 10.4 mM in the segments
illustrated in Figs. 4 and 5. In five experiments (including those
illustrated in Figs. 3 and 4) the insulin release was pulsatile with a
mean periodicity of 0.67/min (±0.12, SEM). Mean electrical burst
periodicity was also 0.67/min (±0.09, SEM). In the experiment illustrated in Fig. 3, the bursts of insulin and electrical activity were
synchronous (in phase). In the experiment illustrated in Fig. 4, the
bursts of insulin release were of the same frequency as the bursts of
electrical activity, but did not show obvious synchrony (out of
phase). In two other experiments, the electrical burst frequency was
higher, it averaged 1.8/min. In these experiments, as illustrated in
Fig. 5, the insulin release did not show regular bursts.

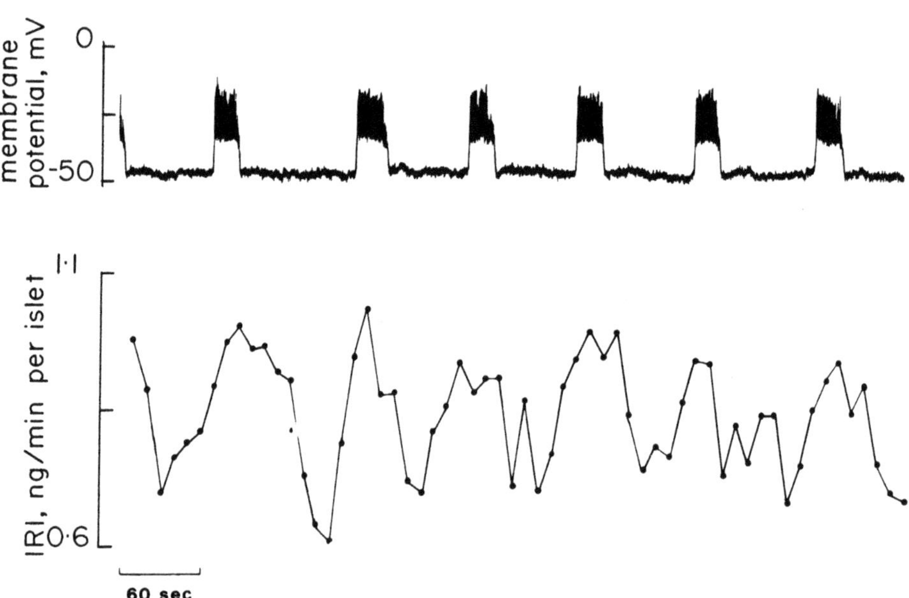

Fig. 4. Oscillations in membrane potential and insulin release of
similar frequency in an ob/ob mouse islet. Upper record, membrane
potential; Lower trace, insulin release measured simultaneously with
membrane potential in 10 sec samples. Concentrations of glucose and
Ca were 16.7 mM and 10.4 mM, respectively. Islet was not exposed to 0
glucose (second protocol described in Methods).

In five experiments, basal insulin release and insulin release induced by 16.7 mM glucose in the presence of several concentrations of Ca were also measured. As reported for normal islets[12] there was a large variation in the absolute levels of insulin release from islet to islet in this study. Basal release ranged from 305 pg/min/islet to 2.13 ng/min/islet (values taken over 5 minutes before exposure to glucose were averaged with the last 5 values taken at the end of the experiment). Average basal release was 1.07 ng/min/islet. The steady state levels of glucose-induced insulin release ranged from 714 pg/min/islet to 5.01 ng/min/islet (values were averaged over 5 minutes taken at least 10 minutes after exposure to glucose). The average release stimulated by 16.7 mM glucose and 2.6 mM Ca was 2.77 ng/min/islet. On average, 16.7 mM glucose induced a 2.6-fold stimulation of insulin release. Similar experiments on normal islets also gave a ratio of about 2.6 for the release at 16.7 mM glucose over the release at 0 glucose.[12]

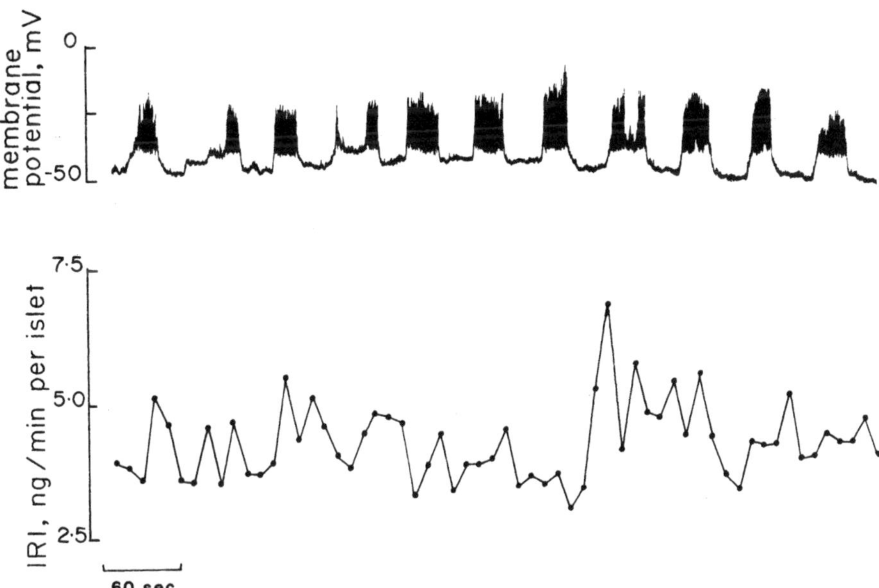

Fig. 5. Irregular oscillations in membrane potential and oscillations in insulin release from an ob/ob mouse islet. Glucose concentration was 16.7 mM and Ca was 10.4 mM. Upper record, membrane potential; Lower trace, insulin levels measured simultaneously with membrane potential in 10 sec samples. Basal release from this islet was 1.72 ng/min (0 glucose) and stimulated release in the presence of 2.6 mM Ca was 4.12 ng/min (16.7 mM glucose). Experiment was performed according to the first protocol described in Methods.

Increasing $[Ca]_0$ to 5.2 mM enhanced release in the presence of
16.7 mM glucose. Further increases in $[Ca]_0$ did not significantly
further enhance release and at $[Ca]_0$ = 10.4 mM, release was significantly inhibited. The maximum amplitude of the oscillations in insulin release in the presence of 16.7 mM glucose and 7.8 mM Ca ranged
from 358 pg/min to 7.58 ng/min (average amplitude = 1.66 ng/min/is-
let). Thus, the maximum amplitude of the oscillations of insulin
release in the presence of 7.8 mM Ca was larger than the difference
between the average basal and stimulated levels in the presence of 2.6
mM Ca. Minimum values (during the trough of the oscillations) ranged
from 414 pg/min/islet to 3.56 ng/min/islet. The minimum levels of
insulin measured during the oscillations in high Ca and glucose were
higher than the basal levels measured in the absence of glucose in all
but one experiment. However, in these experiments, we did not measure
basal levels in the presence of high Ca, so this comparison is not
strictly valid.

DISCUSSION

The main conclusion to be drawn from the experiments shown here
is that insulin release from single islets is usually pulsatile and
that this may be related to the burst pattern of electrical activity.
Thus, in 5 out of 7 of the experiments reported here the bursts of
insulin showed a similar frequency as the bursts of electrical activity, but were not always in phase. The reason for the discrepancy in
burst frequency (2 out of the 7 experiments) is unknown; it may
possibly be attributed to erratic factors inherent to the experimental
procedures used, such as mechanical perturbation of the islets during
microdissection, stretching, or microelectrode impalement, any of
which might dissociate the recording cell from the pattern of the
majority of cells. A number of reasons may explain the lack of
synchrony (or phase shift) observed in many of the experiments between
the electrical and the release events. Restricted diffusion of insulin from the islet extracellular space to the bath may act as a
"filter" that could conceivably delay the pulses of insulin with
respect to the bursts of electrical activity. The diffusion time
constant for insulin to cross 100 μm of islet extracellular space was
calculated to be about 20 sec.[4] This would imply that only islets
exhibiting low electrical burst frequencies could possibly show a

perfect phase relationship between the electrical and secretion
events. This picture may be further complicated by a possible intrinsic lag-time between the influx of Ca^{2+} during a burst and the eventual exocytosis of insulin. For these reasons, the experimental
conditions have been manipulated to induce well-separated, amplified
bursts of electrical activity while maximizing the rate of insulin
release. It is interesting to note that the islet exhibiting close
correlation in both frequency and phase (Fig. 3) was one of the
smaller islets (basal release was only 334 pg/min) while the islet
showing least correlation between electrical activity and insulin
release (Fig. 5) was one of the larger islets (basal release was 1.72
ng/min); both islets responded similarly to a glucose challenge,
showing a 2.1- and a 2.4-fold stimulation, respectively.

There is one other observation which warrants mention here.
Occasionally, insulin release was also observed to be pulsatile at
high glucose concentrations, when spike activity (and presumably Ca^{2+}
influx) is continuous. Further investigation is needed to clarify
this point. Meanwhile, there may be other possible explanations for
the pulsatile release, such as, dynamic variations in the islet extracellular space, where pockets of insulin might be collected and later
emptied towards the bath, or an independent intracellular oscillator
controlling the secretion mechanism.

In this study, we have also shown that the cells in the large
ob/ob islets are electrically synchronous to the same extent as in
normal mouse islets. We have observed that the membrane potential of
the ob/ob β-cells is more sensitive to changes in extracellular Ca.
Furthermore, the electrical response to glucose in the ob/ob islet is
shifted towards higher glucose concentrations by an increased extracellular Ca concentration, tending to normalize its low glucose
threshold. Finally, we have observed that 16.7 mM glucose stimulated
insulin release to about the same extent in these ob/ob islets as in
the large islets taken from normal mice.

ACKNOWLEDGEMENTS

The authors wish to thank Prof. E. Rojas and Dr. J. A. Bangham
for stimulating ideas and critiques and for continued support over the
course of most of this work. This research was supported in part by
the British Diabetic Association and the Wellcome Trust.

REFERENCES

1. I. Atwater, B. J. Frankel, E. Rojas, and G. M. Grodsky, β-cell
   membrane potential and insulin release; role of calcium and
   calcium:magnesium ratio, Quart. J. Exp. Physiol. 68:233
   (1983).
2. I. Atwater, E. Rojas, and A. Scott, Simultaneous measurements of
   insulin release and electrical activity from single microdissected mouse islets of Langerhans, J. Physiol. 291:57P
   (1979).
3. P. M. Beigelman, L. J. Thomas, B. Slavin, M. J. Shu, and S. P.
   Bessman, Insulin from individual isolated islets of
   Langerhans. I: response to glucose, Biochem. Med. 8:392
   (1973).
4. C. M. Dawson, I. Atwater, and E. Rojas, Potassium-induced insulin
   release and voltage noise measurements in single mouse
   islets of Langerhans, J. Membrane Biol. 64:33 (1982).
5. G. T. Eddlestone, A. Goncalves, J. A. Bangham, and E. Rojas,
   Electrical coupling between cells in islets of Langerhans
   from mouse, J. Membrane Biol. 77:1 (1984).
6. J. C. Henquin and H. P. Meissner, Significance of ionic fluxes
   and changes in membrane potential for stimulus-secretion
   coupling in pancreatic β-cells, Experientia 40:1043 (1984).
7. P. Meda, I. Atwater, A. Goncalves, A. Bangham, L. Orci, and E.
   Rojas, The topography of electrical synchrony among β-cells
   in the mouse islet of Langerhans, Quart. J. Exp. Physiol.
   69:719 (1984).
8. H. P. Meissner and M. Preissler, Ionic mechanisms of the glucose-induced membrane potential changes in β-cells, Horm.
   Metab. Res. (Suppl.) 10:100 (1980).
9. E. M. Perez-Armendariz and I. Atwater, Glucose-evoked changes in
   the extracellular concentration of potassium and calcium in
   mouse islets of Langerhans, this book.
10. E. Perez-Armendariz, I. Atwater, and E. Rojas, Glucose-induced
    oscillatory changes in extracellular ionized potassium
    concentration in mouse islets of Langerhans, Biophys. J.
    48:741 (1985).
11. L. M. Rosario, I. Atwater, and E. Rojas, Membrane potential
    measurements in islets of Langerhans from ob/ob obese mice
    suggest an alteration in $[Ca^{2+}]_i$-activated K^+ permeability,
    Quart. J. Exp. Physiol. 70:137 (1985).

12. A. M. Scott, I. Atwater, and E. Rojas, A method for the simulta-
        neous measurement of insulin release and β-cell membrane
        potential in single mouse islets of Langerhans, _Diabetologia_
        21:470 (1981).

COMPARISION OF STIMULUS-SECRETION COUPLING IN NORMAL AND OB/OB
(NORWICH COLONY) MOUSE ISLETS OF LANGERHANS

A.M. Scott and C.M. Dawson

Dept. of Biophysics
School of Biological Sciences
University of East Anglia
Norwich, NR4 7TJ, England

The Norwich colony of ob/ob mice has been compared with albino mice.
The electrical response of the islet membrane to high glucose stimulation
is different in the 2 types of mice, 22.2 mM glucose generating continuous
spike activity in normal but producing bursts of activity in ob/ob mouse
islets. The magnitude and dynamics of the insulin and lactate response to
high glucose are similar in the 2 types of mice, although the absolute
amounts released are higher in the ob/ob mice. Potassium permeability in
islets from normal mice is inhibited by quinine, which potentiates
glucose-induced insulin release. Quinine does not potentiate
glucose-induced insulin release in ob/ob islets but in the absence of
glucose, 100 µm quinine induces electrical activity in such islets.

Cooling from 37°C to 27°C, during steady state glucose stimulation,
reduces both lactate output and insulin release, the temperature
coefficients being similar in both types of mice. The effect of
temperature reduction on electrical activity is more marked in ob/ob islets
than in normal islets. Cooling-induced inhibition of potassium
permeability is greater in ob/ob islets than in normal islets.

Therefore, this new colony of ob/ob mice (Norwich colony) is
comparable to normal mice in terms of the magnitude and dynamics of insulin
release and lactate output in response to glucose stimulation and
steady-state cooling. The response of the β-cell membrane electrical
activity and potassium permeability to glucose stimulation and steady-state
cooling is different in the two types of mice.

INSULIN RELEASE, Ca^{2+} FLUXES AND CALMODULIN CONTENT

OF PANCREATIC ISLETS IN AGING RATS

J.J. Osuna, R. Rubio, E. Rodriguez, and C. Osorio

Departamento de Fisiologia y Bioquimica
Facultad de Medicina
Universidad de Granada
10812 Granada, Spain

Although previous reports have indicated the β-cells from collagenase-isolated islets of aging rats give a smaller secretory response to glucose or leucine stimulation than islets from young rats,[1,2] the underlying mechanism for this phenomenon is still not clear. In the present study we have investigated the effects of age on the calmodulin content, Ca^{2+} movements and insulin release.

Islets were isolated by collagenase digestion from 2 and 24 months old Wistar rats. For measurements of ^{45}CA-uptake, islets were incubated (5 min) in Krebs medium containing ^{45}Ca (20 $\mu C/cm^3$). Perifused islets were used to measure insulin release and ^{45}Ca-efflux. The content of calmodulin of the islets was measured by radioimmuno-assay.

Insulin release in response to a glucose challenge (from 2.7 to 16.7 mM) was significantly lower in islets from old rats compared with islets from young animals. Basal insulin levels (in the presence of 2.7 mM glucose) were 22.3 ± 2.8 $\mu U/cm^3$ per 100 islets in the islets from old rats and significantly higher, 38.0 ± 4.1 $\mu U/cm^3$ per 100 islets, in islets from young rats. Glucose-induced insulin release was reduced by 56% in the islets from old rats and, the second phase of insulin release was complete abolished.

In both, the islets from old and young rats, 16.7 mM glucose produced similar ^{45}Ca-efflux response consisting of an initial fall of the fractional rate of ^{45}Ca-outflow followed a slow rise in the ^{45}Ca-outflow.

Calmodulin content and ^{45}Ca-uptake were significantly lower in islets from old rats compared to islets from young animals.

Since, glucose-stimulated islet ^{45}Ca-efflux was similar in both groups of animals, it is concluded that, alterations in ^{45}Ca-uptake by old islets and old islet calmodulin content might explain the decrease insulin response observed in the islets of Langerhans from old rats.

REFERENCES

1. E. P. Reaven, G. Gold, and G. M. Reaven, Effect of age on glucose-stimulated insulin release by the beta-cell of the rat, J. Clin. Invest. 64:591 (1979).
2. E. P. Reaven, G. Gold and G. M. Reaven, Effect of age on leucine-induced insulin secretion by the beta-cell, J. Gerontol. 35:324 (1980).

PROTEIN CARBOXYL METHYLATION IN RAT PANCREATIC ISLETS: POSSIBLE ROLE IN
β-CELL FUNCTION

J.E. Campillo, P. Mena, S. Alejo and C. Barriga

Departamento de Bioquimica
Facultad de Ciencias
Universidad de Extremadura
06071 Badajoz, Spain

Reversible covalent modifications of proteins play an important role
in cellular function. The phosphorylation and dephosphorylation of
proteins are perhaps the best characterized examples. Recently, another
class of covalent modification, the methylation of protein carboxyl groups,
has been linked to the control of behaviour and to signal transduction in
both prokaryotes and eukaryotes[27,23].

The enzyme protein carboxyl methyltransferase (PCM; EC 2.1.1.24)
transfers methyl groups from S-adenosyl-methionine (SAM) to free carboxyl
groups of protein substrates, the methyl acceptor proteins or MAP. The
second enzyme involved in this system is the protein carboxyl
methylesterase which hydrolyzes protein methyl esters to yield methanol and
the unmethylated protein[19].

Carboxyl methylation of proteins results in the neutralization of
negative charges on the protein substrates which may produce conformational
changes in proteins. These post-translational changes may involve the
activation or inactivation of enzymes, protein hormones or other
biologically active proteins[11]. An alternative function could be the
neutralization of negative charges on intracellular membranes enabling
fusion of different membranes such as occurs in the secretory process[11].

This covalent post-translational modification of proteins has been involved in bacterial and leucocyte chemotaxis[3,27], sperm motility[16]; neural function[13]; repair or degradation of damaged proteins[6] and stimulus-secretion coupling in both exocrine and endocrine secretory cells[12,16,25]. In most cases, these cell functions are associated with ion translocation and at least three proteins which are related to such cell function are good substrates for PCM. These are calmodulin[1,18] the anion transport protein in erythrocytes[6] and the torpedo acetylcholine receptor[14].

The present study was undertaken to investigate as to whether carboxyl methylation of proteins could be implicated in β-cell secretory function. Here, we confirm that rat islets of Langerhans contain PCM activity and endogenous substrates for such activity; we also report the subcellular distribution of PCM activity and MAP capacity in islets as well as the methyl acceptor ability of insulin, and present data consistent with a role for protein carboxyl methylation in β-cell function. Preliminary reports of these findings have been published elsewhere[4,5].

MATERIALS AND METHODS

Reagents

Collagenase (type 1), swine skin gelatin (type 1) were from Sigma. S-adenosyl-L-(methyl-^{3}H) methionine (55-85 Ci/mmol) and L-(methyl-^{3}H)-methionine 80 Ci/mmol) were from New England Nuclear. Insulin RIA kit was from the Radiochemical Center, Amersham. All other chemicals were of the purest grade available.

Isolation and Subcellular Fractionation of Islets

Islets of Langerhans were isolated from pancreases of fed male Wistar rats by a collagenase method[7]. The subcellular fractionation was performed by differential centrifugation at 4°C with resuspension and washing as described in[20]. Succinate cytochrome C reductase activity measured as described in[29] and immunoreactive insulin were used as markers for mitochondrias and secretory granules respectively.

<u>Assay of PCM Activity and MAP Capacity</u>

PCM activity in islet homogenates was assayed using S-adenosyl
L-(methyl-^{3}H) methionine as the methyl donor and pig skin gelatin
(saturating concentration) as MAP, as described in[4,17]. In experiments to
study carboxyl methylation or endogenous islet proteins (MAP capacity), the
gelatin was omitted. Following incubation at 37°C for the times indicated
in the text or figures, the reaction was terminated by the addition of 1 ml
trichloroacetic acid (TCA) 10% (w/v). After centrifugation, the precipitated protein was taken up in 0.4 ml of 1 mM borate buffer pH 11.0 containing
methanol 0.7% (v/v). After 15 min at room temperature to permit hydrolysis
of the protein methyl esters, 0.1 ml aliquots were transferred to 1 ml
plastic tubes and then placed in vials containing 1 ml methanol. The
radioactive methanol in the tubes was selectively recovered by allowing the
vials to equilibrate at 37°C overnight. The radioactive methanol thus
recovered was counted by liquid scintillation spectrometry[4].

<u>PCM Purification</u>

Protein carboxyl methyltransferase was partially purified as described
in[9,21] from fresh bovine adrenal glands obtained from a local
slaughterhouse. The PCM preparation was devoid of MAP capacity when
assayed as described above.

<u>Effect of Incubation of Islets on Protein Carboxyl Methylation</u>

Batches of 25 islets were preincubated at 37°C for 30 min in a
Krebs-Henseleit bicarbonate medium containing 2 mg/ml Dextran (MW 70.000),
3.3 mmol/l glucose, 10 mmol/l L-(methyl-^{3}H) methionine and continuously
equilibrated with 95%O_2/5%CO_2. After the preincubation period, the glucose
concentration was increased in the test tubes by adding 10 µl of medium
containing glucose to allow a final concentration of 20 mmol/l. 10 µl of
medium containing 3.3 mmol/l glucose was added to control tubes. Then the
islets were incubated for 60 min. At the incubation times indicated in
results, the medium was removed and the islets were washed twice in cold
sodium acetate buffer, gently centrifuged and sonicated in the same medium.
Proteins were precipitated by the addition of 1 ml TCA. The protein
carboxyl methyl esters were hydrolysed and quantified as described above.

Electrophoresis of Carboxyl Methylated Proteins

Electrophoresis of carboxyl methylated proteins from subcellular
fractions was performed on 10% polyacrylamide gel (6 x 70 mm) in an acetic
acid urea system[8] according to the procedure described in[17] with minor
modifications. Fractions were incubated at 37°C for 30 min in the presence
of (methyl-^{3}H)-SAM and purified PCM. The reaction was stopped by the
addition of 0.8 M acetic acid, 1.5% N-acetylpyridinium chloride, 3%
2-mercaptoethanol and 5 M urea. The sample was then heated for 5 min at
95°C and applied on polyacrylamide gel. After electrophoresis the gels
were stained, destained and sliced in 2 mm fractions. The protein carboxyl
methyl esters in the two mm gel slices were hydrolysed and quantified as
described above.

Protein Determination

Protein concentrations were determined by the coomassie blue method[26]
using bovine albumin as a standard.

Analysis of Results

The statistical significance of data obtained was evaluated by
Student's t-test for unpaired data.

RESULTS AND DISCUSSION

Protein carboxyl methyltransferase (PCM) is found in most mammalian
tissues but most abundantly in nervous tissue and endocrine glands[12,13].
Methyl acceptor proteins (MAPs) are also widely distributed, the pituitary
gland being the richest source[17,22]. Rat pancreatic islets also contain
PCM activity and endogenous substrates for such activity (MAP activity)[4].
As shown in Figure 1, when islet homogenates were incubated with
methyl-^{3}H-SAM and a saturating concentration of gelatin as exogenous MAP,
label was incorporated into TCA-precipitable material. The methyl-^{3}H group
which was enzymatically transferred to carboxyl groups of proteins was
hydrolysed by alkaline treatment and recovered as methanol-^{3}H as described
under Methods. The presence of MAPs in pancreatic islets was tested by
incubating the islet homogenates in the absence of exogenous substrate.

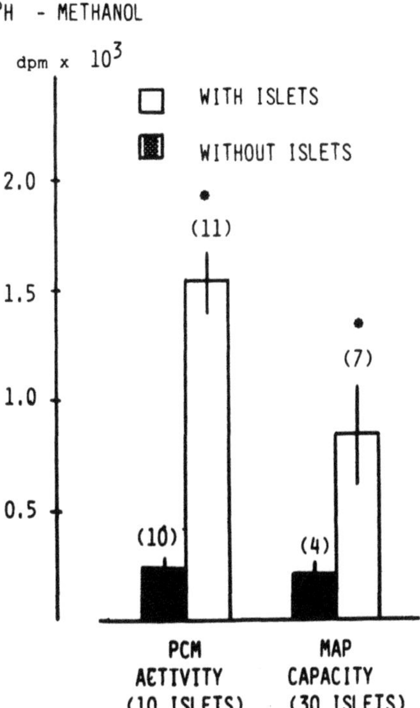

Fig. 1. Protein carboxyl methylation in pancreatic islets. Aliquots of
islet homogenates were incubated with (methyl-^{3}H)-SAM for 30 min.
The extent of protein carboxyl methylation was determined using
endogenous substrate only (MAP capacity) or exogenous gelatin (PCM
activity). Data are given as mean ± SEM for the number of
observations in brackets. (*) corresponds to p <0.01.

Under this condition a significant carboxyl methylation of islet proteins
could be demonstrated although rates were low compared to those seen in the
presence of a saturating concentration of exogenous substrate.

In vivo and in vitro studies in both exocrine and endocrine glands
have suggested that protein carboxyl methylation could play an important
role in exocytotic secretion[12,25,28]. A rapid and reversible increase in
both PCM activity and MAP capacity following stimulation of parotid,
exocrine pancreas, and adrenal medulla has been demonstrated. Specific
inhibitors of secretory activity of these glands inhibited correspondingly
protein carboxyl methylation and secretion[12]. In a previous study[4] we have
reported that no significant differences in protein carboxyl methylation
were found after 90 min incubation of intact pancreatic islets with
L-methyl-^{3}H-methionine to label intracellular SAM under basal (3.3 mmol/l
glucose) and stimulatory (20 mmol/l glucose) conditions. Because of the

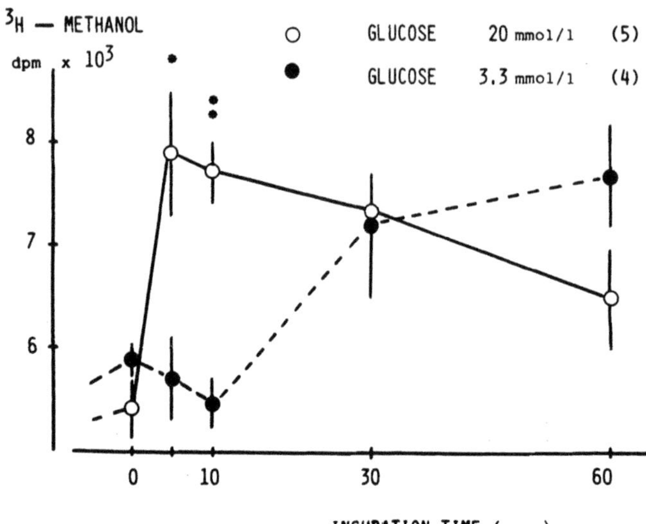

Fig. 2. Effect of high extracellular glucose concentration on protein
carboxyl methylation during the in vitro incubation of intact
pancreatic islets. Batches of 25 islets were preincubated for 30
min with L-(methyl-^{3}H)-methionine to label intracellular SAM under
basal (3.3. mmol/l glucose) conditions. Then the glucose
concentration was acutely raised in the test tubes without
removing the preincubation medium (see Methods). Data are given
as mean ± SEM for the number of observations in brackets.(*) and (**)
correspond to p <0.05 and p <0.01 respectively.

transient and peak-shaped nature of stimulated protein carboxyl methylation
observed in various biological systems[16,24] we have investigated the time
course of protein carboxyl methylation in intact islets incubated with
L-(methyl-^{3}H)-methionine and in the presence of both 3.3 and 20 mmol/l
glucose. As shown in Figure 2, the acute increase in glucose concentration
in the medium resulted in a rapid and transient increase in the protein
carboxyl methyl ester formation which seemed to be maximal at 5 to 10 min.
In the presence of 3.3 mmol/l glucose, islets showed a progressive increase
in protein carboxyl methylation with time. Values obtained after 60 min
incubation were not significantly different under basal and stimulatory
conditions.

The mechanism(s) by which protein carboxyl methylation could act in
the complex sequence of events that constitutes stimulus-secretion coupling
have not been elucidated. In studying the subcellular distribution of MAPs
in bovine adrenal medulla[10,15], it has been found that proteins of the

chromaffin granule membranes were the best substrates. Gagnon et al.[17]
reported that in rat pituitary gland the highest MAP specific capacity was
observed in the granular fraction. From these and other observations[12,16],
it has been hypothesized that carboxyl methylation of granule membrane
proteins would neutralize the negative charges on the membrane and reduce
the electrostatic barrier between the granules and the plasma membrane.
The hydrolysis of the carboxyl methyl esters might initiate or participate
in the retrieval of the granule membrane after exocytosis by increasing its
electronegativity[25].

In a previous study on subcellular distribution of protein carboxyl
methylase and its substrates in rat pancreatic islets[5], we have reported
that specific PCM activity dpm ^{3}H-methanol/µg protein, determined by
incubating subcellular fractions in the presence of methyl-^{3}H-SAM and
gelatin, was higher in microsomal-cytosolic (575) and nuclear (948) than in
granular (154) and mitochondrial (244) fractions. The MAP distribution,
however, was somewhat different. Specific MAP capacity (dpm ^{3}H-methanol/µg
protein) determined as for PCM activity but using purified PCM instead of

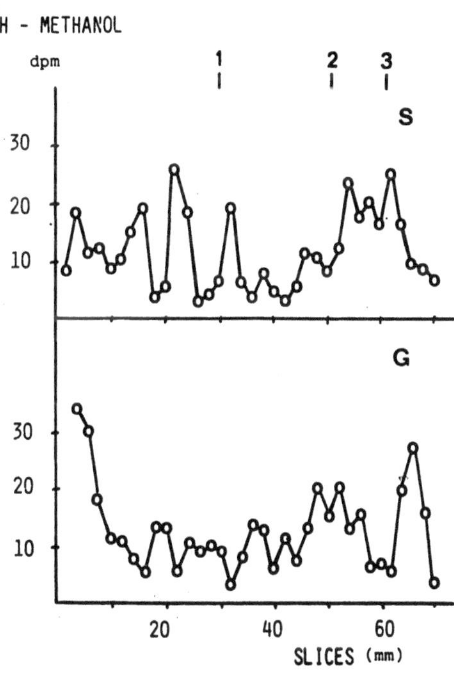

Fig. 3. Electrophoretic profiles of carboxyl methylated proteins in the
granular lysate (G) and microsomal-cytosolic (S) fractions from
pancreatic islets. The vertical lines indicate the position of
the molecular weight markers which were: 1, ovalbumin, 45.000, 2
chymotrypsinogen 25.000 and 3 cytochrome C 12.500.

gelatin, was higher in granular (216) and microsomal-cytosolic (1375) than
in nuclear (13.2) and mitochondrial (99.6) fractions. These results agree
with those reported for subcellular fractions of rat pituitary lobes[17] and
for bovine adrenal medulla[10].

Since the highest concentrations of MAP were found in the granular and
in the microsomal-cytosolic fractions, the MAP from these fractions were
analyzed with an electrophoretic system which prevent the spontaneous
hydrolysis of protein carboxyl methyl esters[17]. As shown in Figure 3,
electrophoretic profiles of MAP from cytosolic and granular fractions were
quite different and showed for each fraction several peaks of radioactivity
suggesting that MAPs in pancreatic islets represent, as for other endocrine
glands[10,17], a heterogenous group of proteins, probably implicated in
different functions. In the granular fraction a significant peak of
carboxyl methylated proteins had a molecular weight smaller than 12.500,
molecular weight of the marker cytochrome C, suggesting that it could
correspond to pancreatic hormones.

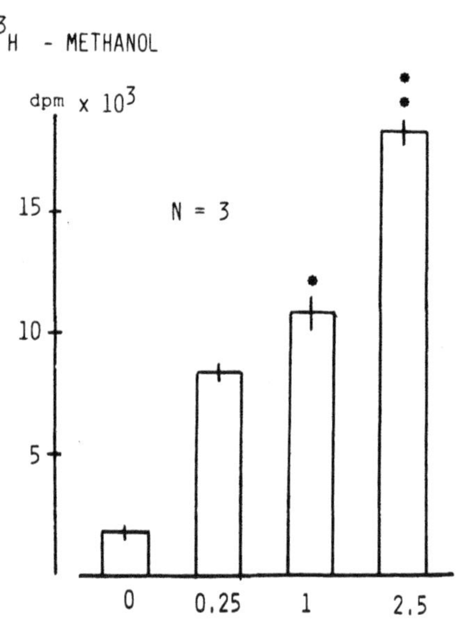

Fig. 4. Carboxylmethylation of porcine purified insulin (Actrapid MC,
NOVO) by incubating the hormone 30 min in the presence of
(methyl-^{3}H)-SAM and purified PCM. The mean value of methyl
acceptor capacity on a molar basis was 5.2 dpm per pmol of
insulin.

In addition to exocytosis, protein carboxyl methylation could play a
role in the storage of secretory proteins. This possibility is supported
by similar observations made on different tissues. In adrenal medulla,
intragranular proteins such as the chromogranins are among the best
substrates known[2,15].

Another intragranular protein neurophysins was shown
to be the major MAP of the posterior pituitary gland[22]. Purified anterior
pituitary peptide hormones are among the best PCM substrates known[18].

These data have prompted us to investigate the ability of insulin to
serve as a substrate for PCM, by incubating porcine monocomponent insulin
(Actrapid MC, Novo) for 30 min in the presence of (methyl-^{3}H)-SAM and
purified PCM. As shown in Figure 4, significant levels of carboxyl
methylation were observed showing a linear relationship with insulin
concentration. Under our experimental conditions the mean value of methyl
acceptor capacity for insulin on a molar basis, was 5.2 dpm per pmol of
insulin, corresponding to 28.8 fmol of methyl-^{3}H incorporated per nmol of
insulin. This value seems to be low if compared with the reported values
for other purified proteins (684 dpm per pmol for luteinizing hormone)[18].
Three important facts must be considered, however, in order to evaluate the
methyl acceptor capacity of insulin: 1 - its low molecular weight, 2 - the
presence in the pig insulin molecule of only four free carboxyl groups
corresponding exclusively to glutamic acid and 3- the sub-stoichiometric
nature of carboxyl methylation reactions in eukaryotes[6].

In conclusion, the presented data raise the possibility that protein
carboxyl methylation may play a role in β-cell secretory activity. The
clarification of this role must await more experimental data regarding
studies on their participation in the complex sequence of events that
constitutes stimulus-secretion coupling.

ACKNOWLEDGEMENTS

We acknowledge the collaboration of Mr. A. Ali in the English text and
the secretarial help of Mrs. A. Romero de Tejada. This work was supported
in part by a grant from the Plan de Formacion del Personal Investigador of
Spain.

REFERENCES

1. M.L. Billingsley, P.A. Velletri, R.H. Roth, and R.J. De Lorenzo,
    Carboxylmethylation of calmodulin inhibits calmodulin-dependent
    phosphorylation in rat brain membranes and cytosol, J. Biol. Chem.
    258:5352 (1983).

2. R.T. Borchardt, J. Olsen, L. Eiden, R.L. Schowen, and C.O. Rutledge,
    The isolation and characterization of the methyl acceptor protein
    from adrenal chromaffin granules, Biochem. Biophy. Res. Comm.
    83:970 (1978).

3. A. Boyd and M. Simon, Bacterial chemotaxis, Ann. Rev. Physiol. 44:501
    (1978).

4. J.E. Campillo and S.J.H. Ashcroft, Protein carboxyl methylation in rat
    islets of Langerhans, FEBS Lett. 138:71 (1982).

5. J.E. Campillo, S. Alejo, P. Mena and C. Barriga, Subcellular
    distribution of protein carboxymethylase and its substrates in rat
    pancreatic islets, Diabetologia 27:262A (1984).

6. S. Clarke and C.M. O'Connor, Do eukaryotic carboxyl methyl-transfer-
    ases regulate protein function? TIBS 8:391 (1983).

7. E. Coll-Garcia and J.R. Gill, Insulin release by isolated pancreatic
    islets of the mouse incubated in vitro, Diabetologia 5:61 (1969).

8. R.H. Davis, J.H. Copenhaver, and M.J. Carver, Characterization of
    acidic proteins in cell nuclei from rat brain by high resolution
    acrylamide gel electrophoresis, J. Neurochem 19:473 (1972).

9. E.J. Diliberto and J. Axelrod, Characterization and substrate
    specificity of a protein carboxymethylase in the pituitary gland,
    Proc. Natl. Acad. Sci. USA 71:1701 (1974).

10. E.J. Diliberto, O.H. Viveros, and J. Axelrod, Subcellular distribution
    of protein carboxymethylase and its endogenous substrates in the

    adrenal medulla: Possible role in excitation-secretion coupling,
    Proc. Natl. Acad. Sci. USA, 73:4050 (1976).

11. E.J. Diliberto and J. Axelrod, Regional and subcellular distribution
    of protein carboxymethylase in brain and other tissues, J.
    Neurochem. 26:1159 (1976).

12. E.J. Diliberto, R.F. O'Dea, and O.H. Viveros, The role of protein
    carboxymethylase in secretory and chemotactic eukaryotic cells, in:
    "Transmethylation", E. Usdine, R.T. Borchardt and C.R. Creveling,
    eds., pp 529-538, Elsevier North-Holland, New York.

13. L.E. Eiden, R.T. Borchardt, and C.O. Rutledge, Protein carboxymethyl-
    ation in neurosecretory processes, in: "Transmethylation", E.
    Usdine, R.T. Borchardt and C.R. Creveling, eds., pp 539-546,
    Elsevier North-Holland, New York.

14. D.D. Flynn, Y. Kloog, L.T. Potter, and J. Axelrod, Enzymatic
    methylation of the membrane-bound nicotinic acetylcholine receptor,
    J. Biol. Chem. 257:9513 (1982).

15. C. Gagnon, O.H. Viveros, E.J. Diliberto, and J. Axelrod, Enzymatic
    methylation of carboxyl groups of chromaffin granule membrane
    proteins, J. Biol. Chem. 253:3778 (1978).

16. C. Gagnon, W. Bardin, W. Strittmatter, and J. Axelrod, Protein
    carboxylmethylation in the parotid gland and in the male
    reproductive system, in: "Transmethylation", E. Usdine, R.T.
    Borchardt, and C.R. Creveling, eds., pp 521-528, Elsevier
    North-Hollard, New York.

17. C. Gagnon and J. Axelrod, Subcellular localization of protein
    carboxylmethylase and its substrates in rat pituitary lobes, J.
    Neurochem. 32:567 (1979).

18. C. Gagnon, S Kelly, V. Manganiello, M. Vaughan, C. Odya, W.
    Strittmatter, A. Hoffman, and F. Hirata, Modification of calmodulin
    function by enzymatic carboxyl methylation, Nature 291:515 (1981).

19. C. Gagnon, D. Harbour, and R. Camato, Purification and
    characterization of protein methylesterase from rat kidney, J.
    Biol. Chem. 259:10212 (1984).

20. S.L. Howell, D.A. Young, and P.E. Lacy, Isolation and properties of
    secretory granules from rat islets of Langerhans III, Studies of
    the stability of the isolated beta granules, J. Cell. Biol. 41:167
    (1969).

21. S. Kim and W.K. Paik, Purification and properties of protein
    methylase II, J. Biol. Chem. 245:1806 (1970).

22. Y. Kloog and J.M. Saavedra, Protein carboxylmethylation in intact rat
    posterior pituitary lobes in vitro, J. Biol. Chem 258:7129 (1983).

23. D.E. Koshland, Jr., Biochemistry of sensing and adaptation in a simple
    bacterial system, Ann. Rev. Biochem. 50:765 (1981).

24. R.F. O'Dea, O.H. Viveros, J. Axelrod, S. Aswanikumar, E. Schiffmann,
    and B.A. Corcoran, Rapid stimulation of protein carboxymethylation
    in leukocytes by a chemotactic peptide, Nature 272:462 (1978).

25. V. Povilaitis, C. Gagnon, and S. Heisler, Stimulus-secretion coupling
    in exocrine pancreas: role of protein carboxyl methylation, Am. J.
    Physiol. 240:G199 (1981).

26. T. Spector, A simple and linear spectrophotometric assay for protein, _Analytical Biochem._ 86:142 (1978).

27. M.S. Springer, M.F. Goy, and J. Adler, Protein methylation in behavioural control mechanisms and in signal transduction, _Nature_ 280:279 (1979).

28. W.J. Strittmatter, C. Gagnon, and J. Axelrod, β-adrenergic stimulation of protein carboxylmethylation and amylase secretion in rat parotid gland, _J. Pharmacol. Exp. Ther._ 207:419 (1978).

29. C. Vallejo, M.A. Gunther-Sillero, and R. Marco, Mitochondrial maturation during Artemia Salina embryogenesis. General description of the process, _Cell. Mol. Biol._ 25:113 (1979).

ROLE OF TRANSGLUTAMINASE IN PROINSULIN CONVERSION AND INSULIN RELEASE

R. Gomis[+], C. Alarcon[*], I. Valverde and W.J. Malaisse[§]

[+]Diabetes Unit, Hormonal Laboratory
Hospital Clinic
Barcelona University
Barcelona, Spain

[*]Fundacion Jimenez Diaz
Universidad Autonoma de Madrid
Madrid, Spain

[§]Laboratory of Experimental Medicine
Brussels Free University
Brussels, Belgium

Among the biophysical events associated with insulin release, the
translocation of secretory granules and their access to the exocytotic site
are thought to be controlled by a microtubular-microfilamentous system.
Only limited information is available on the enzymes involved in these
mechanical events. This report deals with the possible role of
transglutaminase in the secretory process.

Transglutaminase Activity in Islet Homogenates

Transglutaminase catalyzes the cross-linking between endo-γ-glutamyl
and endo-ε-lysyl residues of proteins[2,6]. Transglutaminase activity is
present in islet homogenates. The enzyme is activated by Ca^{2+} with a Ka
for Ca^{2+} close to 90 μM[6]. Hence, transglutaminase could represent a Ca^{2+}-
sensitive target in islet cells[11]. Methylamine, monodansylcadaverine,
N-p-tosylglycine, bacitracin, hypoglycemic sulfonylurea and glycine

methylester inhibit transglutaminase in islet homogenates[5,6,9]. All these
agents were examined for their effect upon islet function. The results of
these studies often failed to provide conclusive evidence for the
participation of transglutaminase in insulin secretion. This is due _inter
alia_ to the poor specificity of these inhibitors, which may for instance,
also exert lysomotropic effects[4], affect cellular pH[8], cause cell damage[6],
interfere with secreted insulin[7], or exert a primary and major action upon
cationic fluxes in the islet cells[5]. In certain cases, the penetration of
the drug into the islet cells could also be questioned[10]. The most
reliable information was obtained, in our opinion, with glycine
methylester.

Effect of Glycine and Sarcosine Methylesters Upon Insulin Release

Glycine methylester (1.0 to 10.0 mM) inhibited, in a dose-related
manner, transglutaminase activity in islet homogenates, decreased
$[^{14}C]$methylamine incorporation into endogenous proteins of intact islets,
and caused a rapid and reversible inhibition of insulin release evoked by
D-glucose, while failing to affect D-$[U-^{14}C]$glucose oxidation. Glycine
methylester also inhibited insulin release induced by other nutrient or
nonnutrient secretagogues[9]. Sarcosine methylester, which was used as a
control to assess the specificity of the primary amino group functionality,
failed to affect transglutaminase activity, $[^{14}C]$methylamine incorporation,
and insulin release. Both methylesters mobilized ^{45}Ca from prelabelled
intact islets, from isolated membranes of islet cells, liver or brain, and
from artificial lipid multilayers, this mobilization of Ca being apparently
unrelated to changes in transglutaminase activity. Taken as a whole, these
data support the view that transglutaminase participates in the machinery
controlling the access of secretory granules to the exocytotic sites[9].

Role of Transglutaminase in Proinsulin Conversion

The results so far discussed refer to the role of transglutaminase in
insulin release. The translocation of secretory granules to the exocytotic
site does not represent, however, the sole functional process dependent on
the integrity of motile events in the islet cells. Thus, the conversion of
proinsulin to insulin also depends on the oriented translocation of
microvesicles from the rough endoplasmic reticulum to the Golgi complex.
We have investigated, therefore, the effect of glycine methylester upon the

rate of conversion of proinsulin to insulin in rat pancreatic islets pulse-labelled with L-[4-^{3}H]phenylalaine[1]. For this purpose, the islets were first preincubated for 90 min, then exposed for 10 min to the labelled amino acid, and eventually incubated for 10 to 85 min in a nonradioactive medium containing unlabelled L-phenylalanine (1.0 mM). D-glucose (16.7 mM) and, as required, either glycine methylester (10.0 mM) or sarcosine methylester (also 10.0 mM) were present in the media throughout the experiment. The rate of insulin release was markedly decreased in the presence of glycine methylester both during the preincubation and incubation periods, whilst being unaffected by sarcosine methylester. Neither of these methylesters affected the incorporation of [^{3}H]phenyl-alanine in islet peptides or the ratio of labelled proinsulin and insulin relative to the total amount of labelled peptides. Glycine methylester, but not sarcosine methylester, caused two anomalies in the rate of conversion of proinsulin to insulin. First, glycine methylester retarded by approximately 20 min the onset of proinsulin conversion. Second, glycine methylester increased by about 50% the true half-life time of the conversion process. These findings support the view that the transgluta-minase-catalyzed cross-bridging of proteins plays a critical role in the control of mechanical events in the insulin-producing β-cell[1].

Regulation of Transgutaminase Activity in the β-cell

Transglutaminase activity in intact islet cells could be increased as a result of a rise in cytosolic calcium activity. In addition, glucose may also increase enzymic activity as a result of the induction of a more reduced state, with subsequent change in thiol-disulfide balance[3]. We indeed observed the transglutaminase activity in islet homogenate was increased after preincubation of the islets at high glucose concentration, and severely decreased after preincubation in the presence of either 1,2-bis-(2-chloroethyl)-1-nitrosurea, which inhibits glutathione reductase in the islets, or 2-cyclohexene-1-one, which lowers the GSH content of several cell types. The enzymic activity was decreased by NAD^+ or $NADP^+$ but not NADH or NADPH, and inhibited by GSSG more than $GSH^{[3]}$. These findings do not rule out the participation of regulatory factors other than Ca^{2+} and redox state in the overall control of transglutaminase activity in the islet cells.

REFERENCES

1. C. Alarcon, I. Valverde, and W.J. Malaisse, Transglutaminase and
    cellular motile events: retardation of proinsulin conversion by
    glycine methylester, <u>Bioscience Reports</u> 5:581 (1985).

2. P.J. Bungay, J.M. Potter, and M. Griffin, The inhibition of glucose-
    stimulated insulin secretion by primary amines. A role for
    transglutaminase in the secretory mechanism, <u>Biochem. J.</u> 219:819
    (1984).

3. R. Gomis, M.A. Arbos, A. Sener, and W.J. Malaisse, Glucose-induced
    activation of transglutaminase in pancreatic islets, <u>Diabetes Res.</u>
    3:115 (1986).

4. R. Gomis, M. Deleers, F. Malaisse-Lagae, A. Sener, P. Garcia-Morales,
    A. Rovira, I. Valverde, and W.J. Malaisse, Metabolic and secretory
    effects of methylamine in pancreatic islets, <u>Cell. Biochem. Funct.</u>
    2:161 (1984).

5. R. Gomis, P.C.F. Mathias, P. Lebrun, F. Malaisse-Lagae, A. Sener, and
    W.J. Malaisse, Inhibition of translgutaminase by hypoglycaemic
    sulphonylureas in pancreatic islets and its possible relevance to
    insulin release, <u>Res. Commun. Chem. Pathol. Pharmacol.</u> 46:331
    (1984).

6. R. Gomis, A. Sener, F. Malaisse-Lagae,and W.J. Malaisse, Transgluta-
    minase activity in pancreatic islets, <u>Biochim. Biophys. Acta</u>
    760:384 (1983).

7. R. Gomis, A. Sener, F. Malaisse-Lagae, and W.J. Malaisse, Corrigendum,
    <u>Biochim. Biophys. Acta</u> 757:141 (1983).

8. P. Lebrun, R. Gomis, M. Deleers, B. Billaudel, P.C.F. Mathias, A.
    Herchuelz, F. Malaisse-Lagae, A. Sener, and W.J. Malaisse,
    Methylamines and islet function: cationic aspects, <u>J. Endocrinol.
    Invest.</u> 7:345 (1984).

9. A. Sener, M.E. Dunlop, R. Gomis, P.C.F. Mathias, F. Malaisse-Lagae,
    and W.J. Malaisse, Role or transglutaminase in insulin release.
    Study with glycine and sarcosine methylesters, <u>Endocrinology</u>
    117:237 (1985).

10. A. Sener, R. Gomis, B. Billaudel, and W.J. Malaisse, Facilitation of
    insulin release by N-p-tosylglycine, <u>Biochem. Pharmacol.</u> 34:2495
    (1985).

11. A. Sener, R. Gomis, P. Lebrun, A. Herchuelz, F. Masaisse-Lagae, and
    W.J. Malaisse, Methylamine and islet function: possible
    relationship to Ca^{2+}-sensitive transglutaminase, <u>Mol. Cell.</u>
    <u>Endocrinol.</u> 36:175 (1984).

INDUCTION OF THE GLUCOKINASE-GLUCOSE SENSOR IN PANCREATIC ISLETS OF
INSULINOMA-BEARING RATS FOLLOWING TUMOR REMOVAL

F.J. Bedoya[1], M.C. Appel[2], R. Goberna[1],
and F.M. Matschinsky[3]

[1]Departamento de Bioquimica
Facultad de Medicina, 41009 Sevilla
Spain

[2]Department of Pathology
University of Massachusetts
Medical Center
Worcester, Massachusetts 01605

[3]Diabetes Center and the Department of Biochemistry
and Biophysics, School of Medicine
University of Pennsylvania
Philadelphia, Pennsylvania 19104

INTRODUCTION

Blood glucose concentrations in mammals are maintained within a narrow
range because the glucose homeostat in the organism controls glucose
storage and usage by a variety of cells.

The islet of Langerhans is the principal endocrine organ in charge of
the glucostat function. Three types of islet endocrine cells, α-cells,
β-cells and δ-cells function coordinately to control glucose storage during
postprandial periods and glucose utilization by the tissues. Although the
organism has several neuroendocrine devices to face decreases in glycemia,
insulin release by islet β-cells is the exclusive mechanism in charge of
correcting hyperglycemia. Furthermore, glucose homeostasis is maintained
under vastly different metabolic situations because of the adaptative
capacity of the β-cells[7,8,9]. Factors that govern β-cells adaptability are
not known in detail but it may be postulated that glucose plays a

preeminent role[14]. Since the enzyme glucokinase is the molecule that confers glucose responsiveness to the β-cell by controlling glycolysis[15] and since glucose controls important functions of the β-cell such as β-cell replication[10], insulin biosynthesis[2], and insulin release[15], we have studied glucose phosphorylating capacity during the adaptative changes that the islet β-cells undergo during starvation-refeeding and also during the transition from a glucose insensitive state to a glucose responsive state. Such a transition takes place in pancreatic islets when a transplantable insulin secreting tumor is removed from a severely hypoglycemic tumor-bearing rat[5].

MATERIALS AND METHODS

For the fast refeeding experiments male albino Wistar rats (250-300 g) were fed _ad libitum_ with standard laboratory chow and then fasted for 48 to 96 h. Some were refed isocalorically either selectively with glucose, or with standard pellet diet or with low-carbohydrate diet. The islets were isolated by the collagenase method[11]. Batches of 250 to 300 islets from three to four rats were suspended in 150 to 300 μl of homogenization medium: 50 mM Hepes pH 7.4, 150 mM KCl, 5 mM $MgCl_2$, 1 mM EDTA and 1 mM 2-mercaptoethanol. The islets were disrupted by sonication at intervals of 5 seconds, three times at position 10 (MSE, MK2 cell disrupter).

Glucose phosphorylation in islet homogenates was estimated by the production of D-(U-^{14}C)glucose-6-phosphate. In brief, islet homogenates were incubated for 20 min at 30°C in 100 μl of (in mM): 50 Hepes pH 7.4, 150 KCl, 5 $MgCl_2$, 5 $MgCl_2$, 10 NaF and 0.5 or 25 D-(U-^{14}C)glucose (specific activities 5.5 mCi/mmol and 0.9 mCi/mmol respectively). The reaction was terminated by the addition of 30 μl of 2 M glucose, 150 mM EDTA. Next, 60 μl aliquots were spotted on DEAE cellulose filters DE 81 (Whatman). Subsequently, filters were washed with 300 ml of deionized water, dried and their radioactivity was counted.

The study of pancreatic islets of insulinoma-bearing rats was performed in Dr. Matschinsky's laboratory in Philadelphia in collaboration with Dr. Appel's laboratory in Worcester. In brief, small fragments of the slow growing line of a radiation-induced insulinoma[5] were implanted subcutaneously into the interscapular region of four months old male inbred NEDH

rats (New England Diaconess Hospital, Boston, MA.). 7-10 weeks after the
tumor was implanted, the animals developed partial limb paralysis and
become moribund. At that point blood was drawn from the tail vein and the
animal was anesthetized with sodium pentobarbital (50 mg/kg) intraperitone-
ally and the pancreas was excised and plunged into isopentane cooled to the
temperature of liquid nitrogen. For another group of animals, the tumor
was surgically removed from etherized rats and the pancreas was excised 24
hours or six days after ablation of the tumor.

All pancreas specimens were shipped packed in dry ice and were stored
upon arrival at -80°C. In order to prepare tissue for quantitative
histochemical analysis, 20 μm thick sections of the pancreases were cut at
-20°C and freeze-dried for 48 h. The dry weight of samples were determined
on a quartz fiber fish pole balance[12]. Sample weights varied between 15
and 80 ng.

For the analysis of hormones in islet tissue, randomly sampled
freeze-dried pieces of islet tissue weighing 20-150 ng were extracted in 1
ml of cold 0.2 M glycine pH 8.8, 0.25% Human Serum Albumin for 30 min; the
extracted samples were stored at -20°C until assayed. Plasma glucose was
determined by a Beckman glucose analyzer. Insulin, glucagon and somatostatin were measured by radioimmunoassay. For the measurement of glucose
phosphorylation in microscopic samples of islet tissue a method based on
the liberation of ^3H_2O from glucose 6-phosphate was used[3b]. For the assay
of glucokinase in tissue samples, 10 mM glucose 6-phosphate was added to
inhibit hexokinase without affecting glucokinase. The sensitivity of the
assay was increased by using small volumes with an oil well procedure.

The actual glucokinase assay was carried out in 0.1 μl of 50 mM Hepes
pH 7.6, 500 mM KCl, 8 mM $MgCl_2$, 10 mM DTT, 0.1% BSA, 10 mM G6P, 0.5 mM or
20 mM D-glucose, and 0.4-0.5 μCi of D-[2^{3}H(N)]glucose, specific activity,
24 Ci/mmol. After 40 min of incubation, the reaction was stopped by adding
1.9 μl of 50 mM Hepes pH 7.6, 1.5 M D-glucose, 110 mM EDTA. The
isomerization of glucose 6-phosphate was accomplished by adding 0.5 μl of
phosphoglucose isomerase (350 U/ml). After 2 hours of incubation at room
temperature the reaction was stopped by adding 5 μl of 0.3 N HCl. ^3H_2O
formed was separated from the radioactive substrate, D-[2^{3}H(N)]glucose by a
diffusion step. For the assay of hexokinase, 0.8 mM D-glucose was used and
10 mM G6P was omitted from the assay mixture; the incubation time of the
tissue with phosphoglucose isomerase was 30 min.

RESULTS

Effect of Fasting and Refeeding on Blood Glucose and Insulin Concentration

Blood glucose fell to 63% after 48 hours of fasting and remained
almost constant for the next 48 hours (Fig. 1). Within 48 h of glucose
refeeding, the blood glucose reached 113% of the basal values in fed rats.
Refeeding with standard diet for 48 hours normalized blood glucose, while
48 hours of low-carbohydrate refeeding increased blood glucose to 86%. The
pattern of blood insulin levels during fasting and refeeding experiments is
shown in Fig. 2; blood insulin fell to 37% after 48 hours of fasting and to
29% after 96 hours of fasting. The different kinds of refeeding restored
blood insulin characteristically: glucose refeeding restored blood insulin
more rapidly reaching 93% of the values in fed rats after 96 hours of
refeeding. Standard refeeding produced a lesser restoring effect raising
blood insulin to 72% after 72 h. However, this effect is more important
than that produced by low-carbohydrate refeeding.

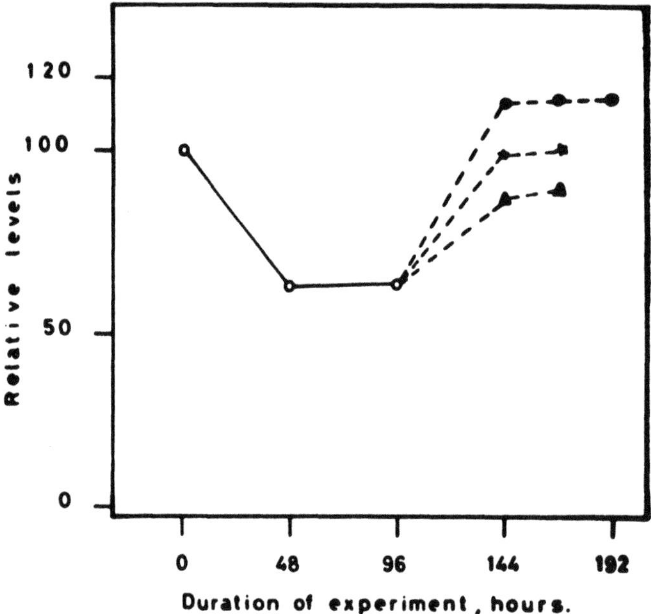

Fig. 1.   Relative levels of blood glucose in rats fasted 48 or 96 hours
(o-o) and refed with glucose (●---●), standard diet (*——*) and
low carbohydrate diet (▲---▲).(Figs. 1-4 taken from Bedoya, et al,
Metabolism, 33:1097 (1984).

450

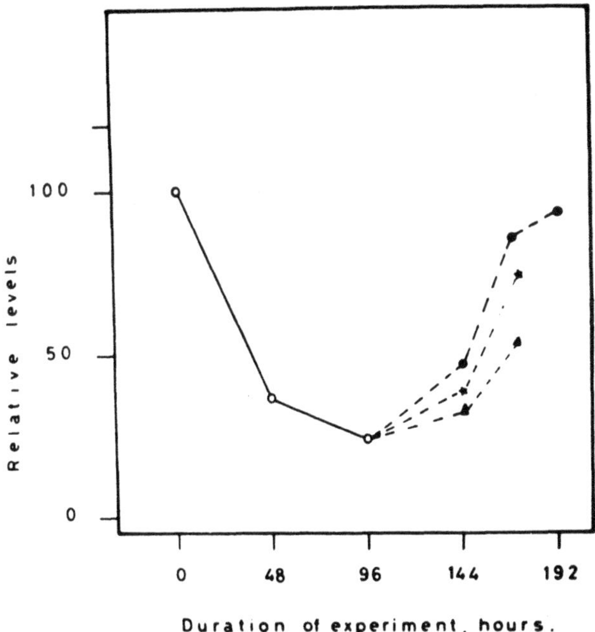

Fig. 2. Relative levels of plasma insulin in rats fasted 48 or 96 hours
(o-o) and refed with glucose (●---●), standard diet (*--- *), and
low carbohydrate diet (▲---▲).

Effect of Fasting and Refeeding on Hexokinase and Glucokinase Activities

Fasting induced an important decrease in islet glucose phosphorylating
activities (Fig. 3). Glucokinase showed a linear fall and its activity
completely disappeared after 96 h of fasting. Refeeding is able to restore
this activity; the most important restoring effect is exerted by glucose
refeeding: after 48 hours of glucose refeeding, the activity is restored by
70% and reaches normal values after 72 hours refeeding. Standard-diet
refeeding exerts a minor inducing effect; after 72 hours of refeeding the
activity is restored by 82%. Finally, low carbohydrate refeeding exerted a
poor inducing effect since the glucokinase activity is restored by 37% only
after 72 hours of refeeding. Hexokinase activity is affected by fasting
less dramatically, reaching 25% of the values of the control group after 96
hours of fasting (Fig. 4). Low carbohydrate refeeding exerts the most
powerful inducing effect on hexokinase activity; refeeding for 48 h with
this type of diet fully normalized hexokinase activity. By contrast,
glucose diet has a slower effect of hexokinase activity in islet cells:
after 4 days of refeeding hexokinase activity is recovered 70%. The dual
system that controls glucose phosphorylation in islet samples is affected
in a very different manner by these experimental manipulations: glucokinase
activity is reduced 90% in islets from 96 h fasted rats. The enzyme is

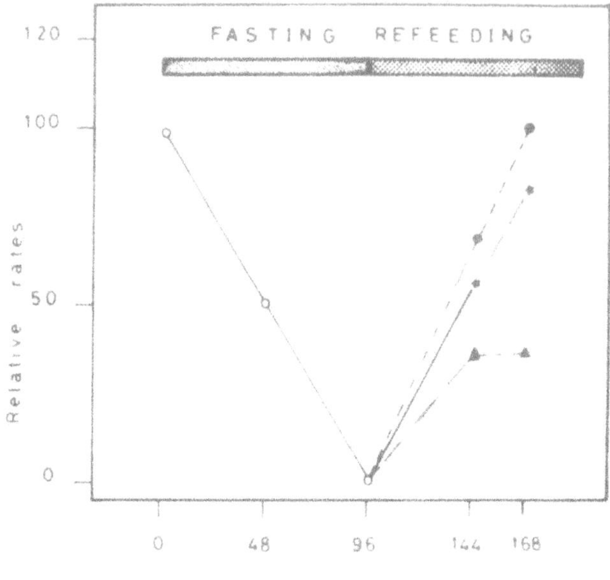

Fig. 3. Relative rates of glucokinase activity in islet homogenates from
rats fasted 48 or 96 hours (0-0) and refed with glucose (●---●)
standard diet (*---*), and low carbohydrate diet (▲---▲).

fully recovered 72 h after refeeding with glucose. Low carbohydrate
refeeding, by contrast, exerts a poor inducing effect. Hexokinase activity
in islets decreases by 75% after 96 h of starvation and low carbohydrate
diet exerts a preeminent role in the recovery of the enzymatic activity
during the period of refeeding.

Studies on β-cell Suppressibility and Recovery

In another experimental model, the implantation of a slow-growing
insulinoma causes in a matter of 2 to 3 months severe hyperinsulinemia
(>10 ng/ml) and extreme hypoglycemia (<20 mg/100 ml) in the host animal.
This causes suppression of endogenous islet β-cells that is documented by a
dramatic decrease of the extractable immunoreactive insulin in islet tissue
(Table 1). α- and δ-cells on the other hand seem to remain unaffected
since neither glucagon nor somatostatin content of the suppressed islet is
diminished.

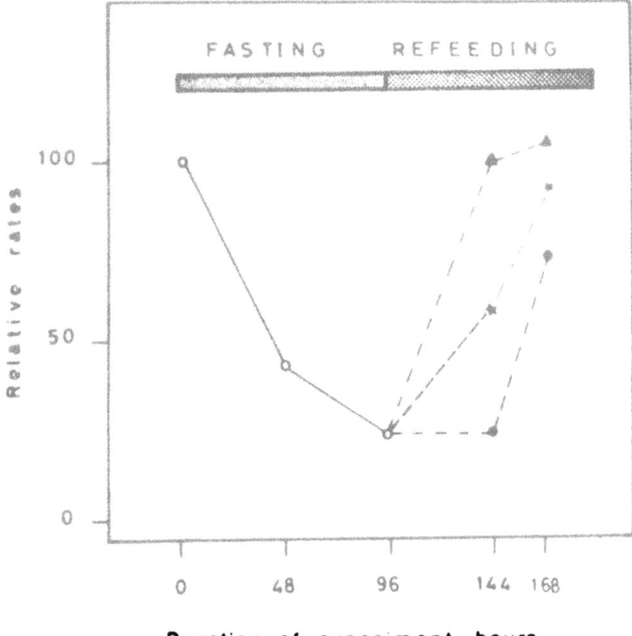

Fig. 4. Relative rates of hexokinase activity in islet homogenates from
         rats fasted 48 or 96 hours (o---o) and refed with glucose (●---●),
         standard diet (*——*), and low carbohydrate diet (▲---▲).

Surgical removal of the tumor causes a transient period of hyperglycemia which peaks 24 h after tumor ablation. The hyperglycemic period is
unable to stimulate the synthesis of insulin sufficiently to replenish the
insulin stores. Serum levels of immunoreactive insulin are only 20% of
normal controls at the 24 h time point following tumor removal. β-cell
function is recovered six days after tumor resection as indicated by normal
blood glucose, serum insulin and insulin content in pancreatic islets.

The data presented in Table 2 show the impact of an insulin secreting
tumor and its surgical removal on the dual system of glucose phosphorylation that exists in the islet cells. Apparent hexokinase activity in
freeze-dried islet fragments is affected neither by the presence nor by the
removal of the insulinoma. By contrast, glucokinase activity is
drastically reduced (78%) in islet samples from insulinoma-bearing animals;
the enzyme fully recovered 24 h after removal of the tumor and remained
high when normal β-cell function is regained 6 days after removal of the
tumor.

Table 1. Effect of implantation and surgical removal of an insulinoma on
          glucose phosphorylating activities on endogenous islet β-cells.

Experimental condition	Glucose Phosphorylation in Islet Tissue	
	Glucokinase	Hexokinase
	Vmax, mmol/kg dry weight/h. (37°C)	
Control	133.5 ± 12.7 (5)	299.9 ± 21.9 (5)
Insulinoma-bearing rat	38.1 ± 4.1 (7)[a]	375.3 ± 42.2 (5)[b]
1 day after resection of insulinoma	120.0 ± 12. (6)[b]	363.7 ± 39.6 (5)[b]
6 days after resection of insulinoma	160.3 ± 25.0 (3)[b]	not analyzed

Small pieces of a slow growing radiation-induced insulinoma were implanted
subcutaneously in NEDH rats. 7-12 weeks after the implantation the tumor
was resected and the pancreas excised 1 day or 6 days after tumor removal.
Pancreas specimens were processed as described in Methods section and
glucose phosphorylation in freeze-dried samples of islet tissue was
determined with radiometric microassay.
Values expressed as means ± SEM with number of animals in parentheses. [a]p
<0.001 compared with control animals; [b]not significantly different from
control animals.

DISCUSSION

In agreement with other authors the present report describes an
important decrease of glucose phosphorylating activities in islet cells
during fasting[4,13]. The fact that the decrease activity reported here is
stronger than previously described may indicate that metabolic factors
generated in islet cells during starvation might inhibit glucose phosphorylating enzymes in such a labile manner that cannot be detected with conventional enzymatic analyses. In this context, it has been reported that

palmitoyl CoA derivatives inhibit allosterically glucokinase purified from liver[16]. The fact that inhibition of fatty acid oxidation in islets from starved animals partially restored insulin secretory response to glucose and glucokinase activity in islets from starved rats speaks in favor of such possibility[3a]. However, there is a lack of direct experimental evidence to support the role of fatty acid derivatives on the regulation of islet cell glucokinase since measurements of free acyl CoA derivatives in islets are not available.

It has been shown that high carbohydrate diet refeeding acts as a signal to unblock secretory response to glucose[1]. We have described that this type of refeeding induces a preponderant restoration of glucokinase activity (Fig. 3). By contrast, low carbohydrate diet restores mainly hexokinase activity and fails to restore insulin secretory response towards glucose (Fig. 4).

On the other hand, there is a preferential decrease of glucokinase activity seen in suppressed islets from hypoglycemic insulinoma-bearing rats. Else, the hyperglycemic peak which is reached 24 hours after removal of the tumor coincides with a complete recovery of glucokinase activity in pancreatic islet cells and precedes all known functional and morphological events that characterize the process of recovery of the β-cells: enhanced insulin biosynthesis and β-cell replication are observed 3 days after tumor removal[1]. Restoration of insulin stores is relatively delayed and is complete only at six days. This coincides with complete recovery of β-cell number, morphology and secretory function.

The collective evidence presented here supports the view that the glucokinase-glucose sensor governs glucose stimulation of insulin release, biosynthesis and storage of the hormone and β-cell replication. It is remarkable that hexokinase is affected in a very different manner under the starvation refeeding manipulations and is not affected by the wide excursions of blood sugar and serum insulin characteristic of the insulinoma-bearing animal model. The role of hexokinase in β-cell metabolism is not understood. The present evidence is consistent with the view which assigns to hexokinase an alternative control of glycolytic flux necessary for the β-cell to respond to secretagogues such as fatty acids, hormones and so forth when the levels of glucose in blood are low.

ACKNOWLEDGEMENTS

Dr. F.J. Bedoya is a Fulbright/Ministerio de Educacion y Ciencia
postdoctoral trainee. This work was supported in part by National
Institutes of Health Grants AM 21122, AM 19525 and by a Grant from the
Comision Asesora de Investigacion Cientifica y Tecnica (0324/81).

REFERENCES

1. M.C. Appel and E.H. Kisklauskis, Effects of glucose infusion in rats
    with a functional insulinoma, Diabetes 33:82A (1984).
2. S.J.H. Ashcroft, J. Bunce, M. Lowry, S.E. Hansen and C.J. Hedeskov,
    The effects of sugars on (Pro) insulin biosynthesis, Biochem. J.
    174:517 (1978).
3a. F.J. Bedoya, R. Ramirez, and R. Goberna, Effects of 2-bromostearate on
    glucose-phosphorylating activities and the dynamics of insulin
    secretion in islets of Langerhans during fasting, Diabetes 33:858
    (1984).
3b. F.J. Bedoya, M.D. Meglasson, J.M. Wilson, and F.M. Matschinsky,
    Radiometric oil well assay for glucokinase in microscopic
    structures, Analyt. Biochem. 144:504 (1985).
4. P.T. Burch, M.D. Trus, D.K. Berner, A. Leontire, K.C. Zawalich, and
    F.M. Matschinsky, Adaptation of glycolytic enzymes, glucose use and
    insulin release during fasting and refeeding, Diabetes 30:923-928
    (1981).
5. W.L. Chick, S. Warren, R.N. Chute, A. Like, V. Lauris, and K.C.
    Kitchen, A transplantable insulinoma in the rat, Proc. Natl. Acad.
    Sci. USA 74:628 (1977).
6. P.R. Flatt, K. Tan, C. Bailey, S.K. Swanston-Flatt, V. Marks and J.D.
    Webster, Plasma glucose and insulin concentrations after
    implantation and surgical resection of a transplantable rat
    insulinoma, Biochem. Soc. Trans. 10:273 (1982).
7. M. Freinkel, On pregnancy and progeny, Diabetes 29:1023 (1980).
8. N.J. Grey, S. Goldring, and D.M. Kipnis, The effect of fasting, diet
    and Actinomycin D on insulin secretion in the rat, J. Clin. Invest.
    49:881 (1970).

9. L. Herberg and D.L. Coleman, Laboratory animals exhibiting obesity and diabetes syndromes, Metabolism 26:59 (1977).

10. D.L. King and W.L. Chick, Pancreatic beta cell replication: Effects of hexose sugars, Endocrinology 99:1003 (1976).

11. P.E. Lacy and M. Kostianowsky, Method for the isolation of intact islets of Langerhans from the rat pancreas, Diabetes 16:35 (1967).

12. D.H. Lowry and J.V. Passonneaw, in: "A Flexible Method of Enzymatic Analysis", pp 237-249, Academic Press, New York (1972).

13. W.J. Malaisse, A. Sener, and A. Levy, Fasting induced adaptation of key glycolytic enzymes in isolated islets, J. Biol. Chem. 251:5936 (1976).

14. F.M. Matschinsky and P.T. Burch, Normal and abnormal pancreatic β-cell function and metabolism, in: "Diabetes Mellitus and Obesity, B.N. Brodorf and S.J. Bleicher, eds., pp 108-116, Williams and Wilkins, Baltimore (1982).

15. M.D. Meglasson and F.M. Matschinsky, New perspectives on pancreatic islet glucokinase, Am. J. Physiol. 246:E1 (1984).

16. P.S. Tippet and K.E. Neet, Specific inhibition of glucokinase by long chain Acyl coenzymes A below the critical micelle concentration, J. Biol. Chem. 257:12839 (1982).

BIOCHEMICAL DESIGN FEATURES OF THE PANCREATIC ISLET CELL GLUCOSE-SENSORY
SYSTEM

F.M. Matschinsky, M. Meglasson, A. Ghosh, M. Appel,
F. Bedoya, M. Prentki, B. Corkey, T. Shimizu,
D. Berner, H. Najafi and C. Manning

Diabetes Research Center and
Department of Biochemistry and Biophysics
University of Pennsylvania
School of Medicine and
Department of Pathology
University of Massachusetts
School of Medicine
Worcester, Massachusetts

Our understanding of the biochemical basis of fuelstat function of the
pancreatic islet cells is still fragmentary in contrast to the comprehen-
siveness of our knowledge about the morphology and physiology of pancreatic
islets. The present account addresses several fundamental issues of β-cell
biochemistry, namely the manner in which these cells keep track of the
blood glucose level and how those blood glucose measurements might be
coupled to insulin secretion. Expressed in the jargon of the field the
presentation will deal with the pancreatic β-cell glucose-sensor-device and
with the problem of stimulus-secretion-coupling.

Research in many laboratories has now firmly established that the
glucose molecule has to be metabolized by the β-cells in order to initiate
insulin secretion[7]. The crucial arguments supporting this notion are:
only metabolizable sugars are effective, the order of potencies of hexoses
for stimulating insulin secretion is the same as the order in which they
are suitable as substrates of metabolism, inhibitors of glycolysis block
hexose and triose induced secretion, both metabolism and secretion are more
effectively activated by the α- than by the β-anomers of glucose and

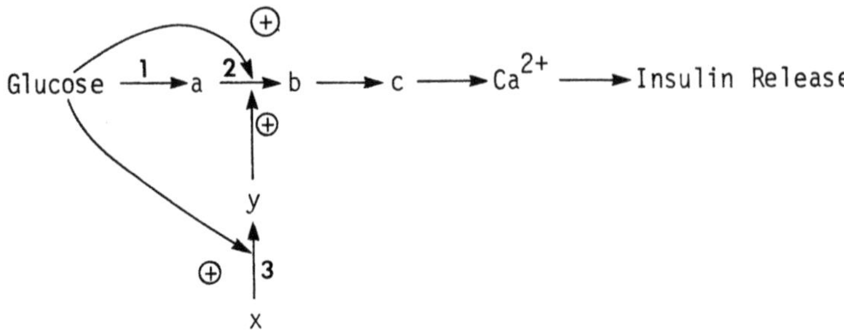

Fig. 1.   Potential glucose-sensor sites in glucose metabolism. Glucose
sensing may occur as glucose is transported into β-cell or at the
glucose phosphorylation step[1]. Glucose may act as an allosteric
activator at a rate limiting step, e.g. phosphofructokinase[2].
Glucose may act as an allosteric activator at an enzyme generating
an activator of glycolysis, e.g. glucose 1,6-diphosphate
synthetase, or an enzyme modifying a rate-limiting step, e.g. by
phosphorylation[3]. (This figure and its legend are taken from
reference 7).

mannose, and the concentration dependencies of glucose induced insulin
secretion and of glycolysis have similar (albeit not identical)
characteristics.

A priori there are many potential glucose sensor sites in glucose
metabolism (Fig. 1). Recent studies have identified the glucose sensitive
element in the glucose metabolizing machinery of the normal pancreatic
β-cells. (It is noteworthy that the α-cell glucose sensor molecule is not
known). It is likely that glucokinase, an enzyme that phosphorylates
glucose, is the glucose sensor molecule[7]. The activity of that enzyme
shows its steepest increase at glucose levels in the physiological range,
from 3-15 mM. The enzyme has a high control strength in the regulation of
glycolysis in pancreatic β-cells. At physiological levels the enzyme
prefers the α-anomer over the β-anomer of glucose. The enzyme is blocked
by certain inhibitors of glucose metabolism which also block glucose
induced insulin secretion and glucose metabolism (e.g. manno-heptulose).

Studies with experimental animals have revealed that small changes of
the enzyme activity (e.g. -30%) have dramatic effects on glucose induced
insulin release. This emphasizes the crucial role of the glucokinase
glucose sensor in β-cell physiology and raises the possibility that
decreases of the enzyme level or of its activity could explain certain
forms of type II diabetes.

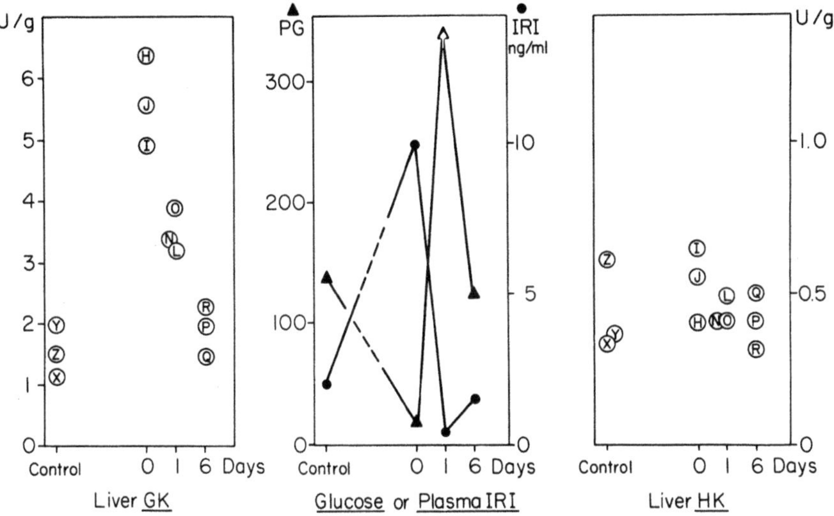

Fig. 2.   The influence of serum insulin and glucose on activities of
glucokinase and hexokinase of liver tissue. Glucose
phosphorylation was measured fluorimetricly in the liver of
insulinoma bearing NEDH rats before (day 0) and after tumor
removal (day 1 and 6) as well as in control NEDH rats. The liver
data were obtained from the same animals which were used for
studies of pancreatic islet glucose phosphorylation and hormone
contents (see Table 1). 1 unit indicates 1 μmol of product
generated per min. Plasma glucose is recorded in terms of mg%.
Results from individual animals are given for the enzyme
measurements.

Strong biological evidence for the pivotal role of glucokinase in
β-cell function was obtained in recent still unpublished studies of islet
from the pancreas of insulinoma bearing rats[3,2]. New England Deaconess
Hospital (NEDH) rats were inoculated with insulinoma tissue and developed
severe hyperinsulinemia and hyperglycemia after 3-4 months of carrying the
tumor. When the tumor was removed a transient diabetes like state ensued.
We measured glucokinase and hexokinase in pancreatic islets and liver of
insulinoma bearing rats when their blood sugar had fallen to about 1 mM,
then 24 hours after tumor removal when insulin had disappeared from the
removal when blood glucose and serum insulin had normalized (Table 1).

The islet glucokinase of the hypoglycemic animal had fallen to 20% of
control. Islet glucokinase normalized within 24 hrs during the hyperglycemic response to tumor resection. Islet hexokinase was not influenced
by these experimental blood sugar or serum insulin excursions. Liver
glucokinase of hypoglycemic animals had reached four times the activity of

Table 1. Effect of implantation and surgical removal of an insulinoma on glucose phosphorylating activities of islet β-cells of the host animal.

Experimental Condition	Glucose Phosphorylation in Islet Tissue	
	Glucokinase	Hexokinase
	Vmax, mmol Kg dry weight/h. (37°C)	
Control	133.5 ± (5)	299.9 ± 21.9(5)
Insulinoma bearing rat	38.1 ± 4.1 (7)[a]	375.3 ± 42.2(5)[b]
1 day after resection of insulinoma	120.0 ± 12 (6)[b]	363.7 ± 39.6(5)[b]
6 days after resection of insulinoma	160.3 ± 25.0(3)[b]	not analyzed

Values expressed as means ± SEM with number of animals in parentheses.
[a] $p \leq 0.001$ compared with control animals; [b] not significantly different from control animals.

Table 2. Results of glucokinase measurements in microdissected human pancreatic islets using a radiometric oil well assay.

Tissue Source	Activity, mmol/kg dry weight/h (37°C)
Human Specimen	
1	118.53 ± 12.96 (5)
2	63.18 ± 19.20 (4)
3	63.23 ± 18.57 (5)
4	47.32 ± 12.00 (4)
5	41.68 ± 7.03 (4)
6	47.09 ± 18.70 (3)

control livers in contrast to islet (Figure 2) and fell to normal levels in
the course of 6 days following tumor removal. These data imply organ
blood and the bloodsugar was about 20 mM and again 6 days after tumor
specific differential regulation of islet and liver glucokinase. The
regulation of the liver enzyme is apparently insulin dependent whereas the
induction of the islet enzyme seems to be governed primarily by glucose.
These are essential design features for a glucose sensor element of the
β-cells.

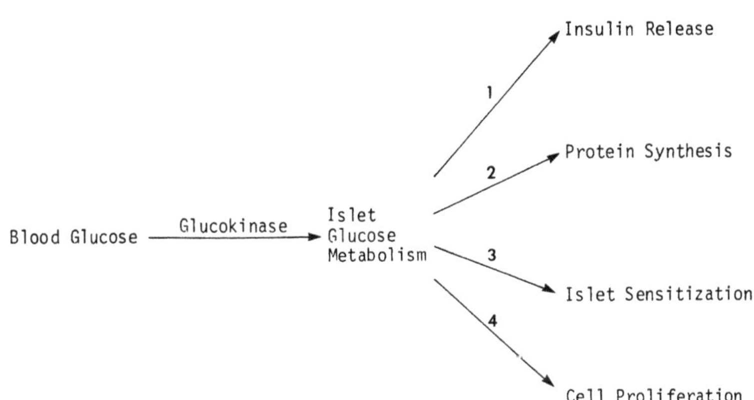

Fig. 3. Pleiotropic effect of glucose on pancreatic β-cells.

In order to assess whether the glucokinase-glucose-sensor is present
in human pancreatic islet cells and, if so, to find out whether its
characteristics are the same as observed in animal tissue, we have applied
a novel microassy for glucokinase to studies of human pancreatic
islets[1,2]. The assay is sensitive enough to detect this macromolecule in
as little as 100 nanogram (1/10,000.000 of 1 gram) of β-cells. This is
about 1/10 of an average pancreatic islet. We found the glucokinase to be
present in human islet tissue (Table 2). Its activity is about half that
found in the rat. Its biophysical constants are also similar to those
found in islets of the rat. The data suggest that the glucokinase glucose
sensor concept has general applicability. Studies of this molecule in the
pancreas from type II diabetes are now possible. Of course, they will be
difficult. Since glucokinase may determine all major β-cell functions
(Figure 3) vigorous attempts should be made to perform such studies with
tissue from diabetic humans.

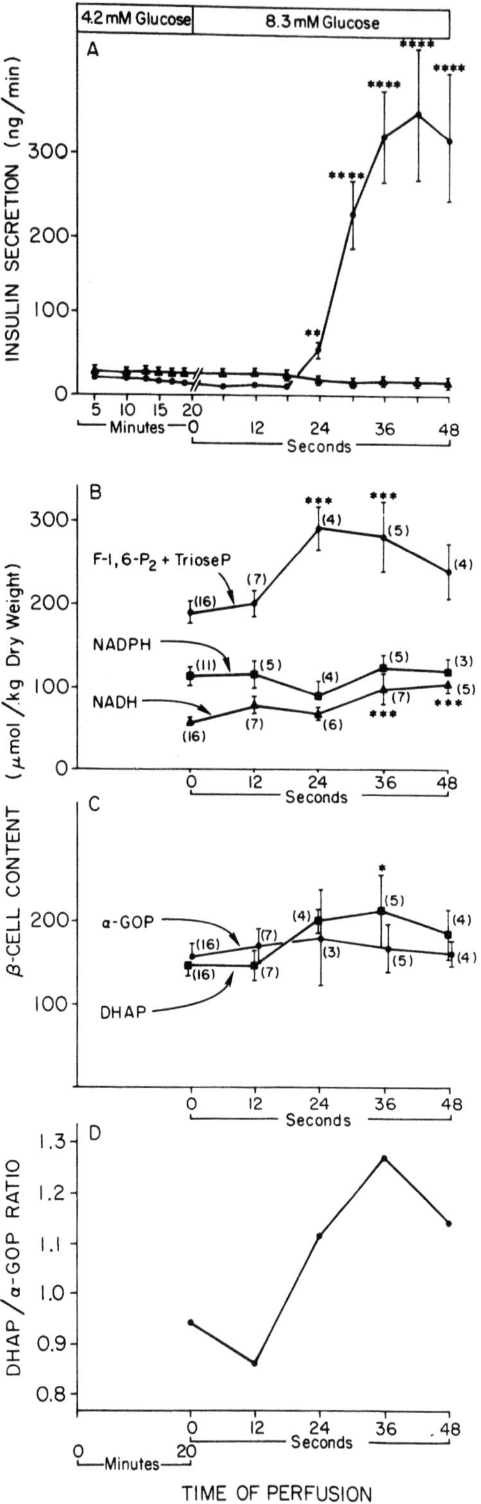

Fig. 4. Studies of the redox potential of pure ß-cells during the first phase of glucose induced insulin release. The isolated perfused rat pancreas was used, and pure ß-cell samples were studied with quantitative histochemical methods. Pure ß-cell samples were obtained by microdissection from freeze-dried pancreas cryotom sections.

The glucose sensor role is apparently not entirely dependent on the
presence of glucokinase. Transmembraneous glucose transport might serve
the glucose sensing role in certain tumorous β-cells. In normal β-cells
glucose transport is extremely fast and intracellular glucose levels
equalize with extracellular levels almost instantaneously over a very wide
concentration range exceeding by far the physiological boundaries.
Transport is not a limiting or regulating step for glycolysis and does not
seem to serve as sensing step in normal β-cells. However in certain clonal
β-cell lines hexose transport may play the role of the sensor[8]. It was
found, for example, that 3-0-methylglucose transport into clonal HIT-T15,
RIN-m5F and RIN-1046-38 cells was much slower than into isulinoma cells
isolated from the transplantable parent tumor (Table 3). HIT-T15 cells
respond to glucose in the hypoglycemic range (<1 mM) whereas the solid
tumor responds in a physiological manner to glucose levels that exceed 5
mM. It is not unlikely that the dose response curve for glucose induced
insulin release in HIT cells is governed by glucose transport in contrast
to the situation in normal β-cells and β-cells in the solid insulinoma.

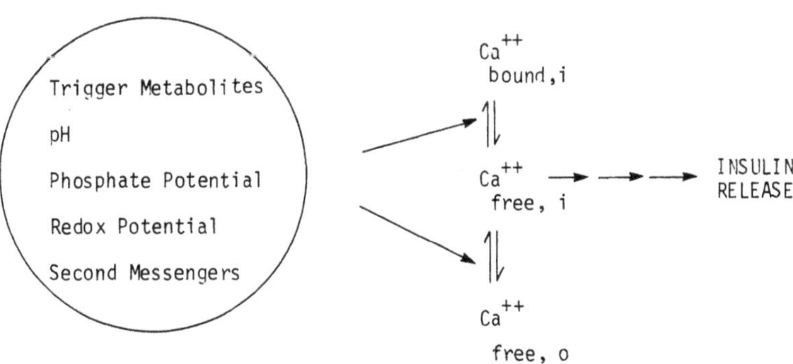

Fig. 5. The central role of Ca^{+2} stimulus secretion coupling in pancreatic
β-cells.

Another distinguishing metabolic design feature of the pancreatic
β-cells is the unusually high activity of mitochondrial α-glycero-P-
dehydrogenase[5]. We have found evidence which suggests that this enzyme
maintains the cytoplasm in an oxidized state during initiation of glucose
stimulation of insulin release. This contrasts with results from previous
studies which indicated an enhanced reduction of the cells following
exposure to high glucose (Figure 4). These new data were gathered from
quantitative histochemical measurements of NADH, NADPH, dihydroxyacetone-P
and α-glycero-P in samples of pure β-cells dissected from freeze-dried

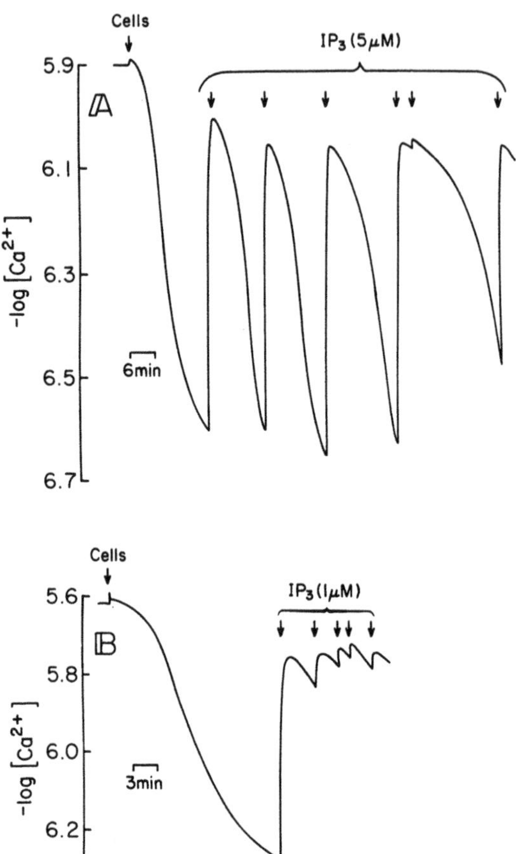

Fig. 6. Effect of pulse additions of IP_3 on the free Ca^{+2} concentration maintained by the nonmitochondrial compartment of permeabilized insulinoma RINm5F cells. A, cells were incubated at 30°C, pH 7.0, in 0.2 ml of buffer containing 110 mM HCl, 10 mM NaCl, 2 mM KH_2PO_4, 1 mM $MgCl_2$, 25 mM Hepes, 1 mM MgATP, 5 mM phosphocreatine, 20 units/ml creatine kinase, 1 μg/ml oligomycin, 0.2 μM antimycin A, and 50 μg saponin. At the points indicated, cells (10^6) or IP_3 (5 μM) were added. B, at the points indicated, cells (10^6) or IP_3 (1 μM) were added. (Figure taken from reference 9).

cryostat sections of the perfused rat pancreas. One plausible explanation is that the α-glycero-P shuttle operates so effectively in islet cells as to prevent accumulation of cytoplasmic NADH which would inhibit glycolysis. Since glucose sensing seems to depend on the unimpeded operation of the glycolytic pathway, NADH removal is an essential design feature of glucose metabolism in the islet cells. It is conceivable that the cytosolic Ca^{2+} level governs the flux through the mitochondrial α-glycero-P dehydrogenase step, since we observed in mitochondria isolated from a glucose sensitive

Table 3. Uptake of 10 mM 3-0-methyl-D-glucose by insulinoma cells and
cultured cells cloned from transformed β-cells or radiation-
induced insulinomas.

Source of Cells (Observations)	3-0-methylglucose space / Water Space × 100
RIN$_s$ Insulinoma (4)	94.5 ± 2.6
HIT-T15 (4)	45.9 ± 3.5
RINm5F (4)	53.4 ± 2.9
1046-38 (4)	37.6 ± 3.3

Dispersed cells were prepared by collagenase digestion of slow-growing,
radiation-induced insulinomas (RIN$_s$) transplantable in NEDH rats. HIT-T15
cells, RINm5F cells, and 1046 cells were grown in monolayer cultures and
prepared as dispersed cells by trypsinization. Dispersed cells were
incubated with ^{3}HOH and [U-^{14}C]-3-0-methyl-D-glucose or [^{14}C]-PEG 4000 for
5 min at 22°C. Cells were separated from incubation medium by centrifuging
through a layer of silicone oil. Intracellular water and 3-0-methylglucose
spaces were computed by correcting for extracellular fluid contaminating
the cell pellet using PEG as a marker of the extracellular space.

Table 4. Oxygen Consumption by Isolated Rat Insulinoma Mitochondria.

Substrate (5 mM)	In the presence of EGTA (nmol O_2/min/mg protein)		In the absence of EGTA	
	State 4	State 3	State 4	State 3
Endogenous	$\leq$1	$\leq$1	1	1
Succinate	39	183	38	185
Glutamate + Malate	19	118	24	117
L-Glycerol-3 Phosphate	31	41	71	153[*]

EGTA: 1 mM
Mean of 4-9 experiments, SE of state 3 respirations <10% of mean values
*, p <0.01 compared to the respiration in the presence of EGTA.

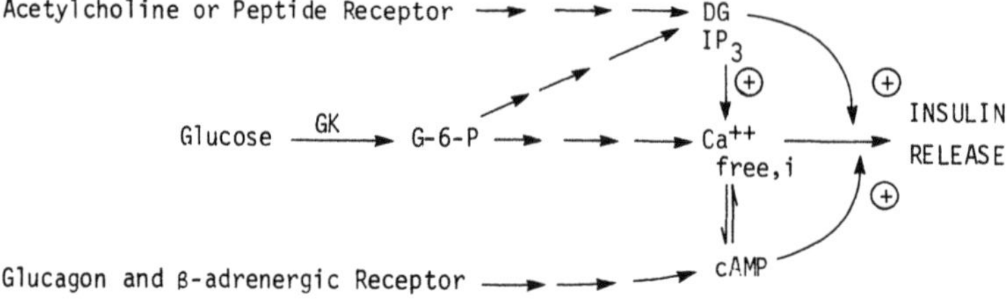

Fig. 7. Intracellular signalling in pancreatic β-cells.
DG: diacylglycerol; IP$_3$: 1,4,5 inositol-trisphosphate; GK;
glucokinase; cAMP: cyclic 3'5'-AMP. Established and putative
signal pathways are indicated.

transplantable insulinoma that the enzyme required micromolar levels of
Ca^{2+} for full activity (Table 4).

The processes involved in stimulus secretion coupling are infinitely
more complex than the relatively simple mechanism of glucose metabolism and
recognition described above. The metabolism of glucose generates a signal
or signals that ultimately change within the β-cell the level of free Ca^{2+}
and of other second messengers with the result being insulin secretion
(Figure 5). Alterations of the intracellular pH, of the phosphate
potential, of the redox potential, of particular trigger metabolites, and
of certain second messengers may be involved. At this point our knowledge
is not sufficient to chose among these and other possibilities. During the
last years a new second messenger candidate has burst on the biochemical
scene: inositol-triphosphate (IP$_3$), which is generated from phospho-
tidylinositol diphosphate, an integral membrane phospholipid, [4,9,10]. It
was tested whether in islet tissue IP$_3$ could mobilize intracellular Ca^{2+},
the ultimate coupling factor causing insulin secretion. Using insulin
secreting tumor cells it was found that IP$_3$ is very effective in islet
cells and that it mobilized Ca^{2+} from permeabilized insulinoma cells
(Figure 6). IP$_3$ together with other second messengers might be involved in
coupling glucose metabolism to insulin secretion (Figure 7). It is implied
in the scheme that phospholipase C and adenylate cyclase might play an
important role in stimulus secretion coupling.

Why is it so important that we understand how, in biochemical terms,
glucose stimulates insulin secretion? Glucose is the preeminent

physiological stimulant of the β-cell and its recognition and metabolic
processing are absolute requirements for other physiological and pharmacological stimuli to influence insulin secretion (e.g. amino acids, ketone
bodies, hormones, neurotransmitters). Impaired glucose sensing by the
β-cell (i.e. type II diabetes) will decrease the response of β-cells to all
known physiological insulin secretagogues and will influence the response
to many known drugs. The importance of the problem is self evident.

REFERENCES

1. F.J. Bedoya, J.M. Wilson, A.K. Ghosh, D. Finegold and F.M.
    Matschinsky, The Glucokinase Glucose Sensor in Human Pancreatic
    Islet Tissue. Diabetes, 35:61-67 (1986).

2. F.J. Bedoya, M. Meglasson, J.M. Wilson and F.M. Matschinsky,
    Radiometric Oil Well Assay for Glucokinase in Microscopic
    Structures, Anal. Biochem. 144:504 (1985).

3. F. Bedoya, M. Appel, T. Shimizu, and F.M. Matschinsky, Differential
    Regulation of Glucokinase Activity in Pancreatic Islets and Liver
    in the Rat, Submitted for publication.

4. S.K. Joseph, R.J. Williams, B.E. Corkey, F.M. Matschinsky, and R.J.
    Williamson, The Effect of Inositol Trisphosphate on Ca^{2+} Fluxes in
    Insulin-secreting Tumor Cells, J. Biol. Chem. 259:12952 (1984).

5. M.J. MacDonald, High Content of Mitochondrial Glycerol-3-Phosphate
    Dehydrogenase in Pancreatic Beta cells During Glucose Stimulated
    Insulin Secretion, J. Biol. Chem. 256:8287 (1981).

6. F.M. Matschinsky, A. Ghosh, M. Meglasson, M. Prentki, and V. June,
    Metabolic Concomitants in Pure, Pancreatic Beta-cells, Submitted
    for Publication.

7. M.D. Meglasson and F.M. Matschinsky, New Perspectives on Pancreatic
    Islet Glucokinase, Am. J. Physiol. 246:E1 (1984).

8. M.D. Meglasson, C. Manning, H. Najafi and F.M. Matschinsky, Glucose
    Transport by a Radiation-induced Insulinome and Clonal Pancreatic
    Beta-Cells, Submitted for Publication.

9. M. Prentki, B.E. Corkey, and F.M. Matschinsky, Inositol
    1,4,5-Triphosphate and the Endoplasmic Reticulum Ca^{2+} Cycle of a
    Rat Insulinoma Cell Line, J. Biol. Chem. 260:9185 (1985).

10. M. Prentki, T.J. Biden, D. Janjic, R.F. Irvine, M.J. Berridge, and
    C.B. Wollheim, Rapid Mobilization of Ca^{2+} from Rat Insulinoma
    Microsomes by Inositol-1,4,5-trisphosphate, Nature 309:562 (1984).

CONTRIBUTORS

Alcron, C.
Hospital Clinico y Provincial
Villarroel 170,
Barcelona 36 (Spain)

Alejo, S.
Departamento de Bioquimica,
Facultad de Ciencias,
Universidad de Badajoz
Badajoz (Spain)

Almers, W.
University of Washington
Department of Physiology and Biophysics
Seattle, Washington 98195 (U.S.A.)

Appel, M.C.
Departamento Bioquimica,
Facultad de Medicina
Sevilla (Spain)

Artalejo, C.R.
Departamento de Farmacologia,
Facultad de Mdeicina,
Universidad de Alicante
Alicante (Spain)

Ashcroft, F.M.
University Laboratory of Physiology
Parks Road, Oxford OX3 9DU (U.K.)

Ashcroft, S.J.H.
Nuffield Department of Clinical Biochemistry,
John Radcliff Hospital
Headington, Oxford OX3 9DU (U.K.)

Ashford, M.L.J.
Department of Pharmacology,
University of Cambridge
Cambridge (U.K.)

Atwater, I.
Laboratory of Cell Biology and Genetics,
NIDDK, National Institutes of Health
Building 8, Room 403
Bethesda, Maryland 20892 (U.S.A.)

Bangham, J.A.
Department of Biophysics
University of East Anglia
Norwich NR4 7TJ (U.K.)

Barriga, C.
Departamento de Fisiologia,
Falcultad de Medicina,
Universidad de Badajoz
Badajoz (Spain)

Bedoya, F.J.
Departamento de Bioquimica,
Facultad de Medicina
Sevilla (Spain)

Berg, M.
Baylor College of Medicine
One Baylor Plaza
Houston, Texas 77030 (U.S.A.)

Bergsten, P.
Department of Medical Cell Biology
Box 571, S-75123
Uppsala (Sweden)

Berner, D.
Diabetes Research Center,
414 Anatomy-Chemistry Building
36th and Hamilton Walk G3
Philadelphia, Pennsylvania 19104 (U.S.A.)

Brocklehurst, K.W.
Laboratory of Cell Biology and Genetics,
NIDDK, National Institutes of Health
Building 8, Room 403
Bethesda, Maryland 20892 (U.S.A.)

Burns, L.J.
Laboratory of Cell Biology and Genetics,
NIDDK, National Institutes of Health
Building 8, Room 403
Bethesda, Maryland 20892 (U.S.A.)

Campfield, L.A.
Northwestern University
303 E. Chicago Avenue
Chicago, Illinois 60611 (U.S.A.)

Campillo, J.E.
Departamento Bioquimica,
Facultad de Ciencias,
Universidad de Badajoz
Badajoz (Spain)

Castellano, A.
Departamento de Fisiologia,
Facultad de Medicina
Avda Sanchez Pizjuan
4, 41009 Sevilla (Spain)

Cervera, T.
Servicio Endocrinologia y Nutricion,
Hospital Sta. Cruz
Antonio Ma Claret 167, 08025
Barcelona (Spain)

Chay, T.R.
Department of Biological Sciences,
Faculty of Arts and Sciences
University of Pittsburgh
Pittsburgh, Pennsylvania 15260 (U.S.A.)

Codina, M.
Servicio Endocrinologia y Nutricion,
Hospital Sta. Cruz,
Antonio Ma Claret 167, 08025
Barcelona (Spain)

Cook, D.L.
Department of Physiology and Biophysics,
University of Washington
Seattle, Washington 98103 (U.S.A.)

Corkey, B.
Diabetes Research Center,
414 Anatomy-Chemistry Building
36th and Hamilton Walk G3
Philadelphia, Pennsylvania 19104 (U.S.A.)

Croghan, P.C.
Department of Biophysics,
School of Biological Sciences
University of East Anglia
Norwich NR4 7NJ (U.K.)

Davidson, H.W.
Department of Clinical Biochemistry
Addenbrooke's Hospital
University of Cambridge
Cambridge (U.K.)

Dawson, C.M.
Department of Biophysics,
School of Biological Sciences
University of East Anglia
Norwich NR4 7NJ (U.K.)

Debuyser, A.
Laboratoire de Physiologie,
Animale Faculte de Sciences
40 Avenue du Recteur Pineau
86022 Poitiers, Cedex (France)

de Leiva, A.
Servicio Endocrinologia y Nutricion,
Hospital Sta. Cruz,
Antonio Ma Claret 167, 08025
Barcelona (Spain)

Dunne, M.J.
Department of Physiology,
University of Liverpool
Brownlow Hill, P.O. Box 147
Liverpool L69 3BX (U.K.)

Ferrer, R.
Departamento de Fisiologia,
Facultad de Medicina
Universidad de Alicante
Alicante (Spain)

Findlay, I.
Department of Physiology,
University of Liverpool
Brownlow Hill, P.O. Box 147
Liverpool L69 3BX (U.K.)

Finol, H.
Escuela de Biologia,
Universidad Central de Venezuela,
Apartado 21201 (Venezuela)

Forsberg, E.J.
Laboratory of Cell Biology and Genetics,
NIDDK, National Institutes of Health
Building 8, Room 403
Bethesda, Maryland 20892 (U.S.A.)

Garcia, A.G.
Departamento Farmacologia,
Facultad de Medicina
Universidad de Alicante
Alicante (Spain)

Garcia-Morales, P.
Laboratorio Metabolismo,
Nutricion y Hormonas
Fundacion Jimenez Diaz, Reyes Catolicos 2
28040 Madrid (Spain)

Garrino, M.G.
Unite de Diabete et Croissance,
Universite de Louvain,
Faculte de Medicine
Avenue Hippocrate 54:B-120
Bruxelles (Belgium)

Gobbe, P.
Laboratory of Pharmacology,
Brussels University,
School of Medicine
Brussels (Belgium)

Goberna, R.
Departamento de Bioquimica,
Facultad de Medicina,
41009 Sevilla (Spain)

Gomis, R.
Hospital Clinico y Provincial
Villarroel 170
Barcelona 36 (Spain)

Gosh, A.
Diabetes Research Center,
414 Anatomy-Chemistry Building
36th and Hamilton Walk G3
Philadelphia, Pennsylvania 19104 (U.S.A.) .

Grimaldi, K.
Department of Clinical Biochemistry
Addenbrooke's Hospital
University of Cambridge
Cambridge (U.K.)

Gylfe, E.
Department of Medical Cell Biology
Box 571 S-75123
Uppsala (Sweden)

Hales, C.N.
Department of Clinical Biochemistry,
Addenbrooke's Hospital
Cambridge CB2 2 QR (U.K.)

Harrison, D.E.
Nuffield Department of Clinical Biochemistry,
John Radcliffe Hospital
Headington, Oxford OX3 9DU (U.K.)

Hashimoto, N.
2nd Department of Internal Medicine
Chiba University School of Medicine
1-8-1 Inohana, Chiba 280 (Japan)

Hellman, B.
Department of Medical Cell Biology
Box 571 s-75123
Uppsala (Sweden)

Henquin, J.C.
Unite de Diabete et Croissance,
Universite de Louvain
Faculte de Medicine
Avenue Hippocrate 54:B-120
Bruxelles (Belgium)

Herchuelz, A.
Laboratoire de Pharmacodynamie et de Therapeutique
Universite Libre de Bruxelles
Faculte de Medicine
Boulevard Waterloo, 115: B-1000
Bruxelles (Belgium)

Hermans, M.
Unite de Diabete et Croissance,
Universite de Louvain
Faculte de Medicine
Avenue Hippocrate 54; B-120
Bruxelles (Belgium)

Hill, R.S.
Department of Medicine and Cell Biology
Baylor College of Medicine
One Baylor Plaza
Houston, Texas 77030 (U.S.A.)

Himmel, D.
Department of Biological Sciences
Faculty of Arts and Sciences
University of Pittsburgh
Pittsburgh, Pennsylvania 15260 (U.S.A.)

Howell, S.L.
Department of Physiology
Queen Elizabeth College
University of London
Campden Hill Road
Kensington, London W87AH (U.K.)

Hutton, P.
Department of Clinical Biochemistry,
Addenbrooke's Hospital
University of Cambridge
Cambridge (U.K.)

Joffre, M.
Laboratoire de Physiologie Animale
40 Ave de Recteur Pineau
86022 Poitiers
Cedex (France)

Jones, P.M.
Department of Physiology
Queen Elizabeth College,
University of London
Campden Hill Road
Kensington, London W8 7AH (U.K.)

Joost, S.
Laboratory of Cell Biology and Genetics
NIDDK, National Institutes of Health
Bethesda, Maryland 20892 (U.S.A.)

Juvent, M.
Laboratory of Pharmacology,
Brussels University
School of Medicine
Brussels (Belgium)

Kanatsuka, A.
2nd Department of Internal Medicine
Chiba University School of Medicine
1-8-1 Inohana, Chiba 280 (Japan)

Lebrun, P.
Laboratoire de Medicine Experimentale,
Universite Faculte de Medicine
Boulevard de Waterloo 115, B-1000
Bruxelles (Belgium)

Lopez-Barneo, J.
Departamento de Fisiologia
Facultad de Medicina
Avda Sanchez Pizjuan 4
4100
Sevilla (Spain)

Lee, G.
Laboratory of Cell Biology and Genetics
NIDDK, National Institutes of Health
Building 8, Room 403
Bethesda, Maryland 20892 (U.S.A.)

Makino, H.
2nd Department of Internal Medicine
Chiba University School of Medicine
1-8-1 Inohana, Chiba 280 (Japan)

Malaisse, W.J.
Laboratoire de Medicine Experimentale,
Universite Libre de Bruxelles
Faculte de Medicine
Boulevard de Waterloo 115, B-1000
Bruxelles (Belgium)

Malaisse-Lagae, F.
Laboratoire de Medicine Experimentale
Universite Libre de Bruxelles
Faculte de Medicine
Boulevard de Waterloo 115, B-1000
Burxelles (Belgium)

Manning, C.
Diabetes Research Center
414 Anatomy-Chemistry Building
36th and Hamilton Walk G3
Philadelphia, Pennsylvania 19104 (U.S.A.)

Marques, A.A.
Departamento de Bioquimica y Fisiologia
Facultad de Medicina
Avda Blasco Ibanez
17, 46010
Valencia (Spain)

Matschinsky, F.M.
Diabetes Research Center
414 Anatomy-Chemistry Building
36th and Hamilton Walk G3
Philadelphia, Pennsylvania 19104 (U.S.A.)

Meglasson, M.
Diabetes Research Center
414 Anatomy-Chemistry Building
36th and Hamilton Walk G3
Philadelphia, Pennsylvania 19104 (U.S.A.)

Mena, P.
Departamento Fisiologia
Facultad de Medicina
Universidad de Badajoz
Badajoz (Spain)

Najafi,H.
Diabetes Research Center
414 Anatomy-Chemistry Building
36th and Hamilton Walk G3
Philadelphia, Pennsylvania 19104 (U.S.A.)

Neher, E.
Max Plank Institut fur Biophysikalische Chemie
Postfach 948 3400
Gottingen (F.R.G.)

Nelson, T.Y.
Department of Medicine and Cell Biology
Baylor College of Medicine
One Baylor Plaza
Houston, Texas 77030 (U.S.A.)

Nenquin, M.
Unite de Diabete et Croissance
Universite de Louvain
Faculte de Medicine
Avenue Hippocrate, 54, B-120
Bruxelles (Belgium)

Niki, A.
Department of Internal Medicine
Aichigakuin University
2-11 Suemori-dori
Chikusa-ku, Nagoya 464 (Japan)

Niki, H.
Department of Internal Medicine
Aichigakuin University
2-11 Suemori-dori
Chikusa-ku, Nagoya 464 (Japan)

Niki, I.
The 3rd Department of Medicine
University of Nagoya
65 Tsurumai Showa
Nagoya 466 (Japan)

Oberwetter, J.M.
Baylor College of Medicine
One Baylor Plaza
Houston, Texas 77030 (U.S.A.)

Osorio, C.
Departamento Fisiologia
Facultad de Medicina
Avda de Madrid s/n. 18012
Granada (Spain)

Osuna, J.I.
Departamento Fisiologia
Facultad de Medicina
Avda de Madrid s/n. 18012
Granada (Spain)

Pasma, A.
Laboratory for Experimental Endocrinology
University of Groningen
1 Bloemsingel
9713 BZ Gronigen (The Netherlands)

Palafox, I.
Facultad de Medicina
Universidad de Alicante
Alicante (Spain)

Paolisso, G.
Unite de Diabete et Croissance
Universite de Louvain
Faculte de Louvain
Faculte de Medicine
Avenue Hippocrate 54; B-120
Bruxelles (Belgium)

Partke, H.J.
Diabetes-forschungs
Institut Auf'm Hennekamp
65. 4000
Dusseldorf (F.R.G.)

Perez-Armendariz, E.
Laboratory of Cell Biology and Genetics
NIDDK, National Institutes of Health
Building 8, Room 403
Bethesda, Maryland 20892 (U.S.A.)

Peshavaria, M.
Department of Clinical Biochemistry
Addenbrooke's Hospital
University of Cambridge
Cambridge (U.K.)

Peterson, O.H.
The Physiological Laboratory
The University of Liverpool
Brownlow Hill, P.O. Box 147
Liverpool L69 3BX (U.K.)

Pogge, Von Strandmann, R.
Department of Clinical Biochemistry
Addenbrooke's Hospital
University of Cambridge
Cambridge (U.K.)

Pollard, H.B.
Laboratory of Cell Biology and Genetics
NIDDK, National Institutes of Health
Building 8, Room 403
Bethesda, Maryland 20892 (U.S.A.)

Pou, J.M.
Servicio Endocrinologia y Nutricion
Hospital Sta. Cruz
Antonio Ma Claret 167 08025
Barcelona (Spain)

Prentki, M.
Diabetes Research Center
414 Anatomy-Chemistry Building
36th and Hamilton Walk G3
Philadelphia, Pennsylvania 19104 (U.S.A.)

Reig, J.A.
Departamento Farmacologia y Terapeutica
Universidad de Alicante
Alicante (Spain)

Rinzel, J.
Mathematical Research Branch
NIDDK, National Institutes of Health
Bethesda, Maryland 20892 (U.S.A.)

Ripoll, C.
Departamento de Fisiologia
Facultad de Medicina
Universidad de Alicante
Alicante (Spain)

Rodriguez, E.
Departamento Fisiologia
Facultad de Medicina
Avda de Madrid s/n. 18012
Granada (Spain)

Rojas, E.
Laboratory of Cell Biology and Genetics
NIDDK, National Institutes of Health
Building 8, Room 403
Bethesda, Maryland 20892 (U.S.A.)

Rorsman, P.
Department of Medical Cell Biology
Box 571, s-75123
Uppsala (Sweden)

Rosario, L.M.
Laboratory of Cell Biology and Genetics
NIDDK, National Institutes of Health
Bethesda, Maryland 20892 (U.S.A.)

Rubio, R.
Departamento Fisiologia
Facultad de Medicina
Avda de Madrid s/n. 18012
Granada (Spain)

Saceda, M.
Laboratorio Metabolismo,
Nutricion y Hormonas
Fundacion Jimenez Diaz, Reyes Catolicos 2
208040 Madrid (Spain)

Sakurada, M.
2nd Department of Internal Medicine,
Chiba University School of Medicine
1-8-1 Inohana, Chiba 280 (Japan)

Sala, F.
Departamento de Farmacologia
Facultad de Medicina
Universidad de Alicante
Alicante (Spain)

Sala, S.
Departamento de Fisiologia
Facultad de Medicina
Universidad de Alicante
Alicante (Spain)

Sanchez-Andres, S.
Facultad de Medicina
Universidad de Alicante
Alicante (Spain)

Santos, R.M.
Laboratory of Cell Biology and Genetics
NIDDK, National Institutes of Health
Bethesda, Maryland 20892 (U.S.A.)

Satin, L.S.
Veterans Administration Medical Center
1660 South Coumbia Way
Seattle, Washington 98108 (U.S.A.)

Scott, A.M.
Department of Biophysics
School of Biological Sciences
University of East Anglia
Norwich NR4 7TJ (U.K.)

Sehlin, J.
Department of Histology and Cell Biology
University of Umea
s-90187
(Sweden)

Sener, A.
Laboratoire de Medicine Experimentale
Universite Libre de Bruxelles
Faculte de Medicine
Boulevard de Waterloo 115, B-1000
Bruxelles (Belgium)

Settle, J.E.
Departments of Physiology and Biomedical Engineering
Chicago and Evanston, Illinois 60611 and 60201 (U.S.A.)

Shimizu, T.
Diabetes Research Center
414 Anatomy-Chemistry Building
36th and Hamilton Walk G3
Philadelphia, Pennsylvania 19104 (U.S.A.)

Smith, P.A.
Departament of Biophysics
School of Biological Sciences
University of East Anglia
Norwich NR4 7TJ (U.K.)

Sohaey, R.
Departments of Physiology and Biomedical Engineering
Chicago and Evanston, Illinois 60611 and 60201 (U.S.A.)

Soria, B.
Departamento de Fisiologia
Facultad de Medicina
Universidad de Alicante
Alicante (Spain)

Sturgess, N.C.
Department of Clinical Biochemistry
University of Cambridge
Cambridge (U.K.)

Siddle, K.
Department of Clinical Biochemistry
Addenbrooke's Hospital
University of Cambridge
Cambridge (U.K.)

Stutzin, A.
Laboratory of Cell Biology and Genetics
NIDDK, National Institutes of Health
Building 8, Room 403
Bethesda, Maryland 20892 (U.S.A.)

Tabares, L.
Departamento de Fisiologia, Facultad de Medicina
Avenida Sanchez Pizjuan 4 41009
Sevilla (Spain)

Tamagawa, T.
Department of Medicine
Aichigakuin University
2-11 Suemori Chikusa
Nagoya 464 (Japan)

Trube, G.
Max Plank Institut fur Biophysikalische Chemie
Postfach 948, 3400
Gottingen (F.R.G.)

Valverde,I.
Laboratorio Metabolismo,
Nutricion y Hormonas
Fundacion Jimenez Diaz
Reyes Catolicos 2
28040 Madrid (Spain)

Van Ganse, E.
Laboratory of Pharmacology
Brussels University School of Medicine
Brussels (Belgium)

Vasseur, M.
Laboratoire de Physiologie Animale
40 Ave de Recteur Pineau
86022 Poitiers
Cedex (France)

Vonk, M.
Laboratory for Experimental Endocrinology
University of Groningen
1 Bloemsingel
9713 BZ Gronigen (The Netherlands)

Wasner, H.K.
Diabetes-forschungs
Institut Auf'm Hennekamp
65. 4000
Dusseldorf (F.R.G.)

Wolters, G.H.J.
Laboratory for Experimental Endocrinology
University of Groningen
1 Bloemsingel
9713 BZ Gronigen (The Netherlands)

Yoshida, S.
2nd Department of Internal Medicine
Chiba University School of Medicine
1-8-1 Inohana, Chiba 280 (Japan)

A-23187, 151, 356

Acetylcholine, 10-14, 15,
16, 18, 139, 143, 345
effects of, 354
evoked ATP release, 19
evoked ATP secretion, 20-24
receptor, 11, 26, 140
release, 25, 140
stimulation, 19

Action potential, 6, 235,
363, 413-427

Active phase, 38

Adrenaline, 298-299

Adenylatecyclase, 404-406,
409-412 (see also cAMP)

Adrenal
medulla, 139
medullary cells, 14, 15

Adrenocortical cells, 125

Adrenocorticotropic hormone
(ACTH), 126

Aging, 429

Agonist, 358

Ag/AgCl electrode, 32

Aminopyridines, 87
4-aminopyridine, 87

AMP
cAMP 403, 409
antagonist (cPIP) 409,

Amplitude distribution, 117

Anion
regulation of insulin
release, 227-234
selectivity, 231

Anti-insulin serum, 161

Arachidonic acid, 373

Apamin, 101

Arsenazo III, 326

Aqueous diffusion, 4

ATP, 7, 9, 14, 15, 16, 17,
21, 23, 24, 267
blockable K-channel, 100,
101
concentration, 25, 248
hydrolysis, 65
inhibited K-channel, 65
release, 7, 10-13, 18, 19-20
secretion, 26
sensitive K-channel, 72-79,
174
sensitivity, 66

Atropine, 13, 350

Basal insulin secretion, 284
cytosolic Ca^{2+}, 286

B-cell
input resistance, 352
effect of atropine on, 352
effect of carbamylcholine
on, 352

granules, 385-394
electrical activity, 265
memory, 347
muscarinic receptor, 352

BAY K 8644, 141, 201
B8-methylene ATP, 58
Bessel filter, 108
Betagranin, 386-390
Bioluminescent reaction, 8
Biphasic
curve, 74
electrical response, 270
Bovine
adrenal medulla, 142
serum albumin (BSA), 294
Burst
kinetics, 56, 57
pattern, 247, 418
frequency, 248, 418
Bursting electrical activity,
94, 265

cAMP, 85, 297-299
production, 74
Caffeine, 243
Ca-activated
K^+ current, 190
K-channels, 64, 97, 89, 183,
190, 257
permeability, 84, 99-100
Ca^{2+}sensitive K-channels,
84, 97, 89
Ca-channel, 37, 139, 159-166,
167-176, 177-188,
189-194, 195-200, 201-206
activation, 142
activity, 356, 364,
antagonist, 143, 195
agonist, 143, 201-206

blocker, 189, 241, 310-311
pharmacology, 139
Ca^{2+}
availability, 403-408
current, 167-175, 180, 237
dependent
ATP release, 7-30
phosphorylation, 313
depletion, 32, 45
efflux, 317, 325, 326
entry, 237, 143
fluxes, 47
induced insulin secretion,
283
regulation of membrane
fusion, 369-384
sensitive electrode, 45
sequestration, 337
threshold for insulin
secretion, 286
Calcium electrode, 11, 12
Calcium spike (see Action
Potential)
Calmodulin, 312, 427, 432
Capsaicin, 84, 89
Carbamylcholine, 329, 354
Carboxymethylation of
proteins, 431
and calmodulin, 432
carboxylmethyltranferase
assay 433
distribution, 434, 436
carboxypeptidase E, 388-389
effect of glucose, 433
methyl aceptor proteins,
432-436
carboxypeptidase E, 388-389
Catecholamine, 10, 26, 142,
secretion, 16, 139, 140,
142, 144

Cation-sensitive

  microelectrode, 31-52

  potential, 36

Cell-attached,

  mode, 66, 110

  patches, 54, 55, 59, 79

Cell-to-cell coupling, 108,
    239, 244, 268

Cell-line, 69-71, 305, 325,
    462

CGP-28392, 202-203

Channel

  activity, 71

  events, 111

  probability, 58

Chaotic burst pattern, 248

Chemizol, 359

Chinese hamsters, 407

Cholinergic

  receptor, 7

  nicotinic receptor, 10

Cholinoceptors, 140

Chord conductance, 212

Chromaffin

  cells, 7 - 25, 139,
      142, 288, 369, 386

  granules, 24

  membrane, 142

Chromogranin A, 390

Cimetidine, 359

C-K model, 253

Closed state,55

Closed time histogram, 113

Coefficients, 4

Collagenase digestion, 9,
    294

Compound exocytosis, 372

Constant-field model, 209

Conductance sub-state, 128

CR1-C1, 71

Collagenase digestion, 84

Culture medium, 54

Current

  injection, 213

  -voltage relation, 55, 112,
    117

Cyclic activation, 31

Cyclic AMP (cAMP), 297-298,
    407-410

Cyclic changes, 37

Cytoplasm, 2

Cytoplasmic

  Ca^{2+}, 1, 212, 325

  side, 71, 100

  signal, 2

Cytosolic

  ATP, 74

  free-Ca^{2+}, 247-264, 317

  Ca^{2+}concentration, 166,
    247-264, 286,293, 305-316

db/db-mouse C57BL/KsJ, 93, 96

Dehydrogenase, 23

Diabetes mellitus, 97

3,4 Diaminopyridine, 88

1,2 Diacylglycerol (DAG), 287,
    298

Diazoxide, 89

Dibutyryl-cAMP 85

Diet, 454

Diffusional access, 4

Diffussional equilibrium, 2

Diffusion limited, 43

Digitonin treatment, 9, 14,
    16, 24, 151, 294-295

Dihydropyridine, 142

Dye, 64

Diolein, 371

D-600, 147

Endoplasmic reticulum, 115
Electron microscopy, 108
Electrical permeabilization, 279
Electrical coupling, 235
Electrical synchrony 416, 419
    effect of increasing Ca2+ on, 421
Electrically coupled cells, 242
Excised
    patches, 79
    inside-out patches, 79
    membrane patches, 79, 127
Excitation-contraction coupling, 229
Exocrine cells, 353
Exocytosis, 19, 81, 287, 377, 431-442
    carboxylmethylation of proteins in, 429-440
    compound, 19
    fusion process, 377 (see also synexin)
Extracellular
    Na+, 319-324
    space, 225

Fasting, 341, 448
Fish islet cells, 109
Fluorescent Ca2+ indicator dyes, 1-6, 305-316
    fura 2, 1-6
    quin-2, 305-316, 332
Flux-ratio equation, 210
Forskolin, 297, 298-299
Frequency-amplitude histograms, 110

Gallamine, 84

Gigaohm resistance seal, 108
Glibenclamide, 96, 98, 99, 100, 101,103
    stimulated activity, 97
Glucagon, 307
Glucokinase, 447-458, 460-463
    assay of microscopic samples, 449
    fasting and refeeding and, 451
    glucose sensor, 460
    human pancreatic islets, 463
    insulinoma transplantation, 461
Glutaraldehyde, 108
Glucose
    homeostasis, 447
    -induced closure, 77
    -induced insulin release, 160
    -6-phosphate, 80
    sensory system, 459, 462, 467
Glyburide, 310
Glyceraldehyde, 77, 177, 181
    D-glyceraldehyde, 78
    dehydrogenase, 80, 463
    phosphate, 80, 465
Glycolysis, 80
Goldman
    equation, 209, 210, 219
    constant field model, 210, 211, 212
Granule, 115, 385

Hepatocytes, 90
Hexokinase activity, 449
High-voltage discharge technique, 279
Hit cells, 306
Histamine, 359
    effect on action potentials,

363

Histaminergic receptor, 358
Hodgkin-Huxley equations, 266
Hyperpolarizing current, 269

IBMX, 298-299
Immunocytochemical, 108
Implantation of an insulinoma,
    452-454
Inositol 1,4,5-trisphosphate
    (IP$_3$), 287, 466
Input resistance, 212, 253, 355
Intracellular
  compartment, 225
  Ca^{2+}, 279, 293, 305, 317,
    319
    distribution, 312
    mobilization, 332
  level, 371
  dialysis, 294
  organelle, 280
Intrinsic noise, 21
Intercellular coupling, 212,
    244, 267
Insulin release, 199, 207,
    279-292, 346-347, 429
  independent of intracellular
    calcium, 293-304
  diffusion time constant, 424
  electrical activity and,
    411-423
  ob-ob mouse and, 413-426
  pulsatile insulin release,
    418, 424
Insulin secreting cells, 77,
    140, 165-167, 177, 201
Insulinoma, 77, 466
  bearing rats, 447
  cell line, 78, 371
Insulinotropic agents, 203

Ion-sensitive electrode, 33,
    43
Isethionate, 16, 227, 228,
    296
Islet of Langerhans, 31, 32
Islet
  ion permeability, 214
  ion conductance, 214
  islet synchrony, 425
  tissue, 34
  ATP content, 73

K+-accumulation, 32, 45
  extracellular oscillations,
    418
K-channel, 37, 53-62, 63-68,
    69-76,
    95-108, 109-124, 125-138,
    167-176, 177-188
  ATP-blockable (G-channel),
  (see also ATP-inhibited
  and ATP-sensitive)
  53-62, 63-68, 69-82, 171
  ATP-blockade of the, 58, 65, 70
  conductance of the, 64, 172
  effects of nucleotides on, 65
  effects of voltage, pH
  and [Ca^{2+}] on
  kinetics of the, 56
  sulphonylureas and, 69
  blockers of the, 190, 242
  Ca-activated, 63-68, 70
    in adrenocortical
      cells, 127
    conductance, 64
  delayed rectifier, 168
  ion fluxes, 45, 47
  fish islet cell, 109-124
  membrane permebability
    and, 37, 53-56, 66, 72,

79, 95, 171

parathyroid cell, 133

pharmacological control of
the, 83-94

K^+-efflux, 45, 47

$[K^+]_i$ oscillations, 38

K^+

current, 258

diffusion coefficient, 35

-evoked release, 141

-induced membrane
depolarization, 144, 160

-induced insulin release,
163

permeability, 37, 66

sensitive micro-electrode,
31-52

Lactate dehydrogenase, 24

Leucine, 72

Linear regression, 113

Lipid vesicles, 1

Loading cells with probes, 1-6
(see also Quin-2)

Lorentzian, 22

Low affinity ^{3}H-QNB binding,
341

Luciferin 25

-luciferase, 7, 10, 14, 59

Luminescence instrument, 7

Mathematical modelling,
247-264, 265-278

Medullary chromaffin
cells, 7, 139

Membrane

channels, 31

conductance, 353

depolarization, 18, 77, 140

dielectric, 24

fusion, 369-384

patch, 2, 218

permeability, 207-224,
247-264

potential, 215, 353

resistance, 73

surface, 57

Melatonin, 367

Metabolic inhibitors, 59

Metallochromic indicator,
319-323

Methylxanthine, 85

Microelectrodes, 1, 31, 34,
35, 40

Microinjection, 2

Mitochondria, 24, 115, 326

Mitochondrial,

ATP, 73

uncoupler (CCCP), 101

Mobilization of calcium,
325-342

Modelling, 247-264, 265-278

Monensin, 397-402

Mouse islets, 34, 83, 95,
159, 225, 235, 351,
359, 413, 427

Murine adrenocortical
cells, 126

Muscarinic

receptor activation, 325-342,

receptor, 13, 143, 351-359,

receptor blocker, 13

Myo-inositol 1,4,5trisphosphate
($InsP_3$), 287

^{22}Na efflux, 225

Na^+ current, 2

Na^+ $2Ca$ exchange, 150,
319, 322

Neonatal rat B-cells, 58

Nernst potential, 212
Neural regulation, 325, 434,
    351, 359, 367
Neurotransmitters, 7, 347
Nicotine, 143
Nicotinic
   receptor, 13, 143, 144
   receptor stimulation, 11
Nifedipine, 26, 171-179, 181,
    195, 197
   blockade induced by, 16, 18
Niludipine, 144
Nitrendipine, 142, 144
Nor-adrenaline, 299, 343
Nutrient, 317
Non-nutrient, 317
N-methyl-D-glucamine, 169

ob/ob mouse, 95, 326, 418, 427
   islet cell electrical
      activity, 96
   glucose and, 97
      quinine and, 101
      sulphonylureas and, 98-101
   basal and glucose stimulated
      insulin release, 425
   glucose threshold, 422
   effect of extracellular
      calcium, 426
   pulsatile insulin release,
      426
Oocytes, 2
Open-time histogram, 57
Open-state probability, 59,
    78, 79, 80
Optical detection, 7-30
Organelles, 24
Oscillations, 32, 39, 40, 41
Oscillatory changes, 37
Oxygen comsumption, 468

Pacemaker neuron, 265
Pancreatic
   acinar cells, 288
   A-cells, 114-118
   B-cells, 7, 167, 187, 195, 201
   D-cells, 114-117
   polypeptide-cell (PP-cell),
      114, 116
Parasympathetic, 343, 351
Parathyroid
   cells, 125
   hormone, 126
Passive fluxes, 48
Patch clamp
   amplifier, 3, 107, 108
   pipettes, 1-6, 108, 189
   technique, 173
Penetration depth, 34, 40
Permeability, 216
   barrier, 16
Permeabilization, 20
Permeabilised islet, 284
Perchlorate, 231
Permeation buffer, 285
Phase plane, 251, 257
Phosphatidylinositol
   4,5-bisphosphate, 287
Phospholipid, 287
Pimozide, 144
cPIP, 410
Pipette
   potential, 111
   resistance, 3
Phorbol ester, 280
Phosphatidylserine, 370
Phosphoglycerate kinase, 80
Photocathode, 7
Photomultiplier, 7, 8, 21
Platelets, 7

Potassium permeability, 53

Power spectrum, 22

Protein,

  carboxyl methylation,

    431-442

  kinase C, 370

  production, 63

Proinsulin, 443-446

Promethazine, 374

Purinergic,

  P_2-receptor, 328

Pulsatile insulin release,

  413-426

Quin-2, 148, 293, 296, 306,

  loading, 308, 311

  leakage, 311

  probe, 317, 326

Quinine, 84, 85, 86, 99,

  100, 101, 235, 240

Quinuclidinyl benzilate

  (^{3}H-QNB), 351

Radioimmunoassay, 161, 228

Rainbow trout, 107

Rb$^+$ efflux, 83-94

Receptor,

  muscarinic, 13, 143,

    325-342, 351-358

  nicotinic, 7-30

  histaminergic, 359-366

    H1, H2, 359

Relaxation time constants, 22

Repolarization, 238, 271, 361

^{86}Rb-efflux, 83, 86-89

RINm5F cells, 180-182, 305, 326

  membrane currents, 183

Secretagogues, 8, 72, 125, 177,

  181, 203

Secretagogue-evoked

  inhibition, 77

Secretory granule, 7, 14,

  369-384,

   385-394

  content in the B-cell, 385

  maturation, 386

  protein formation, 385

  protein localization and

    function, 393

  proton gradient in the, 398

  in the adrenal chromaffin cell,

    387

  pituitary cell, 387

Secretory vesicles, 1

Selectivity ratio, 33

Single channel

  currents, 55, 111

  conductance, 213

  permeability, 213

Skeletal muscle, 73

Smooth muscle, 312

Somatostatin, 235

  secreting cell, 117

Sodium permeability, 218

Spike, 63, 235, 363

  (see also action potential)

  activity, 39, 198, 352

  amplitude, 198, 240

    distribution, 238, 243

  electrogenesis, 235-246

  frequency, 32, 41

Splachnic nerve, 136, 140

Sub-conductance state, 117

Suction pipette, 108

Stimulus-secretion coupling,

  94, 125, 140, 427-428

  ATP-blockable channels and,

    53-62,

    63-68, 69-76, 77-82, 171

adenylatecyclase and,
404-406, 409-412
(see also cAMP)
calcium and, 283, 369-384
carboxymethylation of proteins
and, 431
diacylglycerol (DAG) and,
287, 298
glucokinase, 447-458,
460-463
glucose-sensor, 459, 462, 467
inositol 1,4,5-triphosphate
(IP3),
287, 466
Stochastic model, 272
Sodium-potassium pump, 272
Sulphonylurea (sulfonylurea),
72, 97-100, 214, 297,
307, 311
Sympathetic effect, 343
Synexin,373
liposome agregation induced
by, 376
fusion of liposomes induced
by, 376
arachidonic acid dependent
fusion of
aggregated granules by,
376
[Ca^{2+}]-dependnet binding of
binding to acidic
phospholipids, 376
cloning of, 380
promethazine effect on, 374
synexin II, 378
trifluoroperazine effect on,
374
Synexin-like proteins, 375,
378, 379
Synhibin, 379

Tetraethylammonium (TEA), 84,
170, 180, 182, 190
Tetrodotoxin, 243
Theophylline, 84, 85
Thiocyanate, 228
Tolbutamide, 71, 72, 98, 99,
100, 297
TPA, 370
Transglutaminase, 443-446
activity, 444
inhibition, 444
conversion of proinsulin to
insulin, 445
Trifluoroperazine (TFP), 26,
147, 374
Triton, 9
Triethiodide, 84
Tubocurarine, 84, 90
Turbidity, 373

Unstirred layer, 36

Variance, 22, 356, 363
membrane potential, 354,
361
Verapamil, 147, 178-181,
195-199, 241
Virtual-ground voltage, 8
Voltage
-activated calcium currents,
81, 167-169, 177, 189
-activated Ca^{2+}-channel,
169, 247
-dependent Ca^{2+}-channel,
167, 247, 310
-gated calcium channel, 17,
18, 48, 80, 143-144, 165
-gated potassium-channel,
48

-sensitive Ca -channel,
167

Whole-cell patch-clamp, 2,
177, 189

Yohimbine, 299

GPSR Compliance
The European Union's (EU) General Product Safety Regulation (GPSR) is a set
of rules that requires consumer products to be safe and our obligations to
ensure this.

If you have any concerns about our products, you can contact us on

ProductSafety@springernature.com

In case Publisher is established outside the EU, the EU authorized
representative is:

Springer Nature Customer Service Center GmbH
Europaplatz 3
69115 Heidelberg, Germany